Plato's Problem

Also by Marco Panza

ANALYSIS AND SYNTHESIS IN MATHEMATICS (*co-edited with M. Otte*)

ISAAC NEWTON

NEWTON ET L'ORIGINE DE L'ANALYSE, 1664–1666

NOMBRES: Eléments de Mathématiques pour Philosophes

DIAGRAMMATIC REASONING IN MATHEMATICS (*co-edited with J. Mumma and G. Sandu*)

Also by Andrea Sereni

ISSUES ON VAGUENESS (*co-edited with S. Moruzzi*)

Plato's Problem
An Introduction to Mathematical Platonism

Marco Panza
CNRS (IHPST, Paris), France

and

Andrea Sereni
San Raffaele University, Italy

© Marco Panza and Andrea Sereni 2013

All rights reserved. No reproduction, copy or transmission of this publication may be made without written permission.

No portion of this publication may be reproduced, copied or transmitted save with written permission or in accordance with the provisions of the Copyright, Designs and Patents Act 1988, or under the terms of any licence permitting limited copying issued by the Copyright Licensing Agency, Saffron House, 6–10 Kirby Street, London EC1N 8TS.

Any person who does any unauthorized act in relation to this publication may be liable to criminal prosecution and civil claims for damages.

The authors have asserted their rights to be identified as the authors of this work in accordance with the Copyright, Designs and Patents Act 1988.

First published 2013 by
PALGRAVE MACMILLAN

Palgrave Macmillan in the UK is an imprint of Macmillan Publishers Limited, registered in England, company number 785998, of Houndmills, Basingstoke, Hampshire RG21 6XS.

Palgrave Macmillan in the US is a division of St Martin's Press LLC, 175 Fifth Avenue, New York, NY 10010.

Palgrave Macmillan is the global academic imprint of the above companies and has companies and representatives throughout the world.

Palgrave® and Macmillan® are registered trademarks in the United States, the United Kingdom, Europe and other countries.

ISBN 978–0–230–36548–3   hardback
ISBN 978–0–230–36549–0   paperback

This book is printed on paper suitable for recycling and made from fully managed and sustained forest sources. Logging, pulping and manufacturing processes are expected to conform to the environmental regulations of the country of origin.

A catalogue record for this book is available from the British Library.

A catalog record for this book is available from the Library of Congress.

Contents

Preface vii

Acknowledgements xii

Terminological Conventions xiii

Introduction 1
   Platonism in the Philosophy of Mathematics 1
   Nominalism in the Philosophy of Mathematics 9
   The Indispensability Argument 14

1 The Origins 16
   1.1 Plato as a Platonist? 17
   1.2 Aristotle Between Platonism and Anti-platonism 27
   1.3 Proclus: The Neoplatonic Interpretation of Euclid's Geometry 32
   1.4 Kant: The Transcendental Interpretation of Classical Arithmetic and Geometry 36

2 From Frege to Gödel (Through Hilbert) 45
   2.1 Frege's Logicist Platonism 45
   2.2 Russell and the Separation of Logicism and Platonism 66
   2.3 Set Theory 69
   2.4 The Problem of Foundations 73
   2.5 Gödel's Platonism and the Rise of Mathematical Intuition 90

3 Benacerraf's Arguments 99
   3.1 What Natural Numbers Could Not Be (According to Benacerraf) 101
   3.2 Benacerraf's Dilemma 107
   3.3 A Map of Responses to Benacerraf's Dilemma: Contemporary Solutions to Plato's Problem 110

4 Non-conservative Responses to Benacerraf's Dilemma 112
   4.1 Field's Nominalism: Mathematics Without Truth and Science Without Numbers 112
   4.2 Mathematics as Fiction: Field and Yablo 125

	4.3 Eliminative Structuralism and its Modal Version	136
	4.4 Maddy and the Cognitive Origins of Set Theory	144
5	**Conservative Responses to Benacerraf's Dilemma**	**149**
	5.1 Neo-logicism: A Revised Version of Frege's Programme	149
	5.2 Linsky, Zalta and 'Object Theory': Mathematics and Logic (or Metaphysics) of Abstract Objects	165
	5.3 A First Version of Non-eliminative Structuralism: *Ante Rem* Structuralism	177
	5.4 A Second Version of Non-eliminative Structuralism: Parsons and the Role of Intuition	187
6	**The Indispensability Argument: Structure and Basic Notions**	**196**
	6.1 Four Versions of IA	197
	6.2 The Quine-Putnam Argument and Colyvan's Argument	201
	6.3 (In)dispensability	203
	6.4 Quine's Criterion of Ontological Commitment	210
	6.5 Naturalism	212
	6.6 Confirmational Holism	214
	6.7 The Dispensability of Naturalism and Confirmational Holism	215
7	**The Indispensability Argument: The Debate**	**217**
	7.1 Against Indispensability	217
	7.2 Against Ontological Commitment	224
	7.3 Against Naturalism and Scientific Realism	235
	7.4 Against Confirmational Holism	241
Concluding Remarks		**250**
Notes		255
References		273
Index		296

Preface

The aim of this book, a revised and partially enlarged edition of the Italian original *Il Problema di Platone* published in 2010 (Carocci Editore, Roma), is to offer an introduction to the philosophy of mathematics addressed to all those wishing to familiarize themselves with the subject, and even to those lacking any previous acquaintance with it.

We have done our best to organize the discussion so as to avoid reliance on previous knowledge. An introduction, however, cannot cover everything, and the philosophy of mathematics is a vast domain with a proliferation of close connections to several other domains: mathematics itself and its history, logic, philosophy of language, history of philosophy, just to take some obvious examples. The reader will thus unavoidably encounter remarks for which full understanding requires acquaintance, if only at basic levels, with different disciplines.

Several other introductions to the philosophy of mathematics are available. Mentioning a few of them, and limiting ourselves to books in English, we can point to Brown (1999), Potter (2002), Shapiro (2000a), George and Velleman (2001), Giaquinto (2002), Mancosu (2008a), Bostock (2009), Colyvan (2012). Our book will obviously discuss many topics that are covered also in these introductions, but it is intended to approach them with a particular focus.

Introductions to the philosophy of mathematics have often aimed at offering a comprehensive account of the discussion in the field since at least the end of the nineteenth century. In some cases, the historical background is kept to a minimum, in order to leave room for a theoretical discussion of the main philosophical options available in the contemporary debate. We have followed a different strategy. We have focused on a single problem and a single option that has been offered as a solution of it; we then trace the main historical development of this latter option (Chapters 1–2), discuss the debate it has engendered from the 1960s to the present day (Chapters 3–5) and finally focus in more detail on a currently widely discussed argument that has been advanced in that debate (Chapters 6–7).

The problem we have focused on is that of the ontology of mathematics, the problem that in our title we call 'Plato's problem' (the reasons for this will be clear in § 1.1). The following is a very simple way of presenting it: granted that the statements of mathematics are about something, what are they about? The answer we focus on goes under the name of 'platonism'. It claims, broadly speaking, that these statements are about a domain of abstract objects, which they describe.

We are well aware that this way of proceeding can only offer a very partial representation of past and present philosophy of mathematics, which has dealt with several other problems and answers. Rather than presenting a broad-brush summary of this vast range of topics, we prefer to limit the scope of our introduction so as to make it possible to go into much more detail on the option being discussed and the specific argument in its support. Chapters 6 and 7 are devoted to this latter, the so-called 'indispensability argument'. Our hope is that the reader will be guided through a significant part of present-day discussion in the philosophy of mathematics, possibly paving the way for further studies. In order to achieve this, the book's chapters have been written in three different styles.

Chapters 1 and 2 contain an (inevitably partial) historical reconstruction, with authors and topics considered in chronological order. Connections among specific topics are emphasized. The aim of these chapters is to offer the historical background that seems to us required in order to understand why the contemporary debate is so relevant. Philosophy, even in its most technical and specific regions, is largely motivated by its history and tradition, and no philosophical discussion can really be appreciated when these are wholly disregarded.

Chapters 3, 4 and 5 offer a synchronic reconstruction of several options, some supporting platonism and some opposing it. We have chosen to arrange them depending on which sort of response they offer to the dilemma advanced by Paul Benacerraf in the early 1970s, a dilemma that in our opinion can easily be seen as a modern version of Plato's problem. The aim of these chapters is to present an articulated summary of the current discussion on this problem, one that the philosophy of mathematics, as the previous chapters show, inherits from its tradition.

Chapters 6 and 7 present a systematic reconstruction of a particular argument, aiming to show its assumptions, motivations and difficulties, together with the necessary and/or sufficient conditions for its conclusions. We have opted for a detailed, and often explorative, analysis of the various theoretical ingredients involved in one single argument, an argument that has been suggested during the course of the past fifty years by one of the most authoritative thinkers in the empiricist tradition, Willard van Orman Quine. The contrast between this tradition and the platonic one, that finds a point of intersection in the indispensability argument, is one of its most interesting traits.

We have chosen to devote two chapters to this argument since it involves several crucial issues in contemporary philosophy of mathematics. This is not to say that we intend to appeal to the argument (once appropriately stated) in order to convince the reader of the correctness of the platonist option in the version it supports. Quite the contrary, we believe this argument to have various limitations and difficulties, many of which will be discussed. Our only purpose is to familiarize the reader with the intricacies of contemporary philosophy of mathematics through the consideration of a specific example among the current debate.

Even the choice of treating Plato's problem is partial. Other problems differ from it not only as regards their content, but also as regards their nature. Roughly speaking, we can single out four kinds of problems pertaining to the philosophy of mathematics.

First, there are foundational problems concerning the best way of founding, justifying and organizing the edifice of mathematics or at least some relevant parts of it. Next, there are general interpretative issues relating to the nature of mathematics itself, such as Plato's problem and other problems variously related to it, such as the problem of the nature of mathematical knowledge. (Is there mathematical knowledge, and if so, what kind of knowledge is it?) or that of the logical character of the truths or theorems of mathematics. (If there are any truths in mathematics, what is their source? Are they analytic or synthetic? A priori or empirical? And if there are none, what legitimates the theorems of mathematics? Just to give an example, if '3 + 5 = 8' is not a true statement, in what way does it differ from '3 + 5 = 9'?)

There are also more specific interpretive problems, concerned with particular mathematical theories, or with mathematical practice as it has developed over time. Mancosu (2008a) offers an excellent survey of some of these problems, such as that of the availability of a criterion for acknowledging when arguments are explanatory, or that of visualization, or more generally of diagrammatic reasoning in mathematics, or that of the possibility of offering a principle of purity, selecting some proofs or theories as being better than others.

Last, there are problems relating to the applicability of mathematics, the justification for this applicability and how it takes place, to the role mathematics has in empirical sciences and in our everyday lives.

Many philosophers of mathematics believed, and still believe, that their main task is to celebrate the beauty of mathematics. Carl Jacobi – a great mathematician of the first half of the nineteenth century maintained – against Joseph Fourier – that the only aim of mathematics is "the honour of human mind". Even though we wholeheartedly acknowledge this greatness, and feel the fascination that many mathematical theories exert on us, we believe that this is not the main purpose of the philosophy of mathematics.

We are, then, left with the four sorts of problems mentioned above. They are not certainly unrelated, and indeed many of them, even when they fall under different categories, are so closely intertwined that it is hard to understand how one can be treated in isolation from the others. One of the characteristic traits of the way of pursuing philosophy that is usually labelled 'analytic', however, is that different problems are singled out and treated in their specific aspects in order to deal with them with the required precision, thus avoiding getting lost in overly general considerations. Many of the thinkers we discuss can surely be counted among the practitioners of this kind of philosophy. We ourselves strived to make the subtler distinctions possible, at least so far as the level of generality of an introduction allows. Still, the present book need not necessarily be viewed as an introduction to the analytic philosophy of mathematics.

This is mainly due to the fact that, contrary to what happens in several areas of contemporary philosophy, within the philosophy of mathematics the distinction between the analytic and rival approaches is not that sharp. One reason for this is that the philosophy of mathematics is necessarily receptive to the characters of mathematics itself and to the methodological worries of mathematicians, and this forces some lines of inquiry and naturally excludes others, independently of one's preferred philosophical attitude. Another reason is that mathematics constantly undergoes technical evolutions, and is characterized by an historical development that cannot be openly denied without leading to utterly implausible views. This rules out any approach that might excessively abstract from actual mathematical practice and from historical considerations. A third reason is that the scenery of the philosophy of mathematics has been dominated, between the last decades of the nineteenth century and the early 1970s, by the problem of foundations, and this has inevitably led to a profound confrontation with authors such as Gottlob Frege, Bertrand Russell and Rudolf Carnap, who are unanimously seen as the fathers of the analytic tradition, and others like Henri Poincaré, David Hilbert, Luitzen Egbertus Jan Brouwer, Hermann Weyl and Kurt Gödel, who could hardly be seen as part of that tradition.

These are, in the end, only three different ways of articulating the very same reason, which has resulted in a varied mixture of sensibilities, styles, skills, formative experiences and points of view. For instance, problems falling under the third category described above are now being studied by philosophers that would hardly qualify themselves as analytic, and who nonetheless confront them with methods and assumptions that typically pertain to the analytic tradition. This does not prevent them from grounding many of their arguments on historical considerations or investigations on the technical details of mathematical theories. The philosophy of mathematics thus appears as a privileged field where, contrary to what happens in other areas of philosophy, collaborations and liaisons among different approaches can be the subject of fruitful experimentation.

MARCO PANZA
ANDREA SERENI

Acknowledgements

Many colleagues and friends have, to various extents, made the writing of this book possible. We are of course responsible for any mistakes, omissions and inaccuracies, but we want to thank all who have helped us with comments and suggestions, in particular Andrew Arana, Mark van Atten, Michael Beaney, Jean-Pierre Belna, Consuelo Benfenati, Andrea Bianchi, Claudia Bianchi, Francesca Boccuni, Jacob Busch, Ronan de Calan, Roberto Casati, Riccardo Chiaradonna, Giovanna Corsi, Francesco Berto, Pierre Cassou-Nogues, Karine Chemla, Mario De Caro, Michael Detlefsen, Michele Di Francesco, Jacques Dubucs, Matti Eklund, José Feirrerós, Paolo Freguglia, Maria Carla Galavotti, Massimo Galuzzi, Pieranna Garavaso, Sébastien Gandon, Valeria Giardino, Pierluigi Graziani, Bob Hale, Brice Halimi, Geoffrey Hellman, Paolo Leonardi, David Liggins, Rossella Lupacchini, Paolo Mancosu, Kennet Manders, Daniele Molinini, Gianluca Mori, Sebastiano Moruzzi, Matteo Motterlini, Alberto Naibo, Fabrice Pataut, Richard Pettigrew, Eva Picardi, Matteo Plebani, David Rabouin, Davide Rizza, Ferruccio Repellini, Jean-Michel Salanskis, Marco Santambrogio, François Schmitz, Stewart Shapiro, Ivahn Smadja, Mark Steiner, Jean-Jacques Szczeciniarz, Paolo Togni, Alfredo Tomasetta, Stephen Yablo, Edward Zalta. Special thanks goes to Annalisa Coliva for her continuous support.

During the past years, many of the topics discussed in this book have been presented in seminars and conferences at the universities of Bergamo, Bologna (Cogito), Frankfurt, Modena and Reggio Emilia, Padua, Parma, Milan, Nancy, Paris 1 (IHPST), Rome. We would like to thank all the audiences for their helpful comments and feedback.

Special thanks go to Gianluca Mori at Carocci Editore for having granted us the rights for this translated and revised edition of *Il problema di Platone*, and to Priyanka Gibbons, Melanie Blair, Keith Povey, Rosalind Davies and all those at Palgrave Macmillan who made this book possible, for their support and patience.

Marco Panza
Andrea Sereni

Terminological Conventions

In what follows, we lay down some terminological conventions that are adopted throughout the book. Like all conventions, they are partly arbitrary. They are often required for ease of exposition, and are not meant to rule out different uses that may be equally legitimate. Apart from those listed here, other conventions are adopted where needed, for which separate treatment would have been superfluous, since our use of the relevant expressions will make their intended meaning clear enough (throughout the book we have tried to make these uses as stable as possible). More specific conventions on less frequent words are introduced at specific points in the text.

By 'sentence (of a certain language)' we mean a formula (of that language) as such, as a mere succession of symbols or words subject to certain composition rules. A sentence is often considered for the function it plays (possibly on the basis of an appropriate interpretation of its symbols or words). If this function is to assert or to state something, namely that things stand in a particular way, then we say that it expresses a statement. This statement will then be said to be formulated in that sentence's language. If an appropriate equivalence relation is defined on the sentences of a certain language, the same statement will be expressible by different sentences of that language, each of which will be said to consist in a different formulation of it. Once one of these formulations is given, others are often referred to as reformulations of the statement it expresses. The equivalence relation between sentences expressing the same statements might sometimes be defined also over statements in different languages. We will then say that a given statement, as expressed by a sentence of one of these languages, has a reformulation or can be reformulated in other languages. A statement can thus be understood as an equivalence class of sentences of the same language or of different languages. It is clear, however, that it can only be displayed by one or another among these sentences. In brief, when a sentence is taken as a representative of a certain equivalence class of statements, it will then straightforwardly be called 'statement'.

We distinguish a statement, as a linguistic object, from the linguistic act of asserting something (this act being generally called 'assertion'). We also distinguish a statement from what it states, i.e. its content, which is generally called 'proposition'.

We take the expression 'to make a statement' to have two different senses. In a first sense, which we might call 'propositional', this expression means the act of stating that something is the case, that things stand as the statement (appropriately understood) says they stand. In a second sense, it means the act of performing an utterance with assertoric force: in this sense, 'to make the statement x' means the same as 'to write φ' or 'to utter φ', where 'φ' denotes the sentence expressing x. The context will make clear which of the two senses is relevant on each occasion.

In our book, we shall mainly deal with statements (we will consider sentences and propositions on rare occasions). This is due to the fact that, both in mathematics and in the empirical sciences, sentences are not usually studied as such – contrary to what happens in certain branches of logic or linguistics – but are rather considered for the statements they express.

We will use the term 'theory' to refer to a system of statements that we take, in most cases, to be or have been accepted by someone, generally by a whole scientific community (this simply means that members of this community make or have made these statements or have manifested their proneness to do it). In order to stress the relations that are supposed to hold among these statements and/or in order to be faithful to the linguistic usage of authors, we will sometime speak of bodies of statements – though they will always be taken to form more or less structured theories. Generally, though not necessarily, we will take a theory to be a system of statements deductively closed under appropriate deductive rules (which means that if the theory includes certain statements it also includes all those – generally infinitely many – statements that follow from the former by these rules). Sometimes we will speak of parts or portions of a theory, or of sub-theories, meaning by this to refer to appropriately selected sub-systems of the system of statements forming a given theory (these sub-systems being themselves theories). We will call 'mathematical' those theories that are usually taken to be such, without specifying in general what this adjective exactly means. In accordance with widespread usage, we will on the other hand call 'scientific' those theories usually pertaining to empirical sciences, as distinct not only from mathematical ones, but also from logical, metaphysical, etc.

We will speak of statements of mathematics, of arithmetic, or of geometry in order to refer, respectively, to statements belonging to a given mathematical theory, to a given version of arithmetic, or to a given geometrical theory. These statements are sometimes called 'theorems' or 'consequences (of the relevant theory)'. This holds especially when it is important to stress that these statements follow – according to accepted inference rules – from other statements that will then be called 'axioms'. This happens usually for mathematical and logical theories, which are usually deductively closed. Sometimes we will distinguish pure from impure mathematical theories: the former are meant to include only statements that are formulated in a specifically mathematical language; the latter, on the contrary, are meant to include statements in which, together with the expressions of such a specifically mathematical language, expressions of a non-mathematical, generally scientific, language also occur. It thus follows that an impure mathematical theory can, and usually is, meant to be a scientific theory, even though obviously not all scientific theories are impure mathematical theories.

By 'mathematical statement' we mean, more generally, a statement that is formulated in the language of a given mathematical theory (usually a pure one, but possibly also an impure one, so long as it features expressions of a specifically mathematical language; in these cases we will also speak of 'impure mathematical statements'). By 'nominalistic statement' we mean a statement formulated in a purely nominalistic language (in § 4.1 we will explain what usually counts as a nominalistic language; approximately, it is a language which provides means for speaking only of concrete objects). When a mathematical statement is reformulated in a nominalistic language, we will speak of its nominalistic reformulation. Such a reformulation is often called 'paraphrase'. However, we will speak of paraphrases more generally also to refer to other reformulations of certain statements, not necessarily in a nominalistic language.

According to the foregoing conventions, '5 + 8 = 13' is both a mathematical statement (arithmetical, more precisely) and a statement of mathematics (of arithmetic, more precisely, or of any version of it), whereas the statement '5 + 8 = 12' is a mathematical (arithmetical) statement but is not a statement of mathematics (of arithmetic). The statement 'The acceleration of gravity at terrestrial poles is equal to 9,823 m/s^2' is a statement of an impure mathematical theory which is also a scientific theory, more precisely a version of the gravitational theory. The statement 'Earth poles shift in time' is a non-mathematical statement (save if one wishes to consider 'poles' as part of a geometrical language) belonging to a scientific theory, namely geology. Like statements such as 'Earth poles are always iced' or 'Earth poles are always flowery', it is also a nominalistic statement.

A theory will be said to be true if and only if all of its statements are true (according to a specified conception of truth).

We will use single inverted commas for mention (thus we will say that '3' is a singular term) and double inverted commas for quotation. We will use corners for expressions designating properties, relations or concepts: we will say for example that under the concept ⌜(being) identical to 1⌝ falls only one object.

Introduction

Platonism in the Philosophy of Mathematics

In the philosophy of mathematics, platonism is the thesis according to which mathematical statements, and theorems of mathematical theories in particular, are about abstract objects forming a domain that those theorems describe.

The term 'platonism' obviously traces back to Plato, but it is not meant to denote his philosophy. As we shall see in Chapter 1, Plato can certainly be counted among platonists for his philosophy of mathematics (though not everyone agrees on this) and platonism can be seen as an answer to a problem raised by Plato himself – to Plato's problem, as the book's title says. But platonism can be endorsed within philosophies distant from that of Plato. This expression should thus be regarded as a technical term of contemporary philosophical jargon. Accordingly, we adopt the now widespread convention of writing 'platonism' with non-capital 'p', and we reserve instead the adjective 'Platonic' (with capital 'P') when speaking of Plato's views and works themselves.

Existence of Abstract Objects

A common way of stating the platonist thesis is to affirm the existence of a particular sort of abstract objects, generally called 'mathematical objects', and to claim that mathematical theories are about them, tell us how they behave, which properties they possess, which relations hold among them.

Recourse to the notion of existence does not, however, make things clearer by itself. It does not help in particular to make clear what platonists mean when they speak of abstract objects. On the contrary, it is often through a clarification of the nature of certain abstract objects that one can come to appreciate what their existence comes to.

2 Plato's Problem

Let us consider a simple example. It seems that Italy's national football team should be considered as an object: it is denoted by a proper name apparently endowed with a precise reference; it performs certain actions, such as winning the World Cup; it is distinct from England's national football team or Italy's national basketball team. The status of this object is different, however, from that of the single players that, on a given occasion, enter the field wearing their azure shirts. These we can meet, see and touch. It is hard to do the same with the team they play for. A way of accounting for this difference is indeed to acknowledge that the former are concrete objects, whereas the latter is an abstract object. Now, there is a clear sense in which one might say that Italy's football team exists. It is the same sense in which one can say, for example, that Italy's F1 team does not exist. Here it is clearly by our understanding of the nature of the particular abstract objects under consideration – Italy's football and F1 teams – that we can appreciate in what sense one can say that such objects do or do not exist, and not *vice versa*.

Similarly, it seems that it is only by the understanding of the nature of mathematical objects that we can come to appreciate in what sense one can say that they exist. The main task for a platonist philosopher would thus be to explain in what sense we can plausibly speak of mathematical objects, how they can be identified and thus in what sense it can be said that they exist, and whether there are good reasons for saying so. There are, nevertheless, arguments for platonism that do not presuppose any clarification of the concept of abstract object. One of these arguments – the so-called indispensability argument – leads simply to the conclusion that there are mathematical objects, but leaves it to independent considerations to specify what the nature of these objects is, exactly.

In what follows, we characterize platonism as the claim that abstract (mathematical) objects exist and our mathematical theories are about them; nonetheless, we take it as an open question what their existence comes to in details.

These first remarks already makes clear that platonism is all but an uncontroversial thesis. Not only is its justification hard to obtain, but its formulation is, too.

What Is an Abstract Object?

A similar difficulty emerges with respect to the meaning that should be assigned to the claim that some objects are abstract, and mathematical in particular.

There is no agreement among contemporary philosophers on where the boundary between abstract and concrete objects should be drawn.

This fact might strike as odd, since the distinction is commonly acknowledged and considered of the utmost importance. However, it is however often single cases, such as those of propositions, sets, natural numbers – the non-negative (finite) integer: 0, 1, 2, 3, . . . – or Italy's football team, the *Ulysses* (as distinct from its single copies), that prompt this or that philosopher to think that that distinction should be acknowledged in order to account for certain phenomena. Doubts and divergences of opinion emerge when one leaves these single cases and asks for the general one, pondering how the general notion of abstract object should be understood.

Here, again, it is our understanding of the nature of certain objects that makes us often appreciate in what sense we can say that they are abstract, and not *vice versa*. There are, of course, attempts at clarifying the notion of abstract object in general, among which we recommend Dummett (1973, ch. 14), Zalta (1983) and Hale (1987).[1]

Rather than adopting one proposal among the others, we believe it is appropriate to leave to the single platonist philosopher the burden of elaborating his/her own conception of abstract objects, and mathematical objects in particular. Different conceptions, some of which will be discussed later, can thus originate from different ways of elaborating and defending platonism.

What Is a Mathematical Object?

The difficulties with the formulation of platonism do not end here. It is indeed natural to wonder what makes an object, an abstract object in particular, a mathematical object. Again, it is hard to give a general answer.

This difficulty could be avoided by claiming that it is not the task of a philosopher of mathematics to establish in general what must count as mathematics, or what the nature of the objects of mathematics should be. His/her task would rather be that of accounting for certain accepted mathematical theories and possibly trying to ponder whether they could be conceived of as descriptions of particular domains of objects. A positive answer would *ipso facto* make such a philosopher count as a platonist. This answer would include a characterization of the relevant objects; these will then not be said to be 'mathematical' because they satisfy a given criterion of mathematicality, but simply because it will be established that a given theory, generally considered as mathematical, is about them.

Reasoning in this way means taking mathematics as a fact, or a phenomenon, coming before the philosophy of mathematics, as something given that the latter tries to account for in some way. A philosopher adopting this line of thought might well believe that mathematical objects pre-exist the theories that are about them (as mountains pre-exist the description that geographers make of their profile), but she would leave to the mathematician the task of identifying those objects that are worth being investigated and to which a theory is worth being dedicated. This way of thinking seems rather natural, and indeed this is how the philosophy of mathematics has always been presented: as a reflection on parts of mathematics already given as such. Even those philosophers who have attempted to found mathematics, or believed to accomplish this task by giving new ways of defining certain mathematical objects, have always acknowledged that the new definitions and formulations had to lead to the same results that were obtained within already accepted theories.

Arithmetical Platonism

From this point of view, we can easily conceive of someone endorsing platonism only with respect to a particular mathematical theory, while rejecting it for other theories, or simply withhold judgement about them.

Arithmetic is a case in point.[2] Arithmetical platonism is the thesis according to which arithmetical statements speak of a certain domain of abstract objects, the natural numbers. This form of platonism is widely held. Moreover, it is especially relevant given both the central place of arithmetic in the mathematical edifice, and for the wide-ranging applications arithmetic has. It is widely held since, for reasons that shall become clear later, arithmetic is well suited to be interpreted as a theory about certain abstract objects. It is thus difficult to find philosophers who have maintained platonism for some mathematical theory but have rejected it for natural numbers arithmetic, or even have not taken a stand on arithmetic. Every mathematical platonist seems to be an arithmetical platonist, even though many arithmetical platonists are silent on other mathematical theories or, as happens with Frege (see § 2.1), have different views on some of them (on geometrical theories, in Frege's case). Arithmetical platonism can thus be viewed as a paradigmatic case.

What Is an Object?

A further concern with the formulation of platonism is related to the notion of object itself. How should this notion be made more precise than generalizations and approximations in ordinary discourse allow, if one has to make it the central notion of a philosophical view? Just to make an example (we will come back to this in § 1.4), Immanuel Kant often speaks of objects, but – as his employing in an apparently arbitrary way (Eisler, 1930, p. 391) the two different terms 'Gegenstand' and 'Objekt' shows – he has no explicit theory of objects. Things stand differently, for example, with Frege and Husserl; their theories of objects could be considered by some as unclear or incomplete, but still they are explicitly advanced.

Many philosophers with whom we will be concerned agree with Frege on a crucial point: an object is everything and only what falls under a sortal concept. The notion of sortal concept is usually elucidated by establishing that P is a sortal concept if and only if P is a concept such that if x and y fall under P, then it is possible to establish whether x is or is not identical with y. An object would thus be whatever can be said to fall under a certain concept and for which it can be established what singles it out among anything else falling under this concept. Stars would then be instances of objects: if we observe a star (for example, the first star that we see at sunset) we can, at least in principle, establish whether it is or it is not the same star that we observe on another occasion (for example, the last that we see disappearing at dawn). Things seem to stand differently with emotions: it is not clear whether there are criteria for establishing whether the emotion we experience when we see our star appearing at sunset is or is not the same emotion that we experience when we see it disappearing at dawn. Apparently, thus, emotions would not count as objects.

This way of thinking is not agreed upon by everyone, and it does not solve all of our difficulties: one among those that remains open is the problem of understanding properly what we mean when we speak of concepts. If we adopt the Fregean idea presented above, however, we have reasons for claiming that the identification of a certain domain of objects depends on the definition of an appropriate sortal concept. In order to establish that the objects forming that domain exist, arguments will be needed to show that something falls under that concept, that is, that the concept is not empty, like, for example, the concept of unicorn.

It also follows from this conception that when we speak of objects we refer to individual entities inhabiting worlds that they share with other individuals of the same kind. Obviously, nothing prevents us from thinking of sortal concepts under which only one object falls, like the concept of national football team winner of five World Cups, or even sortal concepts under which necessarily only one object falls (if any does), like the concept of supreme upper bound (*supremum*) of a given sub-set of a totally ordered set. Similar cases, however, usually concern concepts that are obtained through the restriction of other, more general concepts, such as the concept of national football team or that of the elements of a given set (though this is not always the case: think of the concept of God). The fact that under these concepts only one object falls is due to the fact that among the objects falling under those other more general concepts only one has some particular properties. When speaking of mathematical objects, however, one generally intends to refer to individuals inhabiting domains formed by several individuals of the same kind and, in almost all interesting cases, to individuals inhabiting infinite domains. This is what happens with natural numbers, sets, points of $\mathbb{R}^2$, and so on. Some might think that it is appropriate, especially in historical contexts, to speak of the object triangle or the object Riemann integral. But here one would be speaking in a fairly different sense from when we speak of mathematical objects, unless one is speaking with the intention of referring to a given kind of geometrical figure belonging to the domain of geometrical figures (as opposed, for example, to the square or the rectangle) or to a given operator belonging to the domain of functional operators (as opposed, for example, from the Lebesgue integral or the Stieltjes integral, or the Radon integral).

Virtues of Platonism

Many reasons could be offered for doubting that platonism is right: one might think that it has unacceptable consequences, or else that it is not possible to give any satisfactory justification of it. Still, it is hard to deny that platonism is an attractive thesis, since it allows us to give very simple answers to very complex questions about the nature of mathematics. It is easy to see that even if a precise account of what an abstract object is, or of what it means to say that it exists, is lacking, the thesis that a mathematical theory is about abstract objects confers to that theory a structural similarity with other areas of discourse that are about concrete objects.

When we say, for example, that Liverpool is smaller than London we are claiming that a given concrete object, the city of Liverpool, stands in a given relation, the relation of being smaller (relative to the number of inhabitants, let us say), with another concrete object of the same kind, the city of London. This could be true even if we were not in a position to know that it is (maybe because of the complete loss of the archives in the two cities' registry offices). The statement 'Liverpool is smaller than London' has a grammatical structure (the structure of the sentence that expresses it in the particular formulation by which this statement is given to us) that faithfully mirrors the semantic structure that we attribute to it when we understand it as we usually do: the statement is true if the two objects that are referred to by the two proper names 'Liverpool' and 'London' stand in the relation ⌜(being) smaller than⌝ with each other. Its truth does not depend on us in any way, nor in particular on our being able to justify it: it depends on how things stand in the world. Were we able to give an appropriate justification for the statement, as it indeed seems to be the case, this would establish its truth, and we could then deem mistaken anyone who denied it.

If we accepted arithmetical platonism, we could apply a similar interpretation also to arithmetic, with an obvious gain in simplicity. For example, we could claim that when we say that 2 is smaller than 3 we are claiming that a certain abstract object, the number 2, stands in a particular relation, the relation of ⌜(being) smaller than⌝, with another abstract object of the same kind, the number 3. The statement '2 < 3' would thus be understood in such a way that its grammatical structure mirrors its semantic structure faithfully. It would not mean, for example, that in the act of counting we come to count two units before we come to count three units, but rather, and more simply, that the object 2 is smaller than the object 3. Moreover, we could think that this statement is true or false independently from what we could do in order to justify it, namely from our being able to (and, a fortiori, from our way of) proving it. But were we able to give a proof of it, as it is indeed the case, it would establish its truth, and we could then deem mistaken anyone who denied it.

The appeal that platonism exerts is due to the simplicity of this interpretation, and to the fact that it accounts immediately for that feeling of ineluctability that we experience with mathematics and its theorems.

Platonism, Realism and Objectivity

Some clarifications are still in order. The description given above of how arithmetical platonism allows us to interpret the statement '2 < 3' has a gap that, though it might have gone unnoticed, cannot be passed over in silence. For it is one thing to claim that this statement speaks of certain abstract objects, 2 and 3, which could be taken to mean, with the required provisos mentioned above, that these objects exist. It is a different thing to claim that such a statement is true. And in order for this latter claim to hold it is not also required that the truth of that statement is independent from our being able to give a proof of it.

According to the terminology that we adopt here (following the most common usage), the thesis according to which mathematical statements are about abstract objects – or the thesis that there exists abstract objects which those statements are about – is the platonist thesis. Sometimes this thesis is also labelled 'ontological realism'. This thesis should not be confused with another one, usually called 'semantic realism', according to which the statements (theorems) of mathematics are true. This latter thesis could indeed be held even without holding that these statements are true due to the existence of certain objects of which they speak. This second thesis, in its turn, should be distinguished from a third one, known through the works of Michael Dummett,[3] according to which certain statements – mathematical ones, in the present case – possess a definite truth-value independently of our being able to know it. Also this thesis is a version of the philosophical position that is commonly called, without further specification, 'realism'. Despite its relevance for many discussions in the philosophy of mathematics, we will not deal specifically with it in our book. The term 'realism' is also often employed in the philosophy of science to refer to a fourth thesis, called 'scientific realism', with which we will deal in §§ 7.3.2 and 7.4.2 in connection to the indispensability argument. To avoid confusion between these four very different theses, we will always use the term 'realism' followed by the adjectives 'ontological', 'semantic' or 'scientific' in order to refer, respectively, to the first, the second or the fourth thesis.

Difficulties with Platonism

We are now in a position to consider some of the difficulties that are related to platonism.

One of these is that it is difficult to find convincing arguments in support of platonism that are independent from the explanatory benefits delivered from its adoption, and that are not based on assumptions at least as strong as platonism itself. Platonism is often supported by a negative argument: there is no way of accounting for some phenomena concerning mathematics unless we adopt a platonist position. One could rightly feel unsatisfied with an argument of this sort. After all, even the assumption of the existence of a creator God allows us to answer in a simple way many questions concerning the origin of the universe. However, many believe that this is not by itself a sufficient reason for accepting His existence.

There is more than this, however. The way in which platonists sometimes account for our access to the realm of abstract objects about which, in their opinion, mathematics is, is often reminiscent of a religious attitude. For believers, faith gives a means of getting in contact with God, and thus of giving content to their prayers. But it seems unlikely that a similar means exists for getting in contact with mathematical objects and getting information about them. How can we do it, then? The difficulty of answering this question is a second important weakness of platonism.

Nominalism in the Philosophy of Mathematics

There are, of course, many philosophers of mathematics who reject platonism, both for the aforementioned difficulties and for independent reasons. The alternative thesis many adopt is called 'nominalism', and consists in the denial of the existence of the abstract objects mathematical theories would allegedly be about.

Paraphrase

Some adopt a global and radical form of nominalism. They believe that no abstract objects, of any sort, exist, and that it makes no plausible sense to claim that they exist. But what about apparently innocent abstract objects, such as the Italy's national football team? Does it really make no sense to say that Italy's national football team exists, whereas Italy's F1 team does not? In order to defend her/his view, the radical nominalist must find a way of explaining why statements like 'Italy's national football team won the World Cup' or indeed 'Italy's national football team exists, whereas Italy's F1 team does not' seem to be innocuous and intelligible.

A common strategy consists in claiming that statements like these are nothing but abridged reformulations of longer statements about concrete objects, to the effect that the latter provide appropriate paraphrases of the former.

When we say that in 2006 Italy's national football team won the World Cup, what we are really saying is that certain concrete persons – who have signed a contract that binds them and others and that is recognized by a larger community, in accordance with certain rules – acted in a certain way in appropriate circumstances with respect to other persons – also subject to a similar contract and acting in similar ways – and, on the basis of rules that codify that way of acting, have been acknowledged as winners. In this and many other cases the paraphrase is long, tedious, and inconvenient. Still, it is difficult to deny that it really expresses what we really meant with the original statements. The basic idea is clear enough: even though certain statements seem on the surface, that is, on the basis of their grammatical structure, to be about certain abstract objects (football teams, for example), the correct way of reading them is by admitting that, as a matter of fact, they are about other objects, more innocuous from an ontological point of view.

One might think that this holds also for mathematical statements: the way in which they are usually presented is just nothing but an abridged reformulation of longer and more complex statements about concrete objects only. Paraphrasing is a way of reinterpreting. Nominalists who adopt the paraphrase strategy then suggest that the language of mathematics should be interpreted in such a way that it will not be about abstract mathematical entities any more. It will still be possible to express ourselves in the ordinary way, in terms of numbers, classes, functions, etc., but only once it is agreed that this is just a useful expedient, a convenient *façon de parler*.

We will come back to this strategy in Chapters 4 and 7. Here it is worth wondering which alternative strategies are open to a nominalist about mathematics who nevertheless does not believe that the paraphrase strategy works. Such a nominalist should agree with the platonist in believing that mathematical statements are perfectly intelligible when considered at face value, and that they should be understood in accordance with their grammatical structure. But (s)he could not concede that these statements be about abstract objects, even though they appear to be on their surface.

Statements Without Content

A first option for our nominalist is to claim that those statements, while perfectly intelligible, are devoid of content: they do not properly speak of anything, and thus do not describe any object. They would thus be either mere ways of stipulating linguistic conventions or outcomes of these conventions; or else they should be taken as logical truths, like the statement 'Either the Sun is obscured by clouds, or it is not obscured by clouds': this being obviously a statement that is not about today's weather, nor, despite appearances, about the Sun or the clouds. This view has been endorsed, for example, by more orthodox neo-positivists such as Rudolf Carnap. We will come back to it in § 2.4.1.

Empiricism in the Philosophy of Mathematics

An alternative option for our nominalist is to claim that, despite appearances, mathematical statements, even when taken literally, do speak only of concrete objects: '2 < 3' speaks of the objects 2 and 3, but these are concrete objects, like stars and chairs.

This view differs from platonism only insofar as it assigns to concrete objects the role that platonists assign to abstract objects. It might even be suggested that this is nothing but a version of platonism: a sort of empiricist platonism, as it were.

Be this platonism or not, a view on these lines has been repeatedly advanced, most prominently by John Stuart Mill (1843), according to whom numbers are just aggregates of concrete objects and mathematical statements express empirical generalizations; by Penelope Maddy (1990a, 1990b), who labels 'physicalist platonism' a view, close in spirit to that of Mill, that she holds concerning sets (see § 4.4); and by Hartry Field (1980), who seems to suggest that synthetic elementary geometry (namely the geometry of Euclid's *Elements*, presented without appealing to real numbers and sets) is about space–time points conceived of as concrete objects.

Motivations and plausibility aside, these views show that it is possible to distinguish, at least in principle, between two different theses that are normally equated: (*a*) mathematical statements are about a domain of objects; (*b*) mathematical statements are about abstract objects. Some arguments in support of platonism – the indispensability argument among them (see Chapters 6 and 7) – are arguments in support of (*a*) without being at the same time arguments in support of (*b*).

Not every empiricist position in the philosophy of mathematics endorses the claim that mathematical statements, be they paraphrased or not, are about concrete objects (or empirical objects, if you prefer). Some reject this claim, and yet support a form of empiricism only with respect to the justification of those statements. This might be done, for example, by claiming with Charles Parsons (1979–80; 2008, ch. 5) that in order to study their abstract objects, mathematicians need to avail themselves of certain concrete objects that stand in particular complex and important relations with the former. We will come back to this claim in § 5.4. For the time being, let us only notice that other versions of empiricism about justification hinge on the idea that our justification in believing in the existence of abstract mathematical objects is, or should be, of the same kind as that for believing in the existence of concrete objects. This idea has been advanced for example by Imre Lakatos (1976) and is now for different reasons defended by all those who, following suggestions by Quine, adopt the indispensability argument.

Mathematical Statements Are True Only if They Are Vacuously True

Our nominalist could also claim that, since there are no abstract objects, mathematical statements, that nevertheless need be taken literally, are not true, or are true only if vacuously true. This thesis has been held by Hartry Field (1980, 1982) concerning statements of arithmetic and analysis (see § 4.1). Let us here confine ourselves to some brief remarks.

Let us consider the two following statements: (*a*) 'There is an even prime number'; (*b*) 'All prime numbers are odd'.

Let us concede that: (*i*) mathematical statements should be taken literally; (*ii*) these statements can be said to be true only in the same sense in which statements of ordinary language are said to be true, so that, for example, the statements 'There is an honest politician' and 'All politicians are dishonest' are true or false depending on the existence of men who are politicians by profession and are or are not honest; (*iii*) if natural numbers existed, they would be abstract objects, but there are no abstract objects that can be identified with them. Let us also concede, as it seems natural to do, that the grammatical structure of our statements is preserved in the following reformulations in the language of first-order predicate logic: (*a'*) 'There exists an x which is an even prime number'; (*b'*) 'For all x, if x is a prime number, then x is odd'.

From (*i*) it follows that the appropriate semantic structure of (*a*) and (*b*) is the one displayed by (*a'*) and (*b'*). We should then conclude that (*a*) is not true, while (*b*) is. The reason why (*a*) would not be true can be easily appreciated: by clause (*iii*) there are no prime numbers and no even numbers, and thus, a fortiori, there is no x that is an even prime number. Things might appear trickier with (*b*). It suffices however to note that since according to clause (*iii*) there is no x that is a prime number, it follows from (*ii*) that the antecedent 'x is a prime number' is false, whatever x is. Clause (*i*) then forces us to conclude that the whole statement is true: according to the laws of logic, a conditional statement whose antecedent is false is true whatever the truth-value of its consequent is. If (*iii*) is accepted, (*ii*) and (*i*) – that together preclude *ad hoc* solutions – force us to conclude that the truth of our second statement does not depend on its content, but only on its grammatical structure: in this case we say that it is vacuously true.[4]

The two statements (*a*) and (*b*) are quantified statements in which no individual constant occurs that is supposed to denote mathematical objects, like '2', 'the empty set', etc. If clause (*iii*) is accepted, and (*i*) and (*ii*) with it, one should conclude that all mathematical statements of the sort considered behave in the same way, namely they are true only if they are vacuously so. Things could be different for statements in which singular terms occur that are supposed to denote mathematical objects, statements like '2 is an even prime number', 'All prime numbers greater than 2 are odd', '2 + 3 = 5'. If we accept clause (*iii*) the truth of these statements will depend on the way in which we intend to treat statements with empty singular terms, terms like 'Pegasus' or '2' that do not denote any object. There are several options available for dealing with such statements. It should however be clear that, whatever these options are, clauses (*i*)–(*iii*) force us to conclude that these statements can be true only thanks to facts concerning their syntactic structure. For the sake of simplicity, also in this case we say that these statements are vacuously true.

Let us now consider the statements (*c*) '2 is a prime number'; (*d*) '2 is a composite number'. Let us concede again that their grammatical structure is preserved by their reformulations in the language of first-order predicate logic. The problem now is that different reformulations seem possible in this language. This holds in particular for the following: (*c'*) 'For all *x*, if *x* is 2, then *x* is a prime number'; (*d'*) 'For all *x*, if *x* is 2, then *x* is a composite number; (*c''*) 'There exist an *x* which is 2 and is a prime number'; (*d''*) 'There exist an *x* which is 2 and is a composite number'.

Like (*a*) and (*b*), the statements (*c*) and (*d*) seem to contradict each other, so that while (*a*) and (*c*) are usually accepted as consequences of the relevant accepted definitions, (*b*) and (*d*) are usually rejected as conflicting with those definitions. Nevertheless, our nominalist who accepts clauses (*i*)–(*iii*) is not only led to claim that (*a*) is not true, while (*b*) is, but also that (*c*) and (*d*) are both true if they are rendered as (*c''*) and (*d''*). Our nominalist will thus be forced to admit that the acceptance of mathematical statements has little to do with their truth, lest (s)he openly contradicts the most elementary mathematical practice. (S)he could not claim that the virtues, the applications and the utility of mathematics depend on the truth of its theorems. (S)he will rather have to claim the opposite, and will have to explain on what they depend instead. We will see in § 4.1 that this is exactly what Field does.

Fictionalism

Fourth and lastly, our nominalist philosopher might claim that even if we accept clause (*iii*) above, there still is a sense in which we can say that the theorems of mathematics are (non-vacuously) true: we could just reject one of the clauses (*i*) and (*ii*). More specifically, our nominalist could maintain that mathematical theorems are true within mathematics, just as we sometimes say that statements about the characters of a work of fiction are true within the story in which they participate. This view, to which we will return in § 4.2, is called 'fictionalism'. It is held, for instance, by Stephen Yablo (1998, 2001, 2005). Despite appearances, this position is not in contrast with the one presented above, so that even Field (1980; 1989a, *Introduction*) has argued for it.

Fictionalists reject the existence of abstract mathematical objects, but accept that it is possible to make up stories in which these objects are, as it were, characters. Mathematical theories, according to fictionalists, are just pieces of fiction that can nevertheless be highly structured and precise, and abide by definite rules which it is possible to distinguish between correct and incorrect claims. It would thus be perfectly legitimate to speak of the truth of the statements involved in these stories, once it is borne in mind that we are just speaking of truth within that story, and not of truth *tout court*. It makes perfect sense to say that it is true that Sherlock Holmes lives in London, while it is false to say that he lives in Glasgow, even though we do not expect to meet Sherlock Holmes when we visit London. In such contexts, when we speak of truth and falsity we would just be assuming that we are speaking of truth and falsity within the stories told in Conan Doyle's novels. In the same way, it is wholly legitimate to say that '5 + 7 = 12' is true and that '5 + 7 = 13' is false, even without conceding that the numbers 5, 7, 12, and 13 exist outside the story of arithmetic, once we assume that we are speaking of truth and falsity within arithmetic.

The Indispensability Argument

In Chapters 1–5 we will present and discuss several ways of stating and defending platonism, and we will show how platonists strive to give a plausible and precise description of the ontology it implies. We will devote Chapters 6 and 7 to the indispensability argument. This is an argument that has elicited, and as yet stimulates, a vast debate.

One of the reasons why it makes an especially powerful argument is that it is based on considerations that are perfectly available to a nominalist, such as the acknowledgement that certain mathematical theories are employed in the formulation of scientific theories that are true, or that we are justified in believing true. Despite its apparent simplicity, however, the argument hides a number of important assumptions, and involves a number of controversial philosophical theses. Also for this reason, although the most discussed versions of the argument do have some central ideas in common, they are often very different from each other. Before presenting the ongoing debate, we will thus try to clarify, even independently of the usual formulations, what the essential structure of the argument is, and how the notions that it apparently involves should be made precise.

One thing is clear, however: contrary to most of the arguments that we will see in Chapters 1–5, the indispensability argument seems, at least in some of its versions, to promise a justification of the existence of mathematical objects without requiring that the exact nature of these objects is clarified in advance.

This might be seen as a benefit. The argument would give the philosopher of mathematics a solid starting point for her theorizing: once it is established that there exist objects, and that some mathematical theories are about them, we could then proceed and investigate these theories in order to understand the nature of those objects and the sense in which we can speak of their existence. But this could also be seen as the symptom of a difficulty. How can we establish with certainty that certain objects exist if we have not previously made clear what their nature is and in what sense we can speak of their existence?

The difference between these two attitudes is a litmus paper for the centuries-old debate that, with different degrees of elaboration, we set out to reconstruct in our book.

1
The Origins

In a letter to Thomas J. Stieltjes, dated 13 May 1894 (Hermite and Stieltjes, 1905, vol. II, p. 398), Charles Hermite, one of the major analysts of his times, expresses his views about the subject-matter of his own study as follows:

> I believe that numbers and functions of analysis are not the arbitrary product of our mind; I think that they exist outside of ourselves, with the same character of necessity as the things of objective reality, and that we encounter, or discover them, and that we study them like physicists, chemists, zoologists, etc.

In his autobiographical book, *A Mathematician's Apology* (1940), Godfrey H. Hardy, possibly the most influential number theorist of the first half of the twentieth century, gives voice to a similar thought (p. 35):

> For me, and I suppose for most mathematicians, there is another reality, which I will call 'mathematical reality'; and there is no sort of agreement about the nature of mathematical reality among either mathematicians or philosophers . . . I believe that mathematical reality lies outside us, that our function is to discover or observe it, and that the theorems which we prove, and which we describe grandiloquently as our 'creations' are simply our notes of our observations.

It is hard to say whether this is actually the opinion of "most mathematicians". For sure, however, Hermite's and Hardy's point of view has been widely shared among mathematicians of all times. Still recently,

Alain Connes, Field medal in 1982, has declared it without reservation (Changeux and Connes, 1989, p. 28, English translation p. 12):

> Prime numbers . . ., as far as I'm concerned, constitute a more stable reality than the material reality that surrounds us. The working mathematician can be likened to an explorer who sets out to discover the world.

Shortly after the quotation given above, Hardy adds that the same "view has been held, in one form or another, by many philosophers of high reputation from Plato onwards" (1940, p. 35). This is true beyond doubt.

Starting with Plato himself, all those who have shared this view had to face a problem that is inextricably tied to it: if human beings possess the five senses in order to observe the material reality surrounding them, by what means can they observe mathematical reality? Were we able to solve this problem, known today as 'the problem of access', platonism in the philosophy of mathematics would probably be a trivial thesis. But this is not the case. For this reason, platonism, even though it is an ontological view, is always accompanied by an epistemic problem that we could generally formulate as follows: if mathematics is about abstract objects, how can we come to know that which it is about?

Let us begin by examining how Plato deals with it.

1.1 Plato as a Platonist?

In the *Theaetetus*, Plato asks: what is knowledge (*epistēmē*)? Eventually, he will not be content with any of the three answers he proposes, and will reject them all. At the end of the dialogue (210a–b), Socrates summarizes the situation thus: "neither perception . . . nor true opinion, nor reason or explanation [*logos*] combined with true opinion could be knowledge". Many have later endorsed the third proposal, Plato's rejection notwithstanding. According to the most common interpretation, this proposal identifies knowledge and justified true belief. Edmund Gettier (1963) advanced a well-known argument against this way of conceiving knowledge: a true belief might be justified for reasons that have nothing to do with what makes it true: if this were the case, we would hardly take it as knowledge. Plato's argument is not so different from this, *mutatis mutandis*, but it is more radical.

Knowledge is, for Plato, knowledge of something. This something is not a state of affairs, however, but an object: we do not come to know that things are so-and-so, we know things themselves. True opinion consists then, for him, in the identification of an object with what it really is. In terms of Plato's theory of ideas (which is not clearly present in the *Theaetetus*, however), this means that the object is subsumed under the idea of which it is a copy. If knowledge, says Plato (208a–209a), were justified true opinion, it could not be but true opinion together with the "interpretation" (*ermēneia*) of what identifies the relevant object, namely of the "the difference which distinguishes it from other things". But, in order to have knowledge, it would not be enough to have a true opinion about the fact that a given difference distinguishes a given object from others; we should also have knowledge of this difference, that is, we should have a justified true opinion about it. If this is so, the definition is circular or leads to an infinite regress.

In modern terms, Plato's point is the following: even if we were able to identify an object for what it is, we could not be certain that we succeeded in doing so; thus, we could lack the appropriate reasons needed for transforming our true opinion into knowledge. Plato seems then to concede that we can obtain true opinion, and still he denies that this may suffice for having knowledge. To this aim, we also need some ground assuring us that the opinion we have formed is true, and this ground cannot be had. We may reject this argument by claiming that the opinion that must be added in knowledge to true opinion need not provide the warrant Plato asks for: if we concede that knowledge can be either knowledge of something (non-propositional knowledge) or knowledge that something is in a certain way (propositional knowledge), we could indeed claim that one can have knowledge of *p* without having knowledge that one has knowledge of *p*. However, we would not have filled the hiatus Plato deplores, yet. If an opinion is true only insofar as it mirrors the reality of ideas (and the differences distinguishing them from each others), then, Plato seems to maintain, it is not possible to have epistemology without ontology; the former cannot stand alone. Hence the failure denounced in the *Theaetetus*.

Plato, however, also presents milder views compared to those in the *Theaetetus*. The *Meno* outlines a different position. After having acknowledged that "right opinion is just as useful as knowledge" (97c), Socrates must answer *Meno's* question: why, then, do we distinguish among them? And here is Socrates' reply (97e–98a):

> For . . . [true opinions], so long as they stay with us, are a fine possession, and effect all that is good; but they do not care to stay for long,

and run away out of the human soul, and thus are of no great value until one makes them fast with causal reasoning . . . But when once they are fastened, in the first place they turn into knowledge, and in the second, are abiding. And this is why knowledge is more prized than right opinion: the one transcends the other by its bonds.

There is more to it (85d–86a): true opinions are learnt in a time preceding this present earthly life; recollection awakens them, and when they are "awakened by questioning", they "become knowledge". Plato takes geometry by way of example: Socrates aptly questions one of *Meno*'s servants, until he (re)discovers how to double a square. Geometry stems from recollection, which brings up to the soul the true opinions that the soul had acquired before joining the body, and it does this by connecting them with each other, through a system of consequences that makes them abiding and turns them into knowledge. Thus, despite being an earthly matter, geometry manifests knowledge, or, if you wish, is knowledge. The hiatus denounced in the *Theaetetus* is then filled here by the system of premises and consequences binding true opinions together. This system is the justification that is added to true opinion and turns it into knowledge.

While in the *Theaetetus* justification is a criterion for acknowledging truth, in the *Meno* it is a means by which truth is fixed. The *Meno* thus envisages the possibility, denied in the *Theaetetus*, of distinguishing between epistemology and ontology: ontology is concerned with what true opinions are about, what makes them true; epistemology is concerned with the organization of true opinions into a system.

If we want to get clear about the nature of what makes an opinion true, however, we must leave the *Meno* and look elsewhere, at *Phaedrus* for instance. There (246a–249d) Plato describes souls as charioteers driving pairs of winged horses, who "traverse . . . the whole universe" along a "difficult and troublesome" path, up to where the gods are, and "when they reach the top, pass outside and take their place on the outer surface of the heaven", and they contemplate the hyperuranium (*hyperouranios*; literally, 'beyond heaven'). Plato describes what they contemplate as follows (247c–d):

> For the colorless, formless, and intangible truly existing essence, with which all true knowledge is concerned, holds this region and is visible only to the mind, the pilot of the soul. Now the divine intelligence, since it is nurtured on mind and pure knowledge, and the intelligence of every soul which is capable of receiving that which befits it, rejoices in seeing reality [*to on*] for a space of time and by

gazing upon truth is nourished and made happy until the revolution brings it again to the same place.

Those souls that have been capable of following the gods can thus contemplate the true reality during the time in which heaven completes one of its revolutions. They keep the memory of this until they fall back on Earth and join the body. This memory constitutes the true knowledge that only gods and souls that are not made heavy by the body can possess.

Meno's and *Phaedrus*' pictures seem here to come close to each other. Recollection awakens memories of the true reality, but it does not restore true knowledge. The memories it brings to the surface are just true opinions. Since it is now impossible to contemplate again the true reality, only the organization of these opinions in a system can make them abiding. A new sort of knowledge arises then, which is a copy or a trace of the true one. In this way, the pessimistic views in *Theaetetus* and the optimistic ones in *Meno* seem to come to a reconciliation. No earthly justification can restore true knowledge; this, like the reality of which it is contemplation, remains inaccessible to us: this is *Theaetetus*' pessimism. But the organization of true opinions in a system can make them abiding and give rise to a different sort of knowledge: this is *Meno*'s optimism.

We are now in a position to understand what Plato says about mathematics in the *Republic* and in the *Seventh Letter*, which are generally considered the main sources for his philosophy of mathematics.

Let us begin with the *Republic*, namely with the final part of the sixth book (509d–511e), when Socrates asks Glaucon to imagine a line divided into two unequal sections, standing respectively for the realm of the visible and that of the intelligible. For short, let us use symbols, even though Plato does not: let $\mathcal{V}'$ be the first section of this line and $\mathcal{J}$ the second. In the realm of the visible we find physical bodies, their reflected images and their shadows. Thus Socrates suggests that also $\mathcal{V}'$ gets divided into two more sections: one, $\mathcal{V}'_a$, stands for reflected images and shadows of physical bodies; the other, $\mathcal{V}'_b$, stands for these bodies themselves. Socrates adds that $\mathcal{V}'_a$ and $\mathcal{V}'_b$ are to each other in the same ratio as $\mathcal{V}'$ and $\mathcal{J}$, since the intelligible is to the visible as physical bodies are to their reflected images and shadows. No surprise so far. The metaphor's key is in Socrates' further suggestion: he proposes that $\mathcal{J}$ also gets divided according to the same ratio, and demands that its two sections do not any more stand for objects of knowledge, but rather for modes of knowledge.[1] The first section, $\mathcal{J}_a$, stands for a sort of knowledge that makes use of hypotheses. The second section, $\mathcal{J}_b$, stands for a sort of knowledge that makes no use of hypotheses since it directly appeals to ideas. Glaucon has a hard time in understanding all this. Socrates thus helps himself with an example, that of mathematics (510c–511a):

> For I think you are aware that students of geometry and logistic[2] and such subjects first postulate the odd and the even and the various figures and three kinds of angles and other things akin to these in each branch of science, regard them as known, and, treating them as hypotheses, do not deign to render any further account of them to themselves or others, taking it for granted that they are obvious to everybody. They take their start from these, and pursuing the inquiry from this point on consistently, conclude with that for the investigation of which they set out . . . they further make use of the visible forms and talk about them, though they are not thinking of them but of those things of which they are a likeness, pursuing their inquiry for the sake of the square as such and the diagonal as such, and not for the sake of the image of it which they draw . . . And so in all cases. The very things which they mould and draw, which have shadows and images of themselves in water, these things they treat in their turn as only images, but what they really seek is to get sight of those realities which can be seen only by the mind . . . This then is the class that I described as intelligible . . . but with the reservation first that the soul is compelled to employ hypotheses in the investigation of it, not proceeding to a first principle because of its inability to extricate itself from and rise above its hypotheses.

Thus, geometry and logistic pertain to hypothetical knowledge, which, yes, is made of true opinions, but opinions that are justified only by their place in an organized system of hypotheses, be they axioms, postulates, definitions, theorems or solutions of problems. Despite their insurmountable limitations, these disciplines allow speaking of the ultimate reality, of things in themselves, of a realm of ideal objects which we cannot access directly but that geometers represent by means of physical objects bearing to these ideal objects the same relation that reflected images or shadows of physical bodies have with these bodies.

Socrates then suggests (511d–e) associating the four sections of the line to four conditions of the soul: *eikasia* to V'_a, *pistis* to V'_b, *dianoia* to J_a, and *noēsis* to J_b. The correct translation of these terms has posed, and still poses, several interpretative issues; we then prefer to use the original Greek terms. Problems profligate when one considers what Plato writes about geometry a few lines earlier (511d): "And I think you call the mental habit of geometers and their like '*dianoia*' and not '*nous*' because you regard *dianoia* as something intermediate between *doxa* and *nous*."

The term '*doxa*' is what we have so far translated as 'opinion'. Apparently, '*doxa*' and '*pistis*' are almost synonyms or at least semantically close to each other. On the other hand, '*noēsis*' and '*nous*' share the same root word, and it is natural to take them as synonyms. The switch from a distinction among objects of knowledge to a distinction between modes of knowledge that is witnessed in the metaphor of the divided line is then mirrored in the newly introduced distinction: while *eikasia* (that might be rendered as 'conjecture') and *doxa* or *pistis* are conditions of the soul (mental states), *dianoia* and *nous* or *noēsis* are faculties. Even though there is no agreement on the translation of these terms, nor on the proper interpretation of the whole passage, there is enough consensus on the characterization of these faculties: *dianoia* is the faculty of articulating discursive reasons, or to produce arguments; *nous* is the faculty of contemplating ideas.[3] Plato is then telling us that the opinions of which geometry is made of are not only true (what we can only suppose, however, since we have no direct access to ideas), but also justified through the exercise of a special reasoning faculty, *dianoia*. He then adds that this faculty is intermediate between that which produces mere opinions and *nous*, which allows for the contemplation of ideas. Geometry, or mathematics more generally, is then the closest humans can get to true knowledge.

Still, it is not true knowledge. Plato makes this clearer in the seventh book (527a–b):

> This . . . will not be disputed by those who have even a slight acquaintance with geometry, that this science is in direct contradiction with the language employed in it by its adepts . . . Their language is ludicrous and forced, for they speak as if they were doing something and as if all their words were directed towards action. For all their talk is of squaring, applying and extending and the like, whereas in fact the real object of the entire study is pure knowledge . . . of that which always is, and not of a something which at some time comes into being and passes away.

Many have seen a thesis on the ontology of geometry in this passage: geometry would be about ideas or, in modern terms, abstract objects that are eternal, immutable, and existing independently of us. But the text is silent on the nature of these objects. Myles F. Burnyeat (1987) had rather interpreted it as the statement of a problem: that of accounting more precisely for their nature.

Differences in interpretations depend, at least partly, on the translation of *'mala geloiōs te kai anankaiōs'*, which we have rendered as 'ludicrous and forced'. Burnyeat (1987, p. 219) takes *'anankaiōs'* to point to the inevitable recourse to a practical language, and leans towards an interpretation of *'geloiōs'* according to which it indicates the metaphorical nature of this language. Plato would thus be saying that humans can talk about the true objects of geometry – eternal, immutable, and purely intelligible objects – only through the use of terms that apparently refer to other objects – temporary, changeable and sensible objects. But he would not be intending this as the final wold on this matter: rather, he would be posing the problem. Plato would then not be criticizing the language of geometry by contrasting it with its crystal-clear ontology. Quite to the contrary, he would consider this language as inevitable, like the constructive practices to which it refers, and would then be asking how we should properly conceive that ontology.

This interpretation has obvious advantages. First of all, it well accords with the distinction between two modes of knowledge: that of the true knowledge, inaccessible to humans, that we find in the *Theaetetus*, and that of earthly knowledge that we find in the *Meno*. This interpretation would also make Plato's philosophy of geometry compatible with the mathematical practice that a few year after Plato's death would be fixed in Euclid's *Elements*, a practice where not only diagrams are used and apparently treated as the real subject matter of geometrical discourse, but also where geometrical objects are treated not as pure forms (the straight line, the circle, the triangle, etc.) but as particular individuals (straight lines, circles, triangles, etc.) that have these forms, result from appropriate constructions, and are represented by particular diagrams.

The point at issue is that there is no clear ontology that this practice suggests in particular. Hence the problem raised by Plato. According to Burnyeat (1987, pp. 221–2), Plato would accept the following argument:

P.1) The theorems of mathematics (geometry, astronomy, etc.) are true.
P.2) They are not true of physical objects in the sensible world.
[*P*] --------------------------
P.3) They are true of ideal objects, distinct from sensible things.

Premise (*P*.1) would seem to be unassailable for Plato, according to whom it is beyond doubt that the opinions constituting earthly knowledge are true. Premise (*P*.2) would on the other hand hinge on the simple consideration that no physical object can verify the theorems of geometry. Conclusion (*P*.3) would thus be inevitable. Another argument would then follow, this too being accepted by Plato:

P'.1) Mathematicals exist.
P'.2) Mathematicals are different from physical objects in the sensible world.
[*P'*] --------------------------
P'.3) Mathematicals are ideal objects distinct from sensible things.

Conclusion (*P'*.3) would then be inevitable too. The problem raised by Plato is, then, that of clarifying the inevitable conclusions (*P*.3) and (*P'*.3).

The two foregoing arguments are framed in a context where a crucial role is played by the duality of two modes of knowledge and, as a consequence, of two aspects of geometry: on the one side, we have its practice, its constructive language, and the deductive system connecting hypotheses with each others; on the other side, we have the collection of its truths which, pure as they are, tell us about the hyperuranium. The first aspect pertains to human activity; the second to divine contemplation.

Arguments [*P*] and [*P'*] motivate this duality. A different, maybe even stronger, motivation is offered in a passage in the *Seventh Letter* (342a–d):

> Every existing object has three things which are the necessary means by which knowledge of that object is acquired; and the knowledge itself is a fourth thing; and as a fifth one must postulate the object itself which is cognizable and true. First of these comes the name; secondly the definition; thirdly the image; fourthly the knowledge . . . There is an object called 'circle', which has for its name the word we have just mentioned and, secondly, it has a definition, composed of names and verbs . . . And in the third place there is that object which is in course of being portrayed and obliterated, or of being shaped with a lathe, and falling into decay; but none of these affections is suffered by the circle itself . . . Fourth comes knowledge and intelligence and true opinion regarding these objects . . . And of those four intelligence approaches most nearly in kinship and similarity to the fifth, and the rest are further removed.

The presence of the fifth element, transcending geometrical practice, is here motivated by a request for unity, from the need to explain what makes the name, the definition, the image (or diagram) and the propositions of geometry have the strict connection with each other that we attribute to them. Plato's answer to this is simple: they all pertain to a single object, of which they enable us to speak.[4] On the other hand, the fourth element, be it conceived as knowledge, intelligence,[5] or true opinion, coincides with the earthly knowledge we find in the *Meno*, and thus concerns the first three elements. The truth of geometry cannot thus concern these latter elements, and rather lies in the agreement with the fifth one: the object itself, the idea.

Truth and unity: these two requirements for earthly knowledge and geometrical practice lead to postulate a higher realm and to raise the problem of its nature.

So as we have stated it, the problem seems only to concern geometry. But it also arises for the other branch of Greek mathematics: arithmetic. As regards this, Plato suggests a twofold distinction. First, he distinguishes – for instance in *Philebus* (56d–e) and *Theaetetus* (195e–196a) – between the arithmetic that most practise, concerned with unities differing among each others, like two armies or two oxes, and the arithmetic that philosophers practise, according to which unities do not differ among each others. This amounts to distinguishing between the art of counting particular objects and calculating with them on the one side, and pure arithmetic on the other side, concerned with numbers conceived as multitudes of indistinguishable units.[6] The second distinction concerns pure arithmetic. Plato presents this as follows in the *Gorgias* (451a–c):

> . . . Suppose some one asked me . . . : 'Socrates, what is the art of arithmetics?' I should tell him . . . that it is one of those which have their effect through speech. And suppose he went on to ask: 'With what is its speech concerned?' I should say: 'With the odd and the even, however many they are'. And if he asked again: 'What art is it that you call 'logistic'?' I should say that this also is one of those which achieve their whole effect by speech. And if he proceeded to ask: 'With what is it concerned?' I should say . . . that in most respects logistic is in the same case as arithmetic, for both are concerned with the same thing, the odd and the even; but that they differ to this extent, that logistic considers the odd and the even in their multitude not merely in themselves but in relation to each other.

Here Plato apparently distinguishes between (pure) arithmetic proper and (pure) logistic: if we take what he says at face value (according to what we take to be the most faithful translation) Plato seems to claim that the former is confined to the contemplation of numbers in themselves, considering them for their intrinsic property of being even or odd, whereas the latter studies numbers as different multitudes of unities (see the definition VII.2 of Euclid's *Elements*, according to which "A number is a multitude composed of units"), bearing certain relations to each other (both order relations, and additive and multiplicative relations). Another possible interpretation[7] would have it that, when he speaks of odd and even, Plato would intend to refer to numbers as they form a sequence where every even number is followed by an odd number, and would be claiming that arithmetic proper studies this very sequence independently of the fact that each number is, by itself, a given multitude of unities and not another; in modern terms, we would say that it considers natural numbers insofar as they form a progression,[8] whereas logistic studies those relations between numbers that depend on the operations that are defined on these numbers.

If the first interpretation were correct, the distinction in *Gorgias* would amount to the same distinction between the two aspects of geometry we have mentioned above. Arithmetic proper would express true, purely contemplative and humanly inaccessible knowledge; pure logistic (as distinct from calculation effected on particular objects) would express earthly knowledge, and would be what humans practise when they do arithmetic. The truth of the opinions that form this earthly knowledge would then depend on their agreement with the object being contemplated in true knowledge, namely the intrinsic properties of numbers. Their justification, on the other hand, would depend on a system of hypotheses concerning their relations, that would be advanced by considering multitudes of particular objects – even physical ones (for instance, appropriate signs engraved on wax or in clay) – these being pure images of the undistinguished unities that compose numbers as such. The argument [P] would then be easily adapted to the case of numbers: the theorems concerning them are true, but it is not because we come to prove them by arguments relying on the relations with which logistic is concerned that they are so; they are true of ideal objects, distinct from the particular multitudes considered by logistic. The need for unity advanced in the *Seventh Letter* would be present here too: numerical terms, definitions and the particular representations associated with the latter, so as the propositions of arithmetic could find their strict connection in their being about objects of another nature, transcending the intrinsic limitations of a human practice. This would be the same problem Plato raises for geometry: that of understanding the nature of these objects.

When it is generalized to the whole of mathematics, this is what we call 'Plato's problem'.

1.2 Aristotle Between Platonism and Anti-Platonism

Aristotle is not commonly listed among platonists in the philosophy of mathematics. Quite to the contrary, his views on the subject are often regarded as a form of anti-platonism or even nominalism. Nevertheless, they hinge on a proposed solution to Plato's problem. Moreover, this solution can even be viewed, to some extent, as compatible with platonism. This is why it is worth considering it in our discussion.

It must be noticed, to begin with, that Aristotle credits a solution to Plato's problem to Plato himself, though this solution cannot be found as clearly stated in the Platonic dialogues. In *Metaphysics*, I, 6, 987b 15–18, Aristotle claims that, according to Plato, mathematical objects occupy a position intermediate between physical objects and ideas.[9] As opposed to the former, they are "eternal and unchangeable". But of them, Aristotle says, "there are many alike", while the idea itself "is in each case unique". Aristotle seems to be saying that to each idea many intermediate objects correspond.

According to Julia Annas (1976, pp. 19–20), Aristotle is attributing this thesis to Plato as the latter's solution to a problem concerning arithmetic, which Annas calls "Uniqueness Problem". Take the statement '2 + 2 = 4', and let us ask what sort of truth it expresses. For Plato, it will obviously not be a truth about the physical world. Would it then be a truth about ideas? This is also doubtful, since every number would turn out to be, at best, a particular idea, and there would not be two 2s to be summed to each other. One might then have it that it is a truth about the intermediates between ideas and physical objects, and thus argue that there is only one idea of unit, only one of the 2, only one of the 3, and so on, but that to each of these ideas many intermediates correspond: many units, many 2s, many 3s, etc. This solution, however, seems to raise more serious problems than those it could help solving. Indeed, should things stand this way, then either the statement '2 + 2 = 4' is only about three particular intermediates – two particular 2s and a particular 4 –, but then it cannot hold in full generality; or else it is understood as a quantified statement, like 'For every pair of 2s and for every 4, 2 + 2 = 4', but this reading cannot plausibly be credited to Plato. One could argue, of course, that '2 + 2' does not refer to the sum of two 2s, but rather to the sum of 2 with itself. But in this case too, Aristotle's reconstruction of Plato's conception would fall short of being faithful, since it could plausibly be credited to Plato only if he were also credited with the view that the sole ideas to be relevant to arithmetic are those of unit and number, or at best those of the unit, the even and the odd, and that the plurality of different numbers only appears at the level of the intermediates, which is a view that Plato does not endorse, or at least not explicitly. This account applies more easily to geometry, instead, since there is no doubt that this deals, for instance, with a plurality of triangles, whereas it seems plausible to argue that, for Plato, the realm of ideas only includes one idea of triangle. As we shall see in § 1.3, Proclus will bring this conception into a coherent system.

But, be that as it may, it remains that, in *Metaphysics*, XIII–XIV, Aristotle raises several objections against what he takes to be Plato's view and, more generally, against the view that mathematics (be it arithmetic or geometry) is about ideas, or some other entities separate from sensible objects and somehow connected to ideas.

Before coming to the view Aristotle opposes to this, let us see how he formulates (*Metaphysics*, XIII, 1076a, 22–37) the problem which he is supposed to be answering, which is nothing but Plato's problem itself:

> ... We must consider first the objects of mathematics, not qualifying them by any other characteristic – not asking, for instance, whether they are in fact ideas or not, or whether they are the principles and substances of existing things or not, but only whether as objects of mathematics they exist or not, and if they exist, how they exist ... If the objects of mathematics exist, they must exist either in sensible objects ... or separate from sensible objects ... or if they exist in neither of these ways, either they do not exist, or they exist only in some special sense. So that the subject of our discussion will be not whether they exist but how they exist.

Aristotle lists four possible options: mathematical objects (*i*) exist in sensible objects; (*ii*) exist separate from sensible objects; (*iii*) do not exist; (*iv*) exit in some special sense. Scholars agree that (*ii*) is the option Aristotle ascribes to Plato, namely that it corresponds to the thesis that mathematical objects are intermediate between ideas and physical objects (as a consequence of this, Aristotle would not even consider the possibility that mathematical objects could be ideas, nor that mathematics could in the end be about these latter, not even via physical objects taken as their copies). Aristotle's attitude towards options (*i*), (*iii*) and (*iv*) is instead a matter of dispute.

Let us begin with the most cautious interpretation given by Annas (1976, pp. 26–30), according to whom Aristotle's way of dealing with this problem would not be any concession to Plato: despite the fact that Aristotle acknowledges different senses in which mathematical objects can exist, his conclusion would be that in no sense is it literally true that they exist. But he would not be explicit about whether there is any non-literal sense in which they can be plausibly said to exist. The third chapter in Book XIII, where Aristotle offers his own view about the existence of mathematical objects, would just be too vague and approximate, indeed. Aristotle would not say, then, whether option (*iv*) can be made precise enough so as to make it acceptable. Only the following would be clear: (*a*) that Aristotle rejects options (*i*) and (*ii*);[10] (*b*) that for him mathematical objects have something in common with time, place and infinity: they do not exist as such, but rather depend for their existence on the possible existence of physical objects. Aristotle would then be endorsing a reductionist attitude: just like what is said of time, for instance, can be understood only as referring, in the end, to temporal events,[11] so what is said of mathematical objects can only be understood as referring, in the end, to physical objects. In modern terms, this would mean that mathematical statements are not to be taken literally, and that, in the form they are given to us, they are nothing but abbreviated reformulations of longer and more complex statements. Singular terms that occur in them and appear to denote mathematical objects, would thus lack any reference as such, since they are only abbreviations being used in place of more complex expressions in which no such term occurs: and this is typically a nominalist point of view, as we saw in the Introduction.

This interpretation is of no help in explaining the relation that, according to Aristotle, would connect mathematics to physical objects. Both in *Metaphysics*, XIII, 3, and in *Physics*, II, 2, he argues that mathematics originates from a particular way of considering physical objects and their properties, and that this way of considering them is exactly what distinguishes it from physics. Its absence in those chapters notwithstanding, the term 'abstraction' is often used to indicate this particular way of considering physical objects. But how should it be understood? This is the problem with which other interpretations, less cautious than Annas', deal.

According to Burnyeat (1987, p. 222), acceptance of arguments [*P*] and [*P'*] (see § 1.1) is what discriminates between Plato's and Aristotle's views. The former would be trying to properly spell out conclusions (*P*.3) and (*P'*.3); the latter would be trying to resist these conclusions by rejecting premises (*P*.2) and (*P'*.2). According to Burnyeat, Aristotle does indeed accept premises (*P*.1) and (*P'*.1) – by ruling out option (*iii*) above, then – and in *Metaphysics*, XIII, 3 – he gives arguments for the claim that mathematical objects exist and theorems of mathematics are true of them, while arguing that mathematical objects are just physical objects differently considered. Lear (1982) also endorses this reading, and adds that geometry, for Aristotle, is about physical objects in so far as they have certain properties: its theorems would assert that insofar as physical objects have certain (spatial) properties, they also have other properties. Let *F* be one of these properties, for example the property of being a triangle. A theorem of geometry would be asserting, then, that a physical object that has the property *F* also has some other property *G* (for instance, the property of having interior angles adding up to two right angles).

There are at least two problems with this interpretation. First, in the passage quoted above, Aristotle does not even consider the option that mathematical objects are physical objects, but only the option that mathematical objects exist in sensible ones. One could argue that these options are equivalent, after all, or at least that the latter entails the former, if one takes Aristotle as claiming, in *Metaphysics*, XIII, 1076a, that mathematical objects cannot both exist in sensible objects and be distinct from these, and this implies that if they exist in sensible objects, they are not distinct from them. Still it seems quite clear that Aristotle intends these arguments to be against the thesis that mathematical object exist in sensible objects. Here is how he begins, indeed: "That it is impossible for mathematical objects to exist in sensible things, and at the same time that the doctrine in question is an artificial one, has been said already"(*Metaphysics*, 1076a, 37–40; Aristotle refers here to *Metaphysics* III, 2, 997a–998a). Second, it seems that the thesis that mathematics is about physical objects considered differently, or considered only insofar as they have certain properties, can be made precise only by admitting that it is about these very objects regardless of their having other properties. But this is subject to fatal objections, some of which can be made evident by arguments not much dissimilar from those employed by Aristotle. Stewart Shapiro (2000a, pp. 67–8) has recalled some of Frege's arguments, among which is the following (Frege, 1884, § 34):

> The properties which serve to distinguish things from one another are, when we are considering their number, immaterial and beside the point. That is why we want to keep them out of it. But we shall not succeed along the present lines [namely, by claiming that all units are identical to each other] . . . If, for example, in considering a white cat and a black cat, I disregard the properties which serve to distinguish them, then I get presumably the concept ⌜cat⌝. Even if I proceed to

bring them both under this concept and call them, I suppose, units, the whole still remains white just the same, and the black black . . . The concept ⌜cat⌝, no doubt, which we have arrived at by abstraction, no longer includes special characteristics of either, but of it, for just this reason, there is only one.

We might rephrase this argument as follows. If arithmetic concerns Frege's cats independently of some of their properties, it should be concerned with these cats for their remaining properties. But no property of cats seems to have anything to do with arithmetic. At best, arithmetic might be concerned with the collection of those cats only as regards some of the properties of this very collection, in particular as regards those properties that it shares with other suitable collections (namely pairs), such as that formed by a parrot and a tiger. But it would then be hard to claim without circularity that this collection is a physical object, since, for instance, these properties seem shared also by the collection formed by an acute triangle and an obtuse triangle. A similar difficulty seems to arise for geometry too, since it is hard to see which properties of a plot of land could be relevant to geometry, which seems rather concerned with properties that no physical object possesses, such as the property of being two-dimensional. Some might think, after all, that our plot of land has this property; should this be so, however, it would not be possible to dig a hole in it. It is the surface of the plot that is two-dimensional, and this is neatly distinct from the plot itself, just like Frege's two cats are neatly distinct from their collection.

This last remark suggests one more option which, according to the interpretation endorsed for example by Ian Mueller (1970) and Edward Hussey (1991), would be the option chosen by Aristotle. According to this interpretation, Aristotle would accept both premises (P.2) and (P'.2) and conclusions (P.3) and (P'.3), once properly understood. More in particular, he would be arguing that abstraction in mathematics does not merely consist in disregarding some properties of physical objects, nor in a different consideration of the latter, but is rather a process allowing us to access other objects, distinct from physical ones but not separate from them, to which we might only have access through physical objects and via this process (this entails, differently from what Plato held, that men can have access to mathematical objects, even though these are abstract objects).

If this were Aristotle's view, there would be a sense in which he would be a platonist: even though he denies that mathematical objects exist in a realm separate from, and independent of, the one we live in, he would be claiming that mathematics is about abstract objects and that its theorems are true of them. This conclusion contrasts with the commonly accepted opposition between Plato and Aristotle. One might then be willing to amend this conclusion opportunely, as a way of avoiding this result without falling back on the agnosticism prescribed by the first of the three preceding interpretations and without encountering the obstacles the second encounters. Richard Pettigrew (2009) has suggested one way of amending it, arguing that according to Aristotle the objects of geometry are physical objects that nevertheless only exist potentially, until appropriate geometrical procedures allow us to single them out as such.

It still remains to be explained how the objects under consideration can have the relevant geometrical properties and still be physical. Pettigrew's interpretation has nonetheless the merit of drawing our attention to a crucial aspect of Greek geometry, at least as it is presented in Euclid's *Elements*: its intrinsically modal nature. This geometry deals with objects falling under certain concepts that are described in general (the concept of triangle, for instance). These objects, however, are not given to us, *ipso facto*, by the stipulations fixing the relevant concepts, but can only be given to us through codified constructions (the so-called 'ruler and compass constructions'). In suitable circumstances these can be carried out, but still they are not realized just by the mere fact that these circumstances occur. Clearly, these procedures concern physical objects, such as diagrams drawn on suitable supports. Still, it is legitimate to think that they allow the derivation of consequences concerning abstract objects tied to those physical objects by a complex relation that only to some approximation could be conceived as a relation of representation.[12] But the problem remains that there is no direct textual evidence for ascribing this proposal to Aristotle.

1.3 Proclus: The Neoplatonic Interpretation of Euclid's Geometry

Proclus was one of the major figures in the neoplatonic tradition in late antiquity. He was born in Byzantium at the beginning of the fifth century AD, he studied in Alexandria, and moved to Athens when he was still very young, to the then renewed and flourishing Platonic school (the Academy, founded by Plato himself), where he spent the rest of his life. Among his works is a *Commentary on the First Book of Euclid's 'Elements'*. In the *Prologue*, he gives a reading of Plato's philosophy of mathematics that went on to become a classic.

This is based on Aristotle's interpretation, according to which mathematical objects are, for Plato, intermediate between ideas and physical objects. This is crystal clear in the *Commentary*'s opening sentence (Part 1, Chapter 1, 3, 2–8):

> Mathematical being necessarily belongs neither among the first nor among the last and least simple of the kinds of being, but occupies the middle ground between the partless realities – simple, incomposite, and indivisible – and divisible things characterized by every variety of composition and differentiation.

It should not come as a surprise that Platonic ideas are opposed to physical objects, insofar as they are indivisible and simple. Still, it is remarkable that, in pointing to the intermediate nature of mathematical objects, Proclus does not only insist on the fact that to each single idea a plurality of physical objects corresponds, but also emphasizes the opposition between the simplicity and indivisibility of ideas and the composite and divisible character of physical objects. This is due to a crucial thesis that is introduced anew by Proclus: like ideas, mathematical objects are abstract, eternal and immutable, but they differ from ideas not only in so far as to each single idea many of them correspond, but also because they can be composite (and thus divisible) and can have some sort of extension.

According to Proclus (*Prologue*, Part 1, Chapter 2), mathematical objects derive from the two principles generating all things from the first, indescribable and incomprehensible, cause, the One (viewed as a cosmological, absolutely first, principle): the Limit and the Unlimited. The thesis that everything is generated by the One is typically a Platonic one (think of Plotinus, who lived two centuries earlier than Proclus), but it is not what is most relevant here. More relevant is that, according to Proclus, mathematical objects do not just derive from the two principles generating all things from the One, but also from secondary principles, which are in turn derived from the former and generate all intermediate orders of things. In other words, Proclus allows for an indefinite generation of mathematical objects. This presupposes some generative faculty, and here lies the distinctive feature of Proclus' philosophy of mathematics.

Proclus (*Prologue*, Part 1, Chapter 5) comes back to Plato's distinction between *nous* and *dianoia*. In the latter he sees the faculty that is at play in mathematics: *dianoia*, he says (*ibid.*, Part 1, Chapter 5, 2, 4–9) "studies the images of the intelligibles", and produces a form of knowledge that is "dependent on . . . hypotheses", while *nous* "attains to the unhypothetical principle itself". But he goes beyond Plato and asks how *nous* and *dianoia* function, and where the mathematical objects studied by the latter come from. To this, he gives a twofold answer.

First, (*ibid.*, Prologue, Part 1, Chapter 6) he rejects the idea that these objects are derived from the physical ones by abstraction or generalization, and argues that they are generated from the soul, which derives them from itself and from *nous*. This is unclear. Proclus seems to be claiming that the soul has, as if they were engraved on it, traces of the ideas, but he also adds that *nous* is not only at the origin of these traces – since it lets the soul contemplate the ideas – but it also acts on the soul itself, for it lets the soul have access to these traces, and in some sense continues to engrave the soul itself. This continuous activity is what generates mathematical objects. The soul Proclus is speaking of is not the individual soul in each of us, but rather a sort of universal soul. Nevertheless, he does conceive of mathematical objects as being generated, although their existence is prior to the discourse made about them, which is made articulate by *dianoia*.

The second half of Proclus' answer is in the second part of the *Prologue*, devoted to geometry. His new argument (*ibid.*, Prologue, Part II, Chapter 2) seems then concerned with the latter only, and it is hard to apply it to arithmetic.[13]

Arithmetic is about numbers that cannot be divided infinitely, for they are pluralities of indivisible units; these numbers have no extension, despite their being quantities. To the contrary, geometrical magnitudes are extended and infinitely divisible quantities. But objects, even purely intelligible objects, can for Proclus have extension only "through the matter which is their receptacle" (*ibid.*, Part II, Chapter 1, 49, 27–50, 2). Therefore, geometrical magnitudes not only cannot be generated merely by the repeated occurrences of an idea (as seems to happen with numbers, which are generated by multiplying the occurrences of the idea of unit); they have also to possess some sort of matter. In order to account for these magnitudes, Proclus appeals to imagination (*phantasia*). Contrary to Aristotle, who introduced it (in *De anima*, III, 3) as a sort of intermediate faculty between thought[14] and sensibility,[15] Proclus does not see imagination as separate from *nous* and *dianoia*; he rather thinks of it as a tool of these faculties.

Its role hinges on another distinction introduced by Proclus: the distinction between ideas contemplated by *nous* on the one side and, on the other side, other purely intelligible objects, which he calls '*dianoēta*' (the objects of *dianoia*) or '*logoi*' (literally discourses). These are not only unextended, indivisible, and shapeless like ideas; they do not even happen to be many for each single idea. They rather seem to be ideal types, which nevertheless are not only contemplated, like ideas are, but also endowed with reasoning. *Logoi*, however, do not yet provide the objects of geometry. In order to obtain these, *logoi* must multiply and be complemented with extension and infinite divisibility, and thus quantity. It is here that imagination comes into play.

At this point, Proclus' text admits of different interpretations. According to the one we find more convincing, imagination is a sort of passive receptacle, a screen made by extended and yet purely intelligible matter – something we could identify as an ideal space – on which *logoi* are projected, each of them a multitude of times, so as to obtain different projections (*probolai*) which have extension, are infinitely divisible,[16] and can differ as to their dimension and, within certain limits, as to their form (this explains, for instance, why not all triangles are identical or even similar with each other). Projection itself, namely the activity that generates the objects of geometry from *logoi*, seems then to be performed by *nous*; the latter not only does contemplate ideas, it also has the power of generating entities. According to other interpretations, imagination is not purely passive; it is itself an activity that generates the projections of *logoi*. Leaving interpretative issues aside, what is clear is that the extended and infinitely divisible magnitudes with which geometry is concerned – straight lines, triangles, circles, etc., considered in their plurality – are for Proclus the imaginative projections of *logoi*, and that these latter, in turn, are distinct from the ideas *nous* contemplates.

One way of explaining the distinction between *logoi* and ideas is the following. According to Plato, theorems of geometry are true because of their agreement with a domain of inaccessible objects like ideas. Their truth is then independent of their justification. One might think of Proclus as willing to avoid a too sharp separation between the former and the latter. Insofar as no discursive activity, and so no justification, can grant access to the ideas, *dianoia* is kept separate from the contemplative activity of *nous*. But ideas duplicate, and thus generate *logoi*, which can be viewed as our concepts (remember that Platonic ideas are objects, however inaccessible they might be, and not concepts). Since the properties of these *logoi* depend on the nature of ideas, the truth of the theorems of geometry lies in their agreement with the ideas, but they are not directly about ideas, being rather about *logoi* and their various projections. The passage from *logoi* to ideas is left unexplained, but the final picture may seem more suitable than the radical dichotomy we find in Plato.

What Proclus tells us about *logoi*'s projections does not apply to arithmetic. All in all, however, his system of distinctions could be adapted to it. One would begin with the idea of number, or better, with those of unit, even and odd. Appropriate *logoi*, or concepts, would correspond to these ideas. In order to generate the various numbers from these *logoi*, *nous* should help itself with a different form of imagination: this latter would indeed be able to produce the repeated addition of different projections of the *logos* of unit, thus producing alternate projections of the odd and the even. *Dianoia* would be exerted on the *logoi* of unit, odd and even, and would thus lead to the definitions and theorems of arithmetic; these latter would thus be justified by those *logoi*, although they would be properly speaking of *logoi*'s projections and would be made true by their own agreement with the ideas.

In short, Proclus' philosophy of mathematics, to the extent that it might be of interest to our discussion, involves the following elements: (*a*) Platonic ideas (still conceived as inaccessible objects), among which are, for instance, the idea of triangle and unit; (*b*) *logoi*, that could be seen as the concepts associated to those ideas, like the concepts of triangle and unit; (*c*) *nous*, that lets us contemplate ideas, recalls them to the soul when this becomes embodied, and projects *logoi* on the extended (though immaterial) space of imagination, or on some other suitable receptacle, so as to generate appropriate and purely intelligible quantities, such as the different triangles or the different numbers; (*d*) *dianoia*, which is exerted on the *logoi* and lets us formulate definitions, postulates and theorems that concern projections of the same *logoi* and are also justified by the *logoi*, although they are made true by their agreement with the ideas. As regards both geometry and arithmetic, then, *dianoia*, *logoi*, their imaginative projections, and *nous* too – due to its activity of projecting *logoi* – are used for obtaining the theorems of mathematics and the earthly knowledge they express. While ideas, and *nous*, too, again – due to its contemplating ideas – are used for assigning truth to these theorems.

1.4 Kant: The Transcendental Interpretation of Classical Arithmetic and Geometry

Proclus lived eight centuries after Aristotle, Immanuel Kant thirteen centuries after Proclus. But the leap from Proclus to Kant should not come as a surprise to anyone just for those five more centuries it involves in contrast to that from Aristotle to Proclus. Ancient science and philosophy are often perceived as an homogeneous whole, whereas the passage from antiquity (though late antiquity) to Enlightenment is viewed as introducing deep transformations. As a matter of fact, both passages obliterate important distinctions within the complex story of which we are giving but a rough sketch. Nonetheless, there is an intrinsic reason justifying our choice of going directly from Proclus to Kant: Plato, Aristotle, Proclus and Kant were confronted with what was essentially the same sort of mathematics – that which Euclid's *Elements* codified – and their interpretation of it involves similar ingredients.

True, mathematics underwent significant changes in the meanwhile. Even though the *Elements* still contained the backbones of the mathematical edifice in the fifth century, relevant novelties had been introduced by Archimedes, Apollonius, Diophantus and Pappus. In the eighteenth century this backbone had been subject to profound transformations and developed in various directions, often at odds with both the spirit and the letter of the *Elements*. To give some examples, the developments of Arab and Renaissance algebra, the advent of Cartesian geometry, the birth of infinitesimal calculus, and the emergence of Newtonian mechanics, had all introduced deep changes even in the most elementary parts of mathematics, so much so that it was common to talk of a new mathematics, as opposed to the ancient one. Though underlying Kant's transcendental philosophy there is the attempt at explaining the advances of Newtonian mechanics in the understanding of the cosmos, nonetheless his philosophy of mathematics is still anchored to Euclidean mathematics. This is partly due to the way Newton wrote the *Principia* (1678) – the work in which he gave his own explanation of the cosmos – steering clear both of infinitesimal calculus and of Cartesian algebra, and rather relying on an (infinitesimal) extension of Apollonius' geometry. Another reason, however, is Kant's inability to capture the important novelties that the mathematics of his time had with respect to the mathematics of Euclid's time.

This inability goes together with an indisputable historical fact: the mathematical changes which occurred in the two millennia separating Kant from Euclid, and the many philosophical debates which they provoked, did not provide for any substantially new response to Plato's problem, nor for any argument to prove it ungrounded. In the midst of Enlightenment, the platonist solution as formulated by Proclus and the nominalist – or even empiricist – one suggested by Aristotle, were still the major options available.

Kant's philosophy of mathematics introduces something new: not only does it reformulate the problem in an essentially new way, it also connects it to other problems, at least partially new – like that of the logical form of a judgement – and gives an original response to it. This cannot be viewed as a platonist response, to be sure; but it is worth considering it, for some of its ingredients will remain central to the subsequent discussion, and are still today largely in use.

The best known thesis among those Kant held on mathematics is that mathematical judgements are synthetic a priori. We will confine ourselves to this thesis, and will present what we take to be the most convincing interpretation among the many, and often incompatible, available ones.[17]

It is all but easy to establish with some precision what Kant means by 'judgement' (see Martin, 2006; Anderson, 2008). Simplifying on the details, we can say that a judgement is for Kant the assertion (or the denial) that a certain relation among concepts holds, whose nature varies with the form of the judgement involved. For Kant, asserting that certain concepts stand in a certain relation to each other, amounts to representing a state of consciousness where these are given together, so that they are, in some sense, unified in consciousness. This representation presupposes some linguistic means, but concepts can be unified in consciousness by a faculty which acts on them prior to their linguistic representation. Kant calls it 'understanding'. Although this resembles Plato's and Proclus' *dianoia*, it must be kept in mind that concepts are not bound to higher-order objects (ideas), these being in turn contemplated by a different faculty like *nous*. Like Proclus' *logoi*, however, they are general: their function is to collect many contents of consciousness, or representations, under a single matrix. A concept is what lets us think of these contents by subsuming them under their common matrix; intuition is instead that which grants access to these contents. Concepts and intuitions are thus necessarily related to each other: knowledge is not possible if they do not occur together. But understanding can only provide for the concepts and operate with them; it cannot provide for intuitions, and must then be backed by some other faculty.

At this point, things get more complicated. According to Plato, *nous* and *dianoia* are backed by sensibility, which lets men access the earthly appearances of the ideas. Aristotle suggests that imagination is added as an intermediate faculty, while Proclus conceives it as a tool of *nous* and *dianoia* being at play in those cases in which sensibility is not involved. As we have seen, there is nothing like *nous* in Kant, and still sensibility and imagination preserve their role, however differently conceived. Sensibility provides concepts with objects as their content. An object is what acts on us to produce a change in our state of mind, or better, that to which we ascribe the function of acting on us to produce such a change; this we can conceive as the intuition of that object, and thus as the intuition corresponding to the concept under which the object falls. But an object, Kant says, can fall under a concept only if there is a third element correlating the two: some sort of instruction establishing that the former just falls under the latter. This element, called 'schema', is given by imagination.

The objects in question are empirical (or physical), for these are the only objects there are, according to Kant. But concepts under which objects are supposed to fall are not the only concepts there are; the concept of triangle, or number, and mathematical concepts in general, are not of this kind: no object falls, or is supposed to fall, under these concepts. This is not to deny that they have content, but still we need to understand what this content is. This is how Kant reformulates Plato's problem: as the problem of understanding the content of mathematical concepts.

Let us begin by considering Kant's conception of experience, which he does not merely conceive as our relation with a continuous flow of phenomena, but rather as the outcome of a complex construction, culminating in the transformation of these phenomena in a system of objects. Objects are then constructed: they are that to which we assign the function of acting on us to produce an intuition. Therefore, they do not pre-exist the intuition they produce; rather, they are correlates of this intuition. Since intuitions are, in turn, correlates of concepts, the objects producing certain intuitions are given together with the intuitions themselves, since the corresponding concepts are filled by these intuitions. Having an experience is precisely this: having certain concepts and filling them with suitable intuitions.

Now, the first distinction Kant introduces in the *Critique of Pure Reason* (*Introduction*, § 1) is that between *a posteriori* and *a priori* knowledge. The former can be acquired only through experience; the latter we can acquire independently of experience. What has been said so far on the way Kant conceives of experience suffices to understand that *a posteriori* knowledge requires *a priori* one, for it requires concepts and intuitions that it cannot provide by itself. Since we clearly have the former then, we must also have the latter.

Saying that there is *a priori* knowledge amounts, for Kant, to saying that there are *a priori* judgements. To see why this is so, consider that Kant's epistemology makes no room for opinion.[18] Articulating a judgement does not indeed consist in the free association of two concepts, nor to their hypothetical association. If there have to be problematic judgements for Kant – beside assertoric and apodictic ones – their characteristic feature is not to associate two concepts in a conjectural way, but rather to assert (in a non-conjectural way) the possibility of this association. Articulating a judgement is thus an act of unification of two concepts warranted by the nature of these concepts and/or by the corresponding intuition. This is to say that justification comes together with the articulation of the judgement. As a consequence, a judgement is *a posteriori* when in articulating it we need appeal to some experience, while it is *a priori* when in articulating it we need not do so.

According to Kant, there are two sorts of *a priori* judgements: analytic and synthetic ones. This distinction is introduced in the *Introduction* to the *Critique*, section IV (B 10/A 6–7), and is explicitly related only to judgements "in which the relation of a subject to the predicate is thought"; in other words, as it is often said, it is related to judgements in the subject-predicate form. According to Kant, "this relation is possible in two different ways": either the predicate is "contained" in the subject, or it is "outside" it, though it stands "in connection with it". In the former case, the judgement is analytic, in the latter it is synthetic.

Let S and P be two concepts under which objects are supposed to fall. If S is the subject and P is the predicate, a judgement in subject-predicate form tells us that every object falling under S also falls under P. Take, for instance, the judgements expressed by the following statements: 'All husbands are married', and 'All ravens are black'. The former judgement is analytic, the second synthetic. Why does not Kant say, then, that a subject-predicate judgement where S is the subject and P is the predicate is analytic if S is such that the domain of objects falling under it is necessarily contained in the domain of objects falling under P, and otherwise synthetic? A reason could be that he wishes to avoid modal notions like necessity. A different reason may be that he wants to emphasize the very reason why, in analytic judgements, the domain of objects falling under S is necessarily contained in the domain of objects falling under P (which depends of course on the mutual nature of S and P). But a third and more important reason is that such a reformulation in terms of domains of objects falling under S and P would apply only to the case in which S and P are concepts under which objects are supposed to fall, whereas the distinction between analytic and synthetic judgements must be general, and so must depend on a relation between concepts themselves, and not between the domains of objects possibly falling under them.

Kant's definition, however, has often been deemed inadequate, especially when it has been applied in mathematics. Many mathematical judgements do not seem to be of the subject-predicate form. Moreover, this definition seems to rely on a misguided account of concepts, according to which these are arranged in a fixed hierarchy, or at least result from each other by composition, in such a way that any composition of concepts (resulting in a new concept) occurs once and for all, determines the structure of this new concept and locates it in a fixed hierarchical relation to the former concepts. Another interpretation is however suggested by the examples Kant offers in the *Introduction*, section V, in order to illustrate the thesis that "All mathematical judgements, without exception, are synthetic" (B 14). In arguing that, since they are "admitted in mathematics", the judgements expressed by the statements 'The whole is equal to itself' and 'The whole is greater than its part' are synthetic, Kant notices that while the predicate occurring in them is contained necessarily in the respective subject, "it is so in virtue of an intuition which must be added to the concept [namely, to this subject], not as thought in the concept itself" (B 17). The point, obscure as it may seem, makes at least one thing clear: in establishing whether a judgement is analytic or synthetic, what is most important to Kant is whether its articulation requires appeal to intuition, and not that whether the predicate is or is not contained in the subject. This is even more clear when one recalls that according to Kant the articulation of every judgement comes with a justification: what Kant seems to be saying is that a judgement is analytic if its articulation does not depend on the content of the relevant concepts, or, in other words, if it is enough, when articulating it, to consider these concepts as such; while it is synthetic if its articulation requires taking their content into consideration, and thus calls for an intervening intuition. If this is the way things are, the restriction to the case of subject-predicate judgements becomes irrelevant: judgements of any form could be deemed 'analytic' or 'synthetic' according to this definition.

All *a posteriori* judgements are synthetic then, since articulating them requires an intervening intuition of the objects falling under the relevant concepts. Other judgements, such as the one expressed by the statement 'All husbands are married', merely require for their articulation that concepts are considered, independently of their content – or, in the present case, of the objects (supposedly) falling under them. These are clearly analytic *a priori* judgements. A judgement will then be synthetic *a priori* if articulating it requires an intervening intuition which is not intuition of objects, but rather what Kant calls 'pure intuition'. Here is how Kant defines it in the *Critique* (B 34/A 20 – B 35/ A 21), § 1 of the *Transcendental Aesthetic*):

> The pure form of sensible intuitions in general . . . must be found in the mind *a priori*. This pure form of sensibility may also itself be called 'pure intuition'. Thus, if I take away from the representation of a body that which the understanding thinks in regard to it, substance, force, divisibility, etc., and likewise what belongs to sensation, impenetrability, hardness, colour, etc., something still remains over from this empirical intuition, namely, extension and figure. These belong to pure intuition, which, even without any actual object of the senses or of sensation, exists in the mind *a priori* as a mere form of sensibility.

Pure intuition is thus the form of sensibility. It is what, in the intuition of an object, does not depend on its particular nature: what lets us represent an object as an object. Now, for Kant, the intuition of an object is possible only in so far this is located in space and time. Space and time are not to be conceived as mere empty boxes or receptacles existing independently of us and containing us; still they are not objects, nor concepts. Time is the form of inner sense: the frame in which objects are ordered relative to the occurrence of their intuition in us. Space is the form of outer sense: the frame in which objects are placed relative to the occurrence of their intuition as the intuition of what is external to us. Pure intuition is thus intuition of time and space, not in so far as time and space are what is intuited in it, but in so far as it is the form of the intuition of objects which is given by their being ordered in time and located in space.

Objects are not intuited, nor they can be, only in so far they are ordered in time and located in space in a certain way: they are not mere positions in time and space; they are something that is ordered in time and located in space, and the intuition of them must be the intuition of this something. However, they can only be thought of as being ordered in time and located in space in a certain particular way. And thinking of them just for their being so ordered and located gives pure concepts: concepts under which objects are not supposed to fall (since thinking of something ordered and located in a certain way just for its being so ordered and located is not thinking of this something, but merely of a way in which it is necessarily given to us), and whose content cannot thus be delivered by sensibility. These are the concepts of arithmetic and geometry. The former, the concept of the number 5 for instance, are given when we think the mere plurality of a collection of objects ordered in time, but think of no particular plurality of objects. The latter, the concept of triangle, for instance, are given when we think of the mere disposition of an object in space, but think of no particular object. These pure concepts allow us to think of particular objects (that cannot but be, indeed, ordered in time and located in space). Having these concepts, and articulating the corresponding judgements, are thus conditions of possibility of experience, and thus of *a posteriori* knowledge. These are necessary judgements then. Let us see how Kant accounts for them.

We said that no intuition of objects corresponds to mathematical concepts. However, to each of these concepts corresponds a schema with a twofold function (Friedman, 1992, pp. 122–9). First, it allows associating objects with concepts in such a way that those objects are images of these concepts, for they are ordered in time and located in space in the way the concept prescribes. Second, it determines a procedure, associated with the concept under consideration, such that it can be applied in suitable circumstances so as to obtain this image. Coming back to our examples, this image is given by any collection of five objects ordered in time, or by a triangle-shaped diagram drawn on paper, and the corresponding procedures will thus be that of enumeration or of construction of a figure.

By its first function, the schema associates a concept with a content given by the intuition of an object, the latter being a model of the concept even if it does not itself fall under the concept. For instance, it associates the fingers in one hand to the concept of the number 5, even though the hand's fingers, taken together, do not fall under this concept, but only give a model of it. By its second function, the schema gives content to a concept. This content is not an object; it is just the procedure the schema associates to the concept. Kant, however, believes that we do have an intuition of this content, which is given, however, by imagination rather than by sensibility. This intuition is thus *a priori* or pure. Pure intuition, therefore, is not just the form of sensibility: it can also be the intuition of a content.

We can now understand why mathematical judgements are synthetic *a priori*: because pure intuition intervenes in their articulation. This intuition has, in turn, a two-fold function: it warrants that what is proved *via* the consideration of suitable models (for example, the fingers on a hand for the result of a sum, or a diagram for a geometrical theorem) holds in full generality, since it relies on the correspondence between the model and the procedure being determined by the schema; and it also allows drawing a proof without considering any particular model, but rather through the mere consideration of the relevant schemas – in other words, in pure intuition.

One might think of schemas of mathematical concepts as abstract objects produced by imagination, analogous to the projections of Proclus' *logoi*; or of the objects being the models of mathematical concepts as the images involved in the hypothetical knowledge described by Plato in the *Republic* and in the *Seventh Letter*. Both interpretations would make a platonist out of Kant, although one reluctant to use a platonist language and to call the schema of mathematical concepts 'objects'. But this seems to miss the target: despite some analogies, there are indeed crucial differences keeping Kant apart from the platonist tradition. Schemas of mathematical concepts are universal instructions, not particular contents; they are not many for each single concept, like Proclus' projections are many for each single *logos*. There is no plausible sense, therefore, in which they can be seen as objects falling under the corresponding concept.[19] Moreover, concepts, as opposed to Platonic ideas, do not exist in their own realm. They are not transcendental objects to be copied by immanent objects; they are merely ways by which we can think of the objects in our world.

2
From Frege to Gödel (Through Hilbert)

Frege was born in 1848, only forty-four years after Kant's death. Still, Frege's philosophy of mathematics differs from Kant's much more than the latter differs from Plato's, Aristotle's and Proclus'. The main difference lies in Frege's attention to a profoundly renewed way of doing mathematics that had flourished during the nineteenth century, and to his intention to contribute to its foundation, if not to its development. It is often said that Frege was the founding father of analytic philosophy. This, however, does not explain such a concern with mathematics; it is rather a consequence of this concern.

Though nineteenth-century mathematics largely results from previous developments that Kant had disregarded, the state of mathematics in Frege's time was dramatically different from what Kant could have ever imagined.

2.1 Frege's Logicist Platonism

2.1.1 The Development of Analysis Between the Eighteenth and Nineteenth Centuries: The Historical Context of Frege's Programme

To Euclid – as well as to Kant – arithmetic and geometry were separate disciplines, concerned with different domains of objects. Cartesian geometry had already weakened this separation by defining on spatial magnitudes a formalism equivalent to the one applied to numbers. Not only did this make it possible to do geometry by solving and classifying equations, but also, and above all, it diverted attention from the study of numbers and magnitudes to the study of equations themselves. This, together with the rise of infinitesimal methods, gave rise to a new field of inquiry, conceived as more general and more fundamental than both geometry and arithmetic: the theory of functions, also known as analysis.

Euler and Lagrange, two contemporaries of Kant, were the main figures in this turn. Both viewed the theory of functions as the study of abstract quantities. These quantities are individuated only through the relations they bear to other quantities of the same sort, and are expressed by appropriate formulae expressing these very relations (thus, the formula 'x^2' expresses a quantity by expressing the relation that it bears to the quantity x). Functions are identified with these formulae and, via them, with the quantities they express. Arithmetic and geometry are conceived as possible applications of the theory of functions, and the notion of function is thus conceived as the ultimate foundation of mathematics.

This view proved very soon to be problematic. The main difficulty was that the formalism of functions, so understood, yielded entities that did not find a place in the formal universe that this view suggested. During the nineteenth century, mathematicians were involved in the process now known as 'arithmetization of analysis', initiated mainly by Gauss and Cauchy. Central to this process was a new way of conceiving of functions: they were no more taken as formulae or abstract entities, but rather as correspondence laws defined on certain numerical domains. Among the latter, the most important are those of real and complex numbers. Analysis, or theory of functions, then split into two distinct branches: real and complex analysis. But if functions are defined on numerical domains, they cannot provide a ground for mathematics. The need for defining real and complex numbers independently of the theory of functions emerged instead.[1] This need became pressing as long as more problems due to the lack of a clear definition of those numbers were found. These problems led, for instance, to the surreptitious identification of real numbers with points on a line, and real-valued functions with graphs. This produced various misunderstandings, and also subverted the foundational order that was established for mathematics in the previous century, where geometry was conceived as an application of analysis, or, at least, as a theory independent from it.

This is the framework in which Frege's philosophy of mathematics develops. Frege studied as a mathematician in Jena, where he was later to come back as professor, and in Göttingen, one of the most important mathematical centres worldwide. He devoted the main part of his career to the search of an appropriate definition for natural and real numbers. In the second half of the nineteenth century, the mathematical interest towards the definition of natural numbers is motivated by the view that they can and must serve as a basis for the definition of reals. Frege disagrees. In the second volume of his major work, the *Grundgesetze der Arithmetik* (1893–1903), he insists that the domain of real numbers should not be conceived and defined as an extension of the domain of naturals. In §§ 1 and 2 of the *Grundlagen der Arithmetik* (1884), he explains that the "movement in favour of all possible rigour" that was typical of contemporary mathematics should extend to the concept of natural number. The problem, he argues, is not so much to ascertain the "truth" of arithmetical statements, but rather to understand "the dependance of truths upon one another". The request of rigour runs thus parallel, for Frege, to the aim of understanding the proper organization of mathematics. This is for Frege, as it was for Plato, a system of truths. But these are made true, differently from Plato, not so much by their conforming to an inaccessible realm of transcendental objects, but by their dependence from some fundamental truths. The main problem for the philosophy of mathematics is to determine what these truths are.[2]

Among Frege's work are three main treatises (together with a number of shorter essays): the *Begriffsschrift* (1879), and the aforementioned *Grundlagen* and *Grundgesetze*. In the *Begriffsschrift*, he offers a formal system, which, in modern terms, appears as a system of second-order logic. In the *Grundlagen*, he offers a critique of several candidate definitions of natural numbers, replacing them with an alternative one, presented in informal terms. In the first volume of *Grundgesetze*, he updates and extends the system of *Begriffsschrift* so as to obtain a formalization of that definition, or, in other words, to obtain a formal version of arithmetic, to which we will refer as 'GA' (for '*Grundgesetze* Arithmetic').[3] In the second volume, he informally discusses how a definition of real numbers could be attained and begins to formalize it. A third volume was supposed to present the complete formal definition of reals, based on the same system that was used to define naturals. However, Frege's attempt came to a stop when Bertrand Russell discovered a contradiction undermining his formal system.

2.1.2 Frege's Formal System[4]

For us, as for Frege (though with differences that will be stressed below), a formal system is composed of a formal language, some deductive rules that apply to certain formulae of this language – called 'sentences [of that language]' – and those among these formulae that are deduced by means of these rules – called 'theorems [of that system]'.[5] A formal language is given by atomic symbols explicitly introduced and subject to explicit formation rules establishing which strings of these symbols compose a well-formed formula, and which well-formed formulae count as atomic ones and as sentences. A sentence of such a language is thus a string of symbols satisfying some rules. Frege's formal system is axiomatic:[6] some sentences – called 'axioms [of that system]' – are chosen as starting points in deductions, and theorems are obtained from them by the application of the deductive rules (for simplicity it is generally admitted that axioms themselves count as theorems).

Often, a formal system includes explicit definitions, namely stipulations establishing that certain combinations of symbols already introduced can be replaced by new atomic symbols, or by suitable combinations of atomic symbols where new ones occur. This both leads to a simplification of sentences, and allows focusing on some of the system's components. As we shall see, this is crucially the case for Frege's system, since his definition of natural numbers just results from an appropriate system of explicit definitions.

Today, the syntax of a formal system is neatly distinguished from its semantics: the former concerns the formation and deduction of sentences, the latter concerns their interpretation, namely the meaning of the different elements of the language. Semantic stipulations are different from syntactical ones. Hence, the same formal system can admit different interpretations, and it is only under a certain interpretation that the sentences of a formal system, or better the statements expressed by them, can be said in general to be true or false.[7]

Syntactic and semantic stipulations are tied together by the distinction of the atomic symbols of the language and its well-formed formulae into different categories (one of which is that of sentences): this distinction hinges on syntactic stipulations, and semantic ones must respect them.

Some atomic symbols are there to provide, as it were, the fixed fabric of the language. These are its connectives and quantifiers (possibly supplied with parenthesis to be used to separate symbols and formulae in appropriate ways). Other atomic symbols or well-formed formulae work as variables: some of them are supposed to vary on individuals, others (only occurring in second or higher-order languages) on properties, relations or functions possibly belonging to different types (properties of individuals being, for example, of a different type than properties of properties of individuals, properties of properties of properties of individuals, etc.). In axiomatic systems, the mutual syntactical behaviour of logical constants and variables is governed by a number of axioms, which count as logical, and fix a logical system, for example, a system of second-order logic.

To get a non-logical axiomatic system, one has to supply logical axioms with other axioms governing the syntactical behaviour of non-logical (or proper) constants. These are atomic symbols or well-formed formulae that are supposed to denote particular individuals or to design particular properties, relations and functions. Those of them that are supposed to denote particular individuals are usually called 'terms' (usually this label is also conferred to the individual variables). For short, beside variables and non-logical constants, also schematic (or dummy) symbols are often provided. Properly speaking, these are not linguistic symbols, but rather placeholders for linguistic symbols, namely for non-logical constants: a formula including schematic symbols stands indeed for all the formulae that can be obtained from it by replacing these symbols with appropriate non-logical constants. Using schematic symbols allows stating an infinite number of formulae at once.

Semantic stipulations have to respect the foregoing syntactical distinctions. Some of them are intended to fix the way truth and falsity transfer from statements of the relevant language to other such statements independently of any particular interpretation of its variables and non-logical constants. Because of these stipulations, some statements are true or false under every possible interpretation. These are called 'logical truths' and 'logical falsehoods', respectively (of course, a logical axiom is required to be a logical truth). Some other semantic stipulations are intended to fix the individuals, properties, relations and functions that the variables range over and the non-logical constants denote or designate. It is on the basis of these latter stipulations that the statements of the relevant language other than its logical truths and falsehoods can be said to be true or false.

This way of distinguishing and connecting logical and non-logical systems, on one side, and syntax and semantics on the other, originated partly from the problems that were found in Frege's formal system. In the latter, the situation is indeed quite different in both these respects. On one side, the distinction between what counts as logic and what counts as non-logic (namely between logical and non-logical constants and axioms) is drawn in a quite different way than it is today: Frege takes all of his formal system – or at least all of the system he appeals to, in the first volume of the *Grundgesetze*, in order to define the natural numbers, namely GA[8] – to be logical; for us, this is quite controversial, if not unacceptable.[9] On the other side, Frege's distinction between syntax and semantics is much less clear than ours, and it is even possible to doubt that his system admits of any such distinction. Of course, also for him proofs of theorems only result from the application of deductive rules, and are (and have to be) perfectly independent, as such, of the meaning of the symbols involved; and also for him some symbols can be used by leaving their meaning undetermined, at least to a certain extent, and are then subject to different interpretations. Still, his formal language is fixed by attributing a precise meaning to its symbols and formulae – what is done by openly referring to truth and falsehood – and both his axioms and his deductive rules are justified by appealing to this meaning. Moreover, the only symbols whose meaning is left undetermined are the schematic ones, and these are only used within the informal language that Frege appeals to in order to fix the formal one.

These are not however, the only differences between our way of thinking of a formal system and Frege's. His conception of the whole GA as a logical system goes together with the lack of any clear distinction between different sub-systems of it. Looking at it from our modern point of view, we distinguish in it a system of propositional calculus, an extension of it yielding a system of first-order logic, a further extension yielding a system of second-order logic, and finally the whole system GA achieved by a new extension due to the introduction of two axioms concerned with extensions of concepts. No trace of these distinctions can be found in Frege's exposition. The system is directly presented as a whole by appealing to a single fundamental distinction that is taken for granted (though informally elucidated, as far as possible): the distinction between objects and functions.

The notions of object and function are primitive, for Frege. This is reminiscent of Euler's and Lagrange's attitude of considering the latter notion to be the fundamental one for mathematics. As a consequence, functions cannot be conceived by Frege as laws defined on numerical domains. But he no more understands them as formulae or quantities expressed by formulae. There are three reasons for this: (*a*) the notion of a function is, for Frege, prior to the formal language, it is indeed used in fixing the latter; (*b*) it is also prior to the notion of quantity, however, the latter is understood; (*c*) formulae and quantities are objects, while Frege's philosophy is founded on the distinction between objects and functions.

Frege conceives of objects and functions as 'saturated' and 'unsaturated' entities, respectively. Taking a function to be unsaturated has the effect of thinking of it as a schema with empty places to be filled with objects or other functions, these being the function's arguments. Saturating a function consists in filling these places so as to get a value of the function itself, which, insofar it is saturated, is necessarily an object.

This is a metaphorical explanation, but it can easily be rendered in a more precise way. A function with empty places only for objects is a first-level function; a function with empty places only for first-level functions is a second-level function; and so on. There are also functions with empty places for both objects and functions, or for both function of some level and functions of different levels. These functions are unequal-levelled. Let us consider the simplest cases. Let us use 'ξ' and 'ζ' to indicate empty places for objects; 'φ' to indicate an empty place for first-level functions; 'a' and 'b' to denote objects; 'f' and 'g' to designate functions. Then, '$f(\xi)$', '$f(\xi,\zeta)$', '$\varphi(a)$' and '$f(\xi,\varphi(b))$' designate, respectively: a one-place first-level function, a two-place first-level function, a one-place second-level function, and an unequal-levelled function, while '$f(a)$', '$f(a,b)$', '$g(a)$' and '$f(a,g(b))$' denote the values of these functions, which are objects, as already said.

But, one might ask, which objects and functions can be arguments of a given function? If we want to place any constraint on this respect, we should define some restricted domains of objects and functions; but if we are to do this, we must already possess some suitable resources independently of the availability of the notion of function. For instance, if we want to restrict the possible arguments of a function to real numbers, we should be able to define real numbers without using the notion of function. But this is just what Frege is neither able nor willing to do: according to him, the notion of function must be primitive, and should not depend on anything else. He then advances two rather bold claims. First, he claims that all objects and all functions of some appropriate level can be arguments of all functions with suitable empty places. This is tantamount to supposing that functions are defined on the totality of objects and/or functions of the suitable levels. As we now know, without appropriate restrictions and provisos, the notion of totality of objects or functions of a certain level leads to contradictions. Frege was not aware of this, and appealed freely to the two notions. Second, he claims for the existence of two very peculiar objects: the True and the False, which we will denote with 'T' and 'F' respectively.

This enables him to explain in a wholly new way the notion of a concept: a concept of level n is a one-place function of level n that can take only the objects T and F as values. Let us consider the concept horse: it is a one-place first-level function taking value T when its argument is a horse, and F when its argument is any other object. If we can tell horses from other objects, we know how to assign values to this function, and thus possess the concept horse. If we also suppose that we know how to tell any particular horse from all other horses, this concept is *ipso facto* sortal (see Introduction, pp. 4–6). Beside concepts we have relations: n-place functions that can only have the objects T and F as values.

In order to define objects of a given sort, it will be enough that suitable concepts are defined in such a way that it can be established which among them take the value T. This is how Frege proceeds with the natural numbers: if, once we have defined, or established how to define the suitable concepts – those of the individual natural numbers and that of natural numbers – we can establish that there are objects falling under them, then natural numbers exist, and are *ipso facto* objects. In addition, if the relevant concepts have been defined in the appropriate way, or an appropriate way for defining them has been established, they will satisfy all the theorems of arithmetic, and natural numbers will be described by these theorems. The definition of the suitable concepts and the availability of reasons for claiming that objects (whose nature will depend on these very concepts) fall under them, will then provide an argument for arithmetical platonism. This is just Frege's argument.

The two bold claims mentioned above enable him to introduce the fundamental functions upon which his logical system is based. Simplifying the details, these functions can be boiled down to seven.

Five of them correspond, in modern terms, to as many logical constants still in use, though with different notation and underlying conception. Let us consider them first; the others will be introduced in § 2.1.3.[10] They are negation, implication, identity and the two universal quantifiers of first and second order. The negation $\neg\xi$ is the first-level concept taking F as value for T as argument, and T as value for any other argument. The implication $\xi \Rightarrow \zeta$ is the first-level relation taking value F if the first argument is T and the second is not T, and value T otherwise. The identity $\xi = \zeta$ is the first-level relation taking T as value if its two arguments are the same object, and value F otherwise. The first-order quantifier $\forall x[\varphi(x)]$ is the second-level concept taking V as value if the argument is a first-level function $f(\xi)$ such that its value $f(a)$ is T whatever a is, and value F otherwise. The second-order quantifier $\forall P[\mu(P)]$ is the third-level concept taking T as value if the argument whose place is marked by 'μ' is a second-level function $f(\varphi)$ such that its value $f(g)$ is T whatever the first-level function g is, and F otherwise.[11]

Frege's conception of these logical constants as functions, namely concepts and relations, goes together with his conceiving the values of these functions and of those resulting by their compositions as truth-values, namely of the objects T and F, and the formulas denoting these values as names of these objects. In Frege's formal system, then, truth-values are not attributed to statements. They are rather what appropriate components of statements denote. Indeed, a statement is formed by Frege by attaching assertoric force to appropriate names of truth-values – namely to names of values of concepts and relations. This is effected by premising these names by a particular symbol, '⊢', which transforms these names into statements asserting that they are names of T. Each of these statements is then said to be true if and only if this name is actually a name of T.[12] The symbol '⊢' thus indicates both that we are in the presence of a statement and that an act of assertion is being made. The subject performing the assertion is left unspecified, or is generally identified with someone who possibly accepts the relevant theory.[13]

This enables Frege to give a new explanation to the Kantian notion of judgement. He distinguishes between the reference and the sense of a name: the former is the object of which it is a name, the latter is the way in which the object is introduced (and possibly described) through that name. To make an example, then, 'T', '$a \Rightarrow (b \Rightarrow a)$' e '$\forall x f(x) \Rightarrow f(a)$' all have the same reference, since by the definitions given above both the second and the third are names of T whatever the objects a and b and the function $f(\xi)$ are. But they have different senses. The sense of a name of a truth-value is for Frege a thought. And a judgement is "the acknowledgement of the truth of a *thought*" (Frege, 1893–1903, § I.5), namely the acknowledgment that a certain name (other than 'T') is a name of T, and that it can be turned into a true statement by prefixing it with '⊢'.

Using only the five fundamental functions presented so far and composing their values appropriately, we can write statements that are always true: statements in which names for indeterminate objects or functions occur, but that are true whatever these objects and functions are,[14] like the statements '⊢$a \Rightarrow (b \Rightarrow a)$' and '⊢$\forall x f(x) \Rightarrow f(a)$', that are true whatever the objects a and b and the function $f(\xi)$ are. Insofar as in these statements occur only the five functions described above and these functions correspond in modern terms to logical constants, these statements count, in modern terms, as logical truths.[15] This is so also for Frege, though, as anticipated above, he also considers as logical truths other statements in which his two other fundamental functions occur too.

Among the seven axioms of GA, five are statements that only involve the five fundamental functions introduced so far, and count then as logical truths for us too.[16] In order to deduce theorems from his axioms, Frege appeals to three sorts of deductive rules. Substitution rules and transformation rules license the passage from a statement '|—$\mathcal{A}$' to a statement '|—$\mathcal{B}$' such that '$\mathcal{B}$' is, come what may, a name of the same truth-value of which '$\mathcal{A}$' is a name. Inference rules license, instead, the passage from two statements '|—$\mathcal{A}$' and '|—$\mathcal{B}$' to a third statement '|—C' such that if '$\mathcal{A}$' and '$\mathcal{B}$' are names of T, then 'C' is too. It follows that all the theorems that are derived from the five axioms just mentioned count as logical truths for us too.

According to Frege, some of these theorems can be rewritten using new functions introduced from the fundamental ones by explicit definitions. His substitution rules and the conditions under which he believes that it is possible to introduce new functions by explicit definitions, play in his system the same role that we assign to comprehension axioms-schema: they establish that for any name of function that can be written in the theory, there is a function with that name. Hence the system formed by the five axioms mentioned above, the preceding deductive rules and the (often tacit) stipulations regulating the introduction of new functions by explicit definitions, corresponds, in modern terms, to a system of second-order logic with full comprehension.

2.1.3 The Definitions of the Natural Numbers

To define natural numbers, Frege needs more than that. He needs to appeal to his two other fundamental functions, too. To better understand the way he proceeds, it is appropriate to come back to the *Grundlagen*.

In the *Introduction* of this latter work, Frege states three guiding principles of his inquiry. The first and the third are quite well illustrated by his conceptions expounded so far: "always . . . separate sharply the psychological from the logical, the subjective from the objective"; never . . . lose sight of the distinction between concept and object". The second one (also discussed in §§ 60 and 62) is known today as 'context principle': "never . . . ask for the meaning of a word in isolation, but only in the context of a statement" in which it occurs. It is a key principle for Frege's definition of natural numbers.

Clearly, it is not Frege's intention to suggest that the meaning of words changes from statement to statement. Rather, his claim is that this meaning is fixed by the role words have in the context of appropriate statements. When he speaks of meaning (*Bedeutung*), Frege intends to speak of something objective, to be distinguished, by the first among the three principles above, from our representations and intentions. But what are the appropriate statements supposed to be? Certainly, statements like '4 + 3 = 7' are among these, as well as the theorems of arithmetic in general. But there are also statements like 'The King's carriage is drawn by four horses'. It is just because '4' in the first statement has the same meaning of 'four' in the second, that theorems of arithmetic can be applied in everyday life.

Let us consider the second statement first. Frege tells us (§ 46) that when we assert it, we assert of the concept ⌜(being) a horse that draws the King's carriage⌝ that the number four belongs to it. More generally, Frege maintains that every "attribution of natural numbers" (*Zahlangabe*) is an assertion (*Aussage*) about a concept, and should be understood as the assertion that a certain number belongs to a certain concept (in connection to this thesis, Frege, 1919, p. 273, English translation p. 253, refers to Plato's *Hippias Major*; he is likely to have in mind the arguments in 299b–303b; see, in particular, 301d–302b).

This suggests that it should be possible to explain the meaning of the assertion that n is a natural number by picking out a concept to which n belongs: something will be a natural number if and only if it belongs to some concept (in the sense explained above). Frege justifies this idea by showing that if it is appropriately applied, and if natural numbers are defined in accordance with it – each of them as the number belonging to an appropriate concept – then statements like '4 + 3 =7' and all the theorems of arithmetic could be proved by relying on "laws of logic" and suitable explicit definitions alone. Let us see how Frege develops this thought in §§ 55–86.

Whatever exact meaning is attached to the verb 'to belong', it is natural to claim that the number belonging to a concept F is the same that belongs to a concept G if and only if as many objects fall under F as fall under G. There is no circularity here, since it is possible to establish that as many objects fall under F as fall under G without appealing to numbers: all there is to do is to establish that the objects falling under F and the objects falling under G are in one-one correspondence, namely that to every object falling under F one and only one object falling under G corresponds, in such a way that to every object falling under G corresponds one and only one object falling under F. If there is one-one correspondence between the objects falling under the concepts F and G, we say that F and G are equinumerous. We then have the following principle, which is known today as 'Hume's Principle' (henceforth, HP), since Frege introduces it by quoting (in § 63) a passage from Hume's *Treatise of Human Nature* (Book I, Part III, Section I: "When two numbers are so combin'd, as that the one has always an unite answering to every unite of the other, we pronounce them equal"):

$Num.(F) = Num.(G)$ iff F and G are equinumerous,

where F and G are any two concepts, and for every concept X, 'Num.(X)' denotes the number belonging to X.

Let us now stipulate that 0 is the number belonging to the concept ⌜(being) non-identical [*ungleich*] to itself⌝. If we take this concept as well-defined, and accept this definition, we have defined the number 0. We can then consider the concept ⌜(being) identical to 0⌝, under which only 0 falls, and stipulate that 1 is the number belonging to this concept. This would then be a concept under which only one object falls, namely 0, and to which 1 belongs. Moreover, the concept ⌜falling under the concept ⌜(being) identical to 0⌝, but being non-identical to 0⌝ and the concept ⌜(being) non-identical to itself⌝ are equinumerous. Thus, according to HP, 0 belongs to the former, as to the latter. This suggests a way of defining the relation which holds between a number and its successor: n is the successor of m in the succession of natural numbers if there are a concept X and an object x falling under X, such that n belongs to X and m belongs to the concept ⌜falling under X, but (being) non-identical to x⌝. If we now stipulate that 2 is the number belonging to the concept ⌜(being) identical to 0 or to 1⌝, we can conclude that 2 is the successor of 1 in the succession of natural numbers; we can then go on in similar ways, and define all natural numbers.

In order to define the concept ⌜(being) a natural number⌝, a more laborious logical machinery is needed, relying on what is known today as the ancestral relation of a given relation. Suffice here to say that this allows us to define the concept ⌜(being) a natural number⌝ as the concept under which all and only those objects occurring in the succession starting with 0 and then progressing with 0's successor, the successor of this, and so on.

There is more to it, however. For Frege offers the foregoing definitions only after having explained what it means that a certain number belongs to a certain concept. This might seem superfluous, and some think it is indeed. Neo-logicists (see § 5.1), for instance, hold that HP can be taken as an axiom, or as an implicit definition of the operator 'Num.(–)' which, when applied to concepts, gives objects, cardinal numbers in particular.[17] But it is not superfluous for Frege. The reason might strike one as odd, still it concerns a crucial philosophical point. Should we admit HP as a mere stipulation, and take it to suffice for the definition of the natural numbers, we would not be able to know whether Julius Caesar or England are or are not natural numbers. This is known today as the 'Caesar problem'.

There are two faces to this problem. The first and more obvious is this: HP delivers a condition of identity for numbers belonging to concepts, it lays down conditions for when the same number belongs to two distinct concepts; but it is says nothing to explain what a number belonging to a concept is. This situation can be generalized: suppose we have identity conditions for the objects falling under a sortal concept; these conditions tell us when an object a that falls under this concept is the same as an object b that also falls under this concept, but they say nothing to explain what makes an object fall under this concept. These conditions, then, do not seem to suffice for defining our sortal concept, or the objects falling under it. There is no doubt this is so for concepts of concrete objects. We certainly could not define stars by laying down only conditions for when two astronomical observations are observations of the same star. Should we think this possible, we would have probably been misled by the fact that we already know what a star is. But things might stand differently with abstract objects: for one might argue that abstract objects are not substances defined by intrinsic properties, but mere correlates of conditions of identity. The second face of the Caesar problem, the most important for Frege, is made manifest here. If we allow that a sortal concept under which abstract objects are supposed to fall can be defined through the mere stipulation of identity conditions, what could warrant that these objects exist, namely that under that concept actually fall any objects at all? Nothing (apart from contradictions) could prevent us from talking of objects falling under the relevant concept, making as if they existed; but lacking further arguments, this would just be a *façon de parler*, or a mere game with concepts or, even worse, with meaningless linguistic constructions. The solution to the Caesar problem, thus, has a deep significance for Frege's philosophy of arithmetic: according to him, it is just what makes it a platonist philosophy.

The solution is offered in § 68, and relies on a notion Frege borrows from the traditional logical vocabulary: the notion of extension (*Umfang*) of a concept. Kant abundantly employs this notion in the *Critique of Pure Reason*, and in the *Vienna Logic* he defines it as a "*sphaera*", constituted by a "multitude of things that are contained under the concept" (WL, p. 911; LL, p. 354).[18] Quite surprisingly, Frege elaborates no further this notion, and in a footnote to § 68, he even acknowledges that he is assuming "that it is known what the extension of a concept is". This enables him to assert that:

> the natural number which belongs to a concept F is the extension of the second-level concept ⌜(being) a concept equinumerous with F⌝

If we confront this claim with the preceding definitions, we straightforwardly obtain that:

> 0 is the extension of the concept ⌜(being a) concept equinumerous with the concept ⌜(being) non-identical to itself⌝⌝;
> 1 is the extension of the concept ⌜(being a) concept equinumerous with the concept ⌜(being) identical to 0⌝⌝;
> 2 is the extension of the concept ⌜(being a) concept equinumerous with the concept ⌜(being) either identical to 0 or to 1⌝⌝;
> and so on . . .

Similarly, we have that:

> n follows m in the succession of natural numbers if there are a concept X and an object x falling under X, such that n is the extension of the concept ⌜(being a) concept equinumerous with the concept X⌝ and m is the extension of the concept ⌜falling under X, but being not identical to x⌝.

HP then takes the following form:

> The extension of the concept ⌜(being a) concept equinumerous with the concept F⌝ is identical to the extension of the concept ⌜(being a) concept equinumerous with the concept G⌝ if and only if F and G are equinumerous.

If the extension of a concept is defined as Kant suggested, this biconditional can easily be proved to hold, as Frege does in § 73.

This allows solving the Caesar problem. If the second-level concepts that occur in the definitions of natural numbers are well-defined, then they have an extension. Their extensions will be distinct both from Julius Caesar and from England, which clearly are not extensions of second-level concepts. More precisely, says Frege (§ 89), these are objects "which cannot be given to us in sensation". Kant was thus mistaken in thinking that "without sensibility no object would be given to us", and also in holding that mathematical judgements are synthetic, at least as far as arithmetic is concerned. In so far as they follow from the foregoing definitions, the "laws of arithmetic" are indeed analytic (§ 87), at least if Kant's definition of analyticity is opportunely amend by taking a "truth" to be analytic if in giving a proof of it "we come only on general logical laws and on definitions" (§ 3).[19]

This is a quite delicate and crucial point. Frege does not hesitate in thinking that his stipulations, appealing to the Kantian notion of extension of a concept, are logical laws or, at least, innocent explicit definitions. Moreover, he also admits that the concepts formed by their means have extensions, and that these extensions are abstract objects, logical objects in particular, that the theorems of arithmetic describe. But what makes the notion of extension of a concept suitable for being part of logic, as opposed to other Kantian notions like that of a schema of a concept, for example? And why should this notion point to objects, rather than to mere representations? The lack of a clear answer to these questions clearly represents a weakness of the *Grundlagen*. And this, beyond the need of arriving at a formal and more precise definition, seems to be the main reason motivating the new definition offered in the *Grundgesetze*.

Let us come back to this work, then, starting from §§ I.3, I.10, I.11, I.18 and I.20, where Frege introduces his last two fundamental functions[20] and the corresponding axioms. His evident aim is that of giving a new explanation of the notion of extension of a concept – at least of a first-level concept – so to include it in his system. Frege explicitly says (§ I.3) that, granted that a (first-level) concept is nothing but a (first-level) function of a particular kind, its extension is the value-range of this function. Then (§ I.10) he defines the value-range of a first-level function (and then, *a fortiori*, the extension of a first-level concept) as the value taken by another appropriate second-level function for the former function as its argument, and takes this second-level function to be his sixth fundamental function. To designate it, he introduces the new primitive symbol '$\grave{\epsilon}[\varphi(\epsilon)]$', where '$\varphi$' is a placeholder for first-level functions, so that '$\grave{\epsilon}[f(\epsilon)]$' will stand for the value of that function for the function $f(\xi)$ as argument, namely for the value-range of this latter function. No surprise so far. The surprise comes when, rather than introducing the function $\grave{\epsilon}[\varphi(\epsilon)]$ through a stipulation fixing its values for every possible argument, Frege introduces it by laying down its identity conditions, that is by a stipulation with the same logical form as HP:

$$\vdash (\grave{\epsilon}[f(\epsilon)] = \grave{\epsilon}[g(\epsilon)]) = \forall x[f(x) = g(x)].$$

This is the sixth axiom in Frege's formal system, known today as Basic Law V (henceforth, BLV), due to the roman numeral 'V' Frege assigns to it.[21] It stipulates that two functions $f(\xi)$ and $g(\xi)$ have the same value-range if and only if they take the same values for the same arguments.[22] Moreover, since identity is defined as a first-level function, the occurrence of this relation in the first member of the law entails that value-ranges cannot but be objects.

We will later see the problems raised by BLV. Let us now suppose that it is in good standing and suffices to introduce the fundamental function $\dot\varepsilon[\varphi(\varepsilon)]$. It is now easy to introduce the last among Frege's fundamental functions. This is the first-level function denoted by the symbol '\$\xi$', taking the value a, whatever this object is, if its argument is the extension $\dot\varepsilon[a = \varepsilon]$ of the concept $a = \xi$ (the concept ⌜(being) identical to a⌝), and taking the same value of its argument otherwise. Let $C(\xi)$ be a concept under which the sole object a falls. By BLV, its extension is the same as that of the concept $a = \xi$, thus \$\dot\varepsilon[C(\xi)] = a$. It follows that '\$\dot\varepsilon[C(\xi)]$' formally expresses 'the (unique) object falling under the concept C', so that the function \$\xi$ corresponds in modern terms to the definite description operator. From the definition of \$\xi$ also the seventh and last of Frege's axioms, Law VI, follows immediately:

$$\vdash a = \backslash\dot\varepsilon[a = \varepsilon]$$

Adding Laws V and VI to the system described in § 2.1.2 above allows Frege to prove HP (appropriately rewritten in the system's formal language), to define the individual natural numbers by reproducing in the system's language the definitions given in *Grundlagen*, and to prove in the thus extended system Peano axioms: the axioms that Giuseppe Peano (1889) had put forward as an axiomatic basis for arithmetic.[23] It is on this ground that Frege takes himself to have reduced the entire arithmetic to logic. This is the logicist thesis: arithmetic can (must) be reduced to logic. The issue is whether Frege is justified in holding this thesis, and whether a suitable version of arithmetical platonism follows from it.

To see this, let us begin with two remarks. The first is that the role of the operator '*Num.*(–)' occurring in the *Grundlagen* formulation of HP is taken, in *Grundgesetze*, by an appropriate second-level function explicitly defined through the seven fundamental functions. HP, then, neither introduces nor implicitly defines a new operator. The sortal concept of natural number is defined explicitly by appealing to the function $\dot\varepsilon[\varphi(\varepsilon)]$, this being in turn appropriately composed with other fundamental functions in order to define the individual natural numbers, and HP is proved on the basis of the axioms, and becomes then a theorem of GA, whose role is to help in the proof of Peano axioms on the basis of those same axioms and of appropriate explicit definitions. The second remark is that the relation of equinumerosity between concepts is in turn explicitly defined, but no appeal is made for this to the functions $\dot{\epsilon}[\varphi(\epsilon)]$ and $\backslash\xi$. In modern terms, it is then defined in second-order logic alone. HP has thus the effect of reducing the identity of two terms defined in an extension of that logic (an extension obtained through the addition of those functions) to a relation defined within that logic. However, HP is proved by means of Laws V and VI. The envisaged reduction of arithmetic to logic thus hinges on the acceptance of these laws and, *a fortiori*, on the introduction of the functions $\dot{\epsilon}[\varphi(\epsilon)]$ and $\backslash\xi$. As a consequence, arithmetic can be said to have been reduced to logic only if these functions are assimilated to logical constants.

Now, it is clear that the functions $\dot{\epsilon}[\varphi(\epsilon)]$ and $\backslash\xi$ do not directly pertain to numbers, and thus are not arithmetical in character. But is this enough for justifying their assimilation to logical constants? At first glance, it is not. Still, Frege seems to echo the traditional conception of logic as the theory of concepts (a conception partly shared by Kant),[24] to extend it to the theory of objects and functions in its generality, and to take this as a ground for taking the functions $\dot{\epsilon}[\varphi(\epsilon)]$ and $\backslash\xi$ and the corresponding axioms as parts of logic. Moreover, he feels no need to accompany BLV with an explicit identification of value-ranges.

He just rests content with noticing, in § I.10, that nothing stands against assuming that T and F are value-ranges, and in particular that T is the value-range (the extension) of the concept $\xi = (\xi = \xi)$ and F is the value-range (the extension) of the concept $\xi = \neg \forall x[x = x]$. Since $\xi = (\xi = \xi)$ takes T as value only for T as argument, and $\xi = \neg \forall x[x = x]$ takes T as value only for F as argument, this stipulation is in accordance with the Kantian notion of extension of a concept. It also allows Frege to determine which values the four fundamental first-level functions $\neg\xi$, $\xi \Rightarrow \zeta$, $\xi = \xi$, and $\backslash\xi$ take when their arguments are value-ranges. The proof is given in *Grundgesetze*, §§ I.10 and I.31.[25] A natural reading and generalization of the context principle (see Dummett, 1991, pp. 211–12) suggests that names of objects have reference if, when they are used to fill the empty places in the names of a first-level function, we obtain names that themselves have reference (provided of course that the relevant names of function do not fail to name functions, that is, that there are functions named by them, and these names have then a reference, in turn).[26] Frege accepts this, and points to a consequence: that if we ask whether names of value-ranges have reference (in other words, whether there are value-ranges), by considering only the four fundamental first-level functions and other names of objects which surely have reference, we must answer that names of value-ranges (for the moment, only those of the relevant fundamental functions) have reference.

In order to conclude that names of value-ranges in general have reference, it is enough to generalize this argument. For these, three further ingredients are needed: (*i*) a recursive criterion establishing whether names of objects and functions have reference; (*ii*) a base for this recursive criterion;[27] and (*iii*) a warrant that if certain names have reference, then also those resulting from their compositions have reference. Frege explicitly states the condition asked for (*i*) in § I.29. Here, it will be enough to say that the idea underlying this criterion is the same that underlies the aforementioned reading of the context principle: a name of a function has reference if, when we fill some or all of its empty places with names of appropriate arguments having reference, we obtain names of objects with reference; a name of objects has reference if all the names obtained by filling with it some or all empty places for objects in names of functions having reference also have reference. Frege implicitly accepts that the names of the truth-values T and F and those of the fundamental functions can serve as the recursion basis required by (*ii*), and takes for granted the warrant required in (*iii*). He concludes that both the names of natural numbers and that of the concept ⌜(being a) natural number⌝ (this name being obtained in a quite complex way, using suggestions coming from *Grundlagen* expressed in the formal language of *Grundgesetze*) have reference. The natural conclusion is thus that natural numbers exist and are logical objects, and that theorems of arithmetic describe these objects. Should this conclusion prove correct, arithmetical platonism would be established in a version in which natural numbers are logical objects.

2.1.4 Guidelines for a Definition of Real Numbers

Before asking what is wrong in this conclusion and in the grand system conceived by Frege in order to argue for it, let us briefly see how Frege suggests (in the second volume of *Grundgesetze*) dealing with the problem of defining real numbers.

Frege's main point on this respect is that, just as natural numbers provide the cardinality of concepts (namely are things belonging to sortal concepts in relation to the size of the domain of individuals falling under them), so real numbers provide the measures of magnitudes (namely are things belonging to magnitudes, in relation to their respective sizes). Therefore, in the same sense as it is necessary to found the definition of natural numbers on a general theory of objects and functions (among which are concepts), it is necessary to found the definition of real numbers on a general theory of magnitudes.

This analogy has two consequences. The first, often stressed by Frege himself, is that real numbers should not be defined through an extension of the domain of natural numbers. Giving such a definition is not foreclosed; quite to the contrary, this can be done in different ways, and when the second volume of the *Grundgesetze* appeared (in 1903) many of these ways had already been suggested: the two best-known of them are those respectively suggested by Georg Cantor (1872) and Richard Dedekind (1872). According to Frege, definitions like these do not account, however, for the crucial feature of real numbers: their applying to measures of magnitudes. This would just happen to be a lucky consequence of the definition, whereas, Frege argues, it should ground it. The second consequence often goes unnoticed: if Dedekind's and Cantor's definitions are inadequate, this is because there is no way of accounting for the notion of magnitude by relying on our understanding of natural numbers (unless one begins by defining real numbers and then defines magnitudes on their basis, what for Frege would be tantamount to reverting the correct order of the definitions). A deeper problem prevents this: the impossibility of accounting for the notion of magnitude on the sole basis of a theory of functions and objects. This suggests that the definition of real numbers cannot be a matter of pure logic, and that Frege's logicism cannot extend to real numbers unless it is deeply modified.

This does not mean that the definition of real numbers cannot be given in the formal system within which Frege defines the natural numbers, nor that real numbers cannot be thought of as extensions of appropriate concepts. The point is that in order to define real numbers as measures of magnitudes, having some magnitudes at one's disposal will not suffice, for what is needed is an invariant structure understood as the structure of all domains of magnitudes. The reason is clear. Let us suppose we have domains of different sorts of magnitudes, for instance, segments in Euclidean geometry and masses in Newtonian mechanics, and we define real numbers as measures of these magnitudes. We still have to show that we have defined the same numbers in both cases, that the real numbers defined as measures of segments are the same objects as the real numbers defined as measures of masses. And this can be secured, in general, only once a common feature of different domains of magnitudes has been selected in such a way that the definition of their measures can be shown to depend only on that feature.

The crucial point is then that in order to define this invariant structure (actually, in a rather unsatisfactory way, given that Frege only considers a particular kind of magnitude), Frege envisages the need for some axioms that, though expressible in the language of his formal system, are not truths ensuing from the features of his fundamental functions. He is so aware of this that, after pointing to the guidelines for obtaining what he thought was the appropriate definition, he sets himself to prove that some domains of magnitudes exist, that is, systems of objects satisfying the proposed definition. Frege indicates that this proof will rely on the construction of an appropriate domain of objects having natural numbers as its basic elements. Had he given a suitable formal treatment of these ideas, Frege would have offered a purely logical account (according to his notion of logic) of a particular domain of magnitudes. In other terms, he would have identified a domain formed by logical objects satisfying his definition of domains of magnitudes. He could then have defined the measures of these magnitudes. But in order to identify these measures with the real numbers, he should have also proved that they are the same objects as the measures of any other sort of magnitudes whatsoever, possibly of non-logical ones. What is doubtful is that his very notion of logic would have allowed him to consider such a proof, in turn, as purely logical.

2.1.5 The Contradiction in Frege's Formal System

Had Frege been able to argue for this or not, his envisaged definition of real numbers would have collapsed for the same reason that makes his definition of natural naturals and his related argument for arithmetical Platonism ineffective. This reason is that when BLV – which plays an equally crucial role both in Frege's definition of natural numbers, and in his envisaged definition of real ones – is placed into a system like Frege's, it leads to a contradiction, making the whole system inconsistent.

There is no need to appeal to Russell's sophisticated argument in order to realize that something is wrong with BLV. Let us suppose (see Dummett, 1991, p. 133) that we have only three objects, ①, ②, and ③, that can be arguments and values of a one-place first-level function; let $f(\xi)$ be this function. Then, there will be $3^3 = 27$ different combinations of values and arguments for this function: $\{f(①) = ①, f(②) = ①, f(③) = ①\}$; $\{f(①) = ①, f(②) = ①, f(③) = ②\}$ ..., $\{f(①) = ③, f(②) = ③, f(③) = ③\}$. By BLV, there would then been 27 different possible value-ranges for that function. But we have seen that, by BLV, value-ranges cannot but be objects, and we assumed that there are only three objects available that can be arguments and values of a function. Therefore, if we accept BLV, we have either to impose some restriction on the possibility of defining functions on objects, or to accept that our functions are not defined over all objects: otherwise, we will have more value-ranges than objects. This holds for any number of objects grater than 1, since to n objects always correspond n^n possible value-ranges and $n^n > n$, for any n greater than 1. Frege, however, does not place any such restriction on the possibility of defining functions, nor to the domain over which first-level functions are defined and have values. Moreover, he also assumes from the start that there are at least two objects: T and F. It is clear, then, that things cannot work this way.

Russell's argument makes the problem apparent. This argument is raised in a letter from Russell to Frege dated 16 June 1902, which the latter received while completing the second volume of *Grundgesetze*. This volume was published the following year, with an appendix (pp. 253–65) in which Frege tries, in vain, to add local modifications to his system in order to avoid the contradiction. Russell's argument is commonly presented as follows (Russell, 1903, § 100). Take the set of all sets that are not members of themselves, and ask whether this is a member of itself. If it is a member of itself, then, by definition, it is a set that is not a member of itself. If it is not a member of itself, then, again by definition, it is a set that is a member of itself. It follows that this set is a member of itself if and only if it is not a member of itself.

This version of the argument has no direct bearing on Frege's system, since the latter does not involve the notion of set. But it is easy to see how a parallel argument can be used to point to the inconsistency caused by BLV (Boolos, 1986–7, pp. 172–3). Let $R(\xi)$ be a first-level concept under which all and only the extensions of a concept that do not fall under this same concept fall. This concept can be easily defined in Frege's system. Let us then consider its extension $\dot{\varepsilon}[R(\xi)]$, and ask whether it falls under R, namely if it holds that $\vdash R(\dot{\varepsilon}[R(\varepsilon)])$ or rather that $\vdash \neg R(\dot{\varepsilon}[R(\varepsilon)])$. To avoid appealing to our intuitive idea of extension, let us define the concept $R(\xi)$ as the concept under which an object x falls if and only if there is a concept $F(\xi)$ for which it holds both that $\vdash x = \dot{\varepsilon}[F(\varepsilon)]$ and that $\vdash \neg F(x)$. Suppose that $\dot{\varepsilon}[R(\varepsilon)]$ does not fall under $R(\xi)$, namely that $\vdash \neg R(\dot{\varepsilon}[R(\varepsilon)])$. Then there is no concept $F(\xi)$ such that it holds both that $\vdash \dot{\varepsilon}[R(\varepsilon)] = \dot{\varepsilon}[F(\varepsilon)]$ and that $\vdash \neg F(\dot{\varepsilon}[R(\varepsilon)])$. This cannot be the case for $R(\xi)$, then. But if extensions are objects and $R(\xi)$ has an extension, then $\vdash \dot{\varepsilon}[R(\varepsilon)] = \dot{\varepsilon}[R(\varepsilon)]$. It follows that $\vdash \neg[\neg R(\dot{\varepsilon}[R(\varepsilon)])]$, which, in classical logic, is equivalent to say that $\vdash R(\dot{\varepsilon}[R(\varepsilon)])$. Suppose now that $\dot{\varepsilon}[R(\varepsilon)]$ falls under $R(\xi)$, namely that $\vdash R(\dot{\varepsilon}[R(\varepsilon)])$. Then there is a concept $F(\xi)$ for which it holds both that $\vdash \dot{\varepsilon}[R(\varepsilon)] = \dot{\varepsilon}[F(\varepsilon)]$ and that $\vdash \neg F(\dot{\varepsilon}[R(\varepsilon)])$. If this is so, then, by BLV it also holds that $\vdash \neg R(\dot{\varepsilon}[R(\varepsilon)])$. Therefore, according to BLV, both '$\vdash \neg R(\dot{\varepsilon}[R(\varepsilon)])$' and '$\vdash R(\dot{\varepsilon}[R(\varepsilon)])$' imply their own negation. And this is a contradiction.

The contradiction does not only depend on the acceptance of BLV and on the trivial deductive rules involved in the preceding argument.

It also depends on the assumption that $R(\xi)$ is a well-defined function (a concept), or, in other words, on the freedom that Frege's system allows for the formation of names of functions. The contradiction might thus be avoided by limiting this freedom, rather than by renouncing to BLV. Dummett (1991, pp. 217–18) has argued that the source of the contradiction lies in taking for granted that the warrant required by clause (*iii*) of § 2.1.3 above is available, namely in considering that if certain names have reference, then also those resulting from their composition should have reference.[28] More specifically, he argues against the formation of names of functions by means of second-order quantification. If it is not admitted, in general, that names of functions formed in this way have reference, then nothing requires that also the name of Russell's function has reference, and thus that this is a genuine function. This suggests blocking Russell's paradox by constraining this way of forming (names of) functions through appropriate limitations. But then the problem arises of understanding how this can be done without blocking at the same time the possibility of defining natural numbers as proposed by Frege, or in some similar way. *Mutatis mutandis*, this is the problem tackled by Russell himself.

2.2 Russell and the Separation of Logicism and Platonism

Russell discovered the contradiction in Frege's system while he was writing *The Principles of Mathematics* (1903), whose aim was to defend the logicist thesis. At the very beginning of the book's *Preface* (p. 27), Russell presents this thesis as follows:

> all pure mathematics deals exclusively with concepts definable in terms of a very small number of fundamental logical concepts, and . . . all its propositions are deducible from a very small number of fundamental logical principles.

As the title of the book suggests, this seems to be an extension of Frege's programme for arithmetic to the whole of mathematics. Russell's discussion, though, is entirely informal, and his view differs from Frege's under crucial respects. Some of these are apparent from the very first sentence of Chapter 1 (§ 1):

> Pure Mathematics is the class of all propositions of the form '*p* implies *q*', where *p* and *q* are propositions containing one or more variables, the same in the two propositions, and neither *p* nor *q* contains any

constants except logical constants. And logical constants are all notions definable in terms of the following: Implication, the relation of a term to a class of which it is a member, the notion of *such that*, the notion of relation, and such further notions as may be involved in the general notion of propositions of the above form. In addition to these, mathematics *uses* a notion which is not a constituent of the propositions which it considers, namely the notion of truth. [emphasis in the original]

Insistence on theorems of mathematics being implications shifts focus from mathematical truths to their deduction. The way the notion of logical constant is explained brings to light a different conception of what a formal system is. This is made even clearer by the specification that the notion of truth is not itself directly involved in the theorems of mathematics. Surely, not even Frege thought that the names 'V' and 'T' occur in the language of arithmetic; but Russell is saying more than that. He suggests that there is a category mistake underlying the idea of taking truth-values as objects. When this idea is put aside, all the philosophical framework upon which Frege's system is built collapses, and the need emerges to distinguish between functions and predicates, namely between functional letters (letters like 'f', such that '$f(a)$' is a term) and predicate letter (letters like 'F', such that '$F(a)$' – no more '$\vdash F(a)$' – is a sentence).

These crucial changes anticipate modifications in Frege's system that will eventually yield propositional and predicate logic in their modern form. There is yet another change, at least as evident: the set-membership relation occurs among the fundamental logical constants. This signals how relevant in the *Principles* is the notion of set (or class).[29] In Russell's logical system, classes have the role that extensions of concepts had in Frege's. Some attach little philosophical importance to this difference,[30] but from an historical point of view it is crucial, for it points to a shift of focus from the notion of function to the notion of set in searching for a foundation of mathematics.

This shift results in Russell's emending Frege's definition of natural numbers. They are no more identified with extension of concepts, but rather with sets equinumerous with other given appropriate sets (1903, § 111). Though Russell's discussion is informal, the idea is clear, and it immediately suggests the following definitions:

0 is the set of all sets equinumerous to the set of objects not identical to themselves;
1 is the set of all sets equinumerous to the set of objects identical to 0;

2 is the set of all sets equinumerous to the set of objects identical either to 0 or to 1;
etc.

This adaptation of Frege's definition does not by itself suffice to avoid the contradiction. Russell's argument does not only show that Frege's system is inconsistent, but also that the very notion of set is to be handled with care. As Russell notices (1903, § 102), we cannot at the same time accept that stipulating necessary and sufficient conditions for membership to a set suffices for the existence of the set, and that every set can be a member of another set. So, far from avoiding the contradiction, Russell's emendation of Frege's definition brings it even more clearly to light. Though Russell is not explicit on this, one can point to the contradiction even from these very definitions. Suppose that it is always possible to associate to a given set e the set $\{e\}$, of which e is the sole member. Let e be any set. The set $\{e\}$ is clearly equinumerous with the set of objects identical to 0. There are thus so many sets equinumerous with the set of objects identical to 0 as there are sets, and then, according to the definition above, the number 1 is a set with as many members as there are sets. But no set can have as many members as there are sets, on pain of incurring in Cantor's paradox: if that were to be the case, the relevant set should have at least as many members as its subsets are (for these are sets), and this leads to absurdity (see § 2.3 below).

Russell adds a chapter to his book (Chapter x), where he accounts for the contradiction he had communicated to Frege the year before (possibly after he had already arrived at the foregoing definitions), but he argues that his own view can survive it, and adds two appendixes to clarify the matter: Appendix A points to the differences between his and Frege's ideas, while Appendix B sketches his proposal (anticipated in Chapter 10) for avoiding the contradiction. This consists, roughly, in distributing the different sorts of entities named in the language in a hierarchy of types, thus allowing for the formation of names of entities of a given type only by means of names of entities of lower types; this would prevent some complex names from being formed, among which, most notably, is the one involved in Russell's paradox.

In the *Principles*, this idea is only sketched its broad outlines, and Russell leaves it to future work to investigate how it affects the possibility of elaborating a logical system powerful enough to include mathematics (or a large portion of it). This further work will lead to the publication of a new monograph, written together with Alfred North Whitehead, that was intended by its authors to offer a formal version of the reduction of mathematics to logic that had been envisaged in the *Principles*. The book was published some years later, in three volumes, with a title reminiscent of Newton's masterpiece: *Principia Mathematica* (1910–13). The ideas outlined in the earlier work are here presented through a sophisticated theory of types (in his "ramified" version, anticipated in Russell, 1908; see Coquand 2010, Landini, 2011; for an introduction to the theory of types see for instance Andrews, 2002). But the idea that well-formed names of sets refer to objects whose existence is warranted by the way such names are introduced is now abandoned. These names are rather treated as "incomplete symbols": symbols that occur in meaningful formulae but have no meaning if considered in isolation.[31] This view goes together with the abandonment of the idea that mathematics is about abstract objects, sets in particular, in favour of a conception that is known as 'no-class theory'. This conception stands against Frege's platonism, that was still apparently endorsed in the *Principles* instead. Moreover, Russell and Whitehead require their construction to satisfy the so-called Vicious Circle Principle, banning recourse to impredicative definitions, namely those definitions that appeal to a set while defining elements of this very set (see Chihara, 1973). This forces them to introduce *ad hoc* axioms, which, quite independently of one's views on the reference of names of sets, frustrate any attempt to offer anything similar to Frege's logical argument for platonism. Today, it is generally accepted that these axioms do indeed frustrate also Russell's and Whitehead's own attempted reduction of mathematics to logic. What is more important for us here, is that *Principia* witnesses, *contra* Frege, a separation of the logicist program from platonism.

2.3 Set Theory

In giving a central place to the notion of set in the *Principles*, Russell is following one of the main trends in the mathematics of his time, pursued mainly in the works of Dedekind and Cantor.[32] Some years earlier, the former had published a short treatise (1888) in which he had shown, while steering clear of any explicitly philosophical remarks, that the succession of the natural numbers can be defined as a set satisfying appropriate conditions involving a function with values and arguments in this set. In showing this, he made use of notions that can be seen as precursors of a theory of sets. In the same year, this theory was being developed in various essays by Cantor (1872, 1874, 1879–84, 1883, 1895–7), initially devoted to issues in reals analysis, and later progressively focused on problems raised by the autonomous development of the theory. More specifically, Cantor defined two new kinds of numbers that, as one might say, go beyond the naturals. On the one side we find transfinite cardinals, extending the naturals to give the cardinality of infinite sets of any sort; on the other hand, we find transfinite ordinals, yielding various kinds of ordering that infinite sets can take. Cantor did not merely define these numbers, he also elaborated transfinite arithmetics for them, differing from each other and from the arithmetic of natural numbers. The development of set theory between the end of the nineteenth century and the beginning of the twentieth is strictly connected to the attempt to find an appropriate deductive framework within which these arithmetics could be framed and results on those numbers proved. Two examples might help in understanding what is at stake here.

Bernard Bolzano (1851) had already foreseen the possibility of defining an infinite set as a set a proper subset of which can be put into one-one correspondence with the set itself. A clear case is given by natural numbers: there are as many natural numbers as there are even numbers, odd numbers, multiples of 3, etc. This definition was finessed by Dedekind (1888), and a set satisfying this condition is then said to be 'Dedekind-infinite'. A interesting question is thus: can we prove that a set is Dedekind-infinite if and only if it is infinite in the usual sense (namely it cannot be put into one-one correspondence with any initial segment {0, 1, 2, . . ., n} of natural numbers)? The answer to this question depends on which axioms are accepted: the proof which is common today requires the so-called Axiom of Choice, which allows forming a new set from every set of non-empty sets by picking, in an arbitrary way, one and only one element from each of these latter sets.

Let us see the second example. A set that can be put into one-one correspondence with the set of natural numbers is said to be 'denumerably infinite'. Cantor proved that this is, for instance, the case for the set of rational numbers, but not for that of real ones. He also proved that no set can be put into one-one correspondence with the set of all its subsets, and that the set of real numbers can be put into one-one correspondence with the set of all the subsets of the set of natural numbers. Insofar as two sets have the same cardinal number if and only if they can be put into one-one correspondence, this means that the sets of natural numbers and that of rational ones have the same cardinal, but this differs from the cardinal of the set of reals, which is in turn the same of those of all the subsets of the sets of natural and rational numbers. More generally, the cardinal of any set is different from the cardinal of the set of all its subsets. Suppose we have defined a total order on cardinals – namely, a strict ordering relation, <, such that for any two cardinals α and β either $\alpha = \beta$, or $\alpha < \beta$, or $\beta < \alpha$ – and that this has been done in such a way that if α is the cardinal of a given set, and β is the cardinal of the set of all subsets of the latter, then $\alpha < \beta$. It follows that if $\aleph_0$ (read: *aleph* zero) is the cardinal of the set of natural numbers, the cardinal of the set of reals is[33] $2^{\aleph_0}$, and that $\aleph_0 < 2^{\aleph_0}$. We can ask, then: is there any cardinal between $\aleph_0$ and $2^{\aleph_0}$? The negative answer to this question is usually called 'Continuum Hypothesis'. Today, thanks to Kurt Gödel (1939a, 1939b, 1940) and Paul Cohen (1963), we know that the usual axioms of set theory with the addition of the Axiom of Choice do not suffice to decide between the Continuum Hypothesis and its negation.

These axioms have been put forward in a series of essays published between 1908 and 1930 by Ernst Zermelo and Abraham Fraenkel, and form a theory called 'ZF' after them. When the Axiom of Choice is added to this theory, one gets a stronger theory quite naturally called 'ZFC'. These theories do not require stratification in types in order to avoid Russell's paradox, since no set like Russell's paradoxical one can be proved to exist on the basis of their axioms (and no other paradox is known to arise in them, so that they are generally regarded as consistent). These theories offer an appropriate formal framework in which a vast part of modern mathematics can be interpreted, including the theories of natural and real numbers.

There is more than one way of defining natural numbers, within ZF or ZFC. The simplest is surely to pick an appropriate family of sets and define on it the usual arithmetical relations and operations. A way of doing this, suggested by Frege's definition, has been proposed by John von Neumann (1923, p. 347). It consists in first identifying zero with the empty set, and then using a recursive clause associating a given set to the union of this set with the set of which it is the sole member (that is, given a set e, one gets the set $e \cup \{e\}$). One thus gets a progression ordered according to this recursive clause:

$$\emptyset; \{\emptyset\}; \{\emptyset, \{\emptyset\}\}; \{\emptyset, \{\emptyset\}, \{\emptyset, \{\emptyset\}\}\}; \ldots$$

Another, equally simple way has been proposed by Zermelo (1908). Zero is still identified with the empty set, but the recursive clause now associates to a given set the set of which the former is the sole member (from *e* one gets {*e*}). The resulting progression, ordered by that recursive clause, is then the following:

$$\varnothing; \{\varnothing\}; \{\{\varnothing\}\}; \{\{\{\varnothing\}\}\}; \ldots$$

With these definitions, we are nevertheless still far from having solved the problem of the foundation of arithmetic, and still less we have offered any argument for platonism. What we have shown is just that within set theory it is possible to construct models of Peano axioms, that is progressions whose elements are sets.[34] This warrants that it is possible to identify the natural numbers with the elements of one of these models (it does not matter which one),[35] but nothing warrants that their theory is properly founded, or that natural numbers, having been thus identified, exist as objects, in some appropriate sense. In order to obtain this, we must look at set theory, and see whether it is properly founded and whether sets definable in it exist as objects, in some appropriate sense. Both the problem of foundations, and that of giving an argument for or against platonism, are then transferred to set theory. This holds not just for arithmetic, but also for any branch of mathematics that can be expressed and we want to express within set theory.

Generally, speaking of set theory, one has ZFC (or ZF) in mind. ZFC, however, differs from other available versions of set theory, involving different, or differently formulated, axioms. Moreover, the fact that a decision within ZFC of some hypothesis (the Continuum Hypothesis, for instance) has proven impossible, has led many to search other axioms to be added to those of ZFC in order to increase its deductive power, just as the Axiom of Choice is added to ZF.[36]

This makes evident by itself that the axioms of ZFC, like those of any other available version of set theory, not only cannot be taken as logical truths, but do not even have the sort of immediacy and topic-neutrality we find in the axioms of first-order logic, or even, according to someone, of second-order logic. It is utterly implausible to think that they could ever ground a neutral framework that could be used in intellectual or empirical inquires while having no substantial influence on what we have to say about the investigated phenomena. Quite to the contrary, set theory gives rise to a realm of idiosyncratic entities, whose nature, existence and appropriateness are still a matter of dispute: whatever our idea of logic is, set theory is not a branch of logic, but is itself a mathematical theory.

2.4 The Problem of Foundations

2.4.1 Carnap: Between Logicism and Logical Empiricism

We have seen that set theory offers no autonomous solution to the problem of foundations. It is then natural that, while the theory was being further developed, several attempts at a solution were put forward. One of these was offered by Carnap, the third great logicist philosopher of the first half of the twentieth century. With Carnap, the separation between logicism and platonism became even stronger than it was with Russell. Carnap was one of the major figures of logical empiricism (see Ayer 1946; Jacob, 1980; Coffa, 1991; Stadler, 1997; Richardson and Üebel, 2007), whose main tenet he wholly accepted:[37] a statement to which factual content is to be assigned is meaningful if and only if there is, at least in principle, a method for its verification; and since what decides of factual matters is experience, the method must rely on it. If they are to be meaningful, therefore, mathematical statements must either be experimentally verifiable, or lack factual content. For Carnap, however, mathematics is doubtlessly *a priori*: as a consequence, its theorems are meaningful only if they lack factual content, and more precisely if there is no fact of the matter deciding for their truth or falsity. Nor is there a fact of the matter, then, about the existence of the objects that are possibly described by these theorems, nor about what is the right way of founding mathematics. This does not entail that every existence problem concerning mathematics is ill-formed, nor that the problem of its foundations is. These are meaningful problems, but they must be properly understood and formulated.

If mathematics has no factual content and its theorems are meaningful, it cannot but reduce to a linguistic framework, in which some of our questions and statements concerning our experiences can be expressed, to be used in the process of verifying other statements: it only concerns the structure of our language. The problem of its foundation will thus be the problem of finding the most appropriate way of shaping this framework. Being logicist will thus amount to claiming that shaping this framework is a matter of logic: this is just what Carnap means when he claims that "mathematics is reducible to logic" (Carnap, 1931, pp. 91–2). This statement is specified as follows (*ibid.*):

> We split the logicist thesis into two parts . . .: *i*) the *concepts* of mathematics can be derived from logical concepts through explicit definitions; *ii*) the *theorems* of mathematics can be derived from logical axioms through purely logical deduction. [emphasis in the original]

Carnap justifies thesis (*i*) by appealing to the system of Russell's and Whitehead's *Principia Mathematica*, and by advancing some simplifications that are made possible by his acceptance of impredicative definitions. The price to be paid for these simplifications is a sort of definitional vice, to be seen as innocent, however, as long as the aim is that of merely devising a linguistic framework (Shapiro, 2000a, p. 125).[38] Thesis (*ii*) is where the real problem lies. Carnap does nothing to hide the problem posed to the system of *Principia* by the presence of axioms that he himself takes as non-logical. Still, he notices that they are needed in order to prove certain theorems only if these are conceived in the usual way. Once they are rewritten as implications whose antecedent is given by these very axioms (possibly together with others), and whose consequent is given by their usual formulations, the problem is avoided: the controversial axioms are no more needed for a proof of the rewritten theorems.

This is acceptable only if one also accepts that the logicist thesis is not concerned with the content of mathematics, but only with its mere form. Since mathematics lacks content for Carnap, this is obvious to him. Saying that mathematics is reducible to logic, then, is nothing more than saying that it can be written in a logical language and that its proofs can be expressed in the form of deductive chains. Clearly, the logicist thesis, thus understood, has little to do with Plato's problem, to which Frege tried to answer; but Carnap wants to be very explicit on this. He insists on the existential character of the disputed axioms, claiming that a "purely formal science" can "make only conditional . . . statements about existence" (1931, p. 96). Some years later, Carnap (1950, § 2) endorses an even more radical view, introducing a distinction that will be often used in the subsequent debate: that between internal and external existence, or better, between internal and external existence questions. The former concern "the existence of certain entities . . . *within the framework*", the latter concern "the existence or reality *of the system of entities as a whole*" (emphasis in the original). The former are well-formed and have answers that depend on the relevant theory. In arithmetic, for instance, one might ask whether there are infinite prime numbers. The proof of this is simple, and lets one conclude that the statement 'There are infinite prime numbers' is true. From this, however, it does not follow that there are prime numbers (nor, thus, natural numbers) outside arithmetic, and that they are described by the theorems of arithmetic. Claiming this would be answering to an external existence question and, Carnap says, such questions are meaningful only when they are taken as pragmatic questions, concerning the opportunity of the adoption of the relevant framework.

It is the clear that for Carnap the truths of mathematics are always analytic, even when their logical form (but not their content) is that of an existential statement. This is so since they are nothing but logical consequences of the definitions introducing the relevant concepts: they are analytic truths since they are true in virtue of their meaning, this being fixed by those definitions. But are we sure that every internal question can be given an answer (within the relevant theory)? Shapiro (2000a, pp. 131–3) considers Fermat's last theorem as an example. This answers to an internal existence question: are there triples $\{p, q, r\}$ of natural numbers such that, for some natural number n greater than 2, $p^n + q^n = r^n$? Thanks to the proof given by Andrew Wiles (1995), we now know that the answer is in the negative. But that proof employs resources going well beyond arithmetic. Should then the question be considered internal or external? And if it is internal, should Wiles' proof be considered appropriate? Is the theorem proved by Wiles an analytic truth, is it true in virtue of its meaning? A positive answer to these questions, in defence of Carnap's thesis, could only be had in two cases: either (*i*) the meaning of the theorem is taken to rely on an extremely vast family of definitions, transcending the limits of arithmetic and very hard to delimit; or (*ii*) it is claimed that by investigating certain statements on natural numbers in a wider context, we can better understand what actually only depends on their meaning and is fixed by the sole definitions of arithmetic. Both options seem implausible. Similar problems do not only affect Carnap's logicism and logical empiricism as a whole; they also put pressure on his very distinction between external and internal existence. This is why the distinction is often rejected.[39]

2.4.2 Hilbert and the Axiomatic Foundation of Geometry

In §§ 143–4 of the second volume of the *Grundgesetze*, Frege discusses what he calls "creative definitions": stipulations introducing new entities by laying down that they are to satisfy certain conditions. He offers two examples from Otto Stolz's *Vorlesungen über allgemeine Arithmetik* (1885). Here is the second of them. Suppose a commutative operation $*$ is defined on a given domain of objects such that, given two elements a and b of this domain, there is no element x such that $a * x = b$. Then, Stolz says, we can introduce this element by denoting it by a new symbol, say '$b\sim a$', and stipulating that $a * (b\sim a) = (b\sim a) * a = b$. Once this is done, we can assign to $b\sim a$ all the properties we like, provided we avoid contradictions. Definitions of this sort are fairly common in mathematics: think of the way negative integers are often introduced from naturals: for any natural n, the negative integer $-n$ is introduced by stipulating that

$n + (-n) = -n + n = 0$. But Frege mounts a fierce criticism against them: the only way of establishing that some properties are not mutually inconsistent is to pick an object satisfying them all; creative definitions are thus in good standing only if they are useless, namely if we already are able to point at the objects that they are supposed to introduce. Given that for Frege a property is what makes an object fall under a given concept, and that concepts compose with each other, his argument is based on a general principle that might be stated as follows: "the only way to show the consistency of a concept or collection of concepts is to exhibit some thing that falls under it (or them)" (Blanchette, 1996, p. 328).

The two examples given by Frege are too simple to show the full bearing of this principle. They both concern definitions introducing new objects from others taken as given; in other words, they lead to extensions of the already given domain. His point is much more general, however, and he must have been aware of this, though he does not explicitly says so.

Four years before the second volume of the *Grundgesetze* appeared, David Hilbert, perhaps the most authoritative mathematician of the time, had published a treatise whose impact on mathematics, philosophy of mathematics, and logic will prove enormous: *Die Grundlagen der Geometrie* (1899). His aim is to suggest a certain way of understanding and founding Euclidean geometry (and also showing its relation with projective geometry). The basic idea was not new: geometry is not about spatial objects to be accessed by some sort of intuition, as Kant thought and Frege still held; it is not about objects endowed with intrinsic properties individuating them singularly; it is rather about a relational structure characterizing its elements only in so far as they satisfy the conditions the structure imposes on them. It is then neither necessary nor useful, in order to found Euclidean geometry, to establish what a point is, or what an Euclidean straight line or plane are; rather, one must axiomatically fix the conditions that points, straight lines and planes must satisfy if they are to possess the mutual relations that are ascribed to them in this geometry (notice that Hilbert defined circles as *loci* of points). Even if we concede that the content of geometry, as presented by Euclid, can be given in intuition, this would provide, at best, a particular instance of a system of conditions that can be fixed independently of the consideration of this very instance. It is this system of conditions that should be seen as the proper subject matter of Euclidean geometry, instead of its (alleged or real) intuitable instances.

When it is thus understood, Euclidean geometry is a system of consequences drawn from a finite number of axioms, involving primitive terms defined by these very axioms. In a certain sense, these axioms work like Frege's BLV: they state conditions that the entities named by certain terms must satisfy, without saying anything else about them and without appealing to any prior definition. According to Hilbert, the issue is not, however, that of introducing a certain entity in a system of entities that have been already explicitly defined, nor that of describing entities that can somehow be conceived as logical.

Though he is not interested in a highly formalized exposition of his system – which he expounds through natural language – it is clear that its axioms, however stated, require using a somehow neutral language, a language that is not specifically geometric in character and that must be taken as given, or at least as understandable independently of our understanding of geometry. *Mutatis mutandis*, we can think of this as the language of the logic underlying Hilbert's system, including only the logical constants involved by this system. But what is crucial is not how this language is thought of, but the fact that, whatever it might be, it cannot suffice for the formulation of the relevant axioms. This requires that new elements are added (working as non-logical constants) whose meaning is fixed by the axioms themselves.

Among these, not only terms denoting points, straight lines or planes, or predicate constants denoting the properties of being a point, a straight line or a plane, must occur. Also predicate constants denoting relations are required. Let us take axiom II.2 as an example: 'If A and C are two points of a straight line, then there exists at least one point B lying between A and C and at least one point D so situated that C lies between A and D'. The words 'if . . . then', 'and', 'there exists at least one', 'so . . . that' are clearly part of the neutral language underlying geometry. But this is clearly not the case for 'point', 'straight line', 'being a . . . of . . .', and 'lying between . . . and . . .'. This is less clear for 'A', 'B', 'C', and 'D', but a little reflection will convince us that they merely work as individual variables whose range is specified by the words 'points of a straight line'. The words 'point', 'straight line', 'being of . . .', and 'lying between . . . and . . .' are predicate constants instead, of which the first two are monadic, the third is dyadic and the fourth is triadic (the word 'two' is clearly used only for ease of exposition and is actually superfluous). The axiom can then be written as follows, in a first-order predicate language:

$$\forall x,y[[Pnt(x) \land Pnt(y) \land \exists z[StrtLn(z) \land Of(x, z) \land Of(y, z)]] \Rightarrow$$
$$\Rightarrow \exists v[Pnt(v) \land Of(v, z) \land Btw(v, x, y)] \land \exists w[Pnt(w) \land Of(w, z)$$
$$\land Btw(y, x, w)]]$$

The predicates '*Pnt*(–)', '*Strln*(–)', '*Of*(–,–)', and '*Btw*(–, –, –)' are constants that have a meaning only in so far as they occur in this and other axioms of the system. These and the other geometrical constants occurring in Hilbert's axioms are not thus defined each independently of the others by a single definition; they are (implicitly) defined together, through various stipulations that are such that in each of them more than one constant can occur, and that each constant is allowed to occur in several stipulations: they are defined as elements of a structure that is fixed as a whole by all of the stipulations taken together.

Soon after the publication of the *Grundlagen der Geometrie*, and before that of the second volume of *Grundgesetze*, Frege had an intense correspondence with Hilbert focused on Hilbert's (implicit) definitions (Frege, WB, pp. 39–65; PMC, pp. 31–52). Frege denies these are in good standing, with arguments similar to those he mounted against Stolz. He now concludes that Hilbert's axioms cannot define the concepts of point, straight line and plane, since they do not secure the existence of the corresponding objects. Nothing warrants, as a consequence, that the system they form is consistent. If these objects are available to us independently of the axioms (thanks to our spatial intuition, for instance), then, Frege seems to claim, we have grounds for believing that these axioms are consistent, but then, they are useless; if these objects are not available to us independently of the axioms, nothing warrants that these axioms are consistent.

Frege's argument does not only affect Hilbert's axioms and his claim that these provide a foundation for Euclidean geometry. What is more generally at stake are the arguments Hilbert offers in the second chapter of his treatise, which is devoted to prove the consistency and independence of his axioms. It begins with the following claim (Hilbert, 1899, p. 17, English translation p. 4):

> The axioms, which we have discussed in the previous chapter . . . are not contradictory to one another; that is to say, it is not possible to deduce from these axioms, by any logical process of reasoning, a fact which contradicts any of them. To demonstrate this, it is sufficient to provide [*anzugeben*] a geometry where all of . . . [them] are fulfilled.

In the remaining part of the chapter, Hilbert proves this through a construction of the relevant "geometry". He then proves the independency of some axioms from others, through a construction of different "geometries" satisfying the system formed by the latter and the negation of the former. Hilbert apparently endorses the following criterion: if we can provide (or there is) a "geometry" satisfying the given system of axioms, then this system is consistent. At first sight, Hilbert seems to agree, *mutatis mutandis*, with the general principle adopted by Frege concerning the consistency of concepts; at least, he seems to acknowledge that the condition stated in this principle is sufficient for the consistency of axioms systems. But things stand very differently.

To see this, we must first understand what Hilbert means when he says that to prove the consistency of his axioms it suffices "to provide a geometry" where they are "fulfilled". Should this "geometry" happen to be the usual Euclidean geometry, the consistency proof would make its axioms useless, as Frege suggests. But the indeterminate article before 'geometry' suggests things must stand differently. Hilbert is speaking here of any domain of objects satisfying the axioms which can be produced independently of these (in modern terms, he is speaking of a model of the axioms). In particular, in his consistency proof he considers the domain Ω of "those algebraic numbers which result by beginning with the number 1 and applying to it a finite number of times the four arithmetical operations (addition, subtraction, multiplication, and division) and the fifth operation $|\sqrt{1+\omega^2}|$, where ω represents a number arising from the five operations already given" (1899, p. 34; English translation p. 17). Algebraic numbers form a proper subset of real ones. Those belonging to Hilbert's domain Ω form in turn a proper subset of algebraic numbers (including all rational and some irrationals). The way they are generated from 1 allows one to define on them the operations that are usually defined on all reals, and to show that, like all the reals, they form a field: they satisfy some algebraic properties that the set of real numbers satisfy. Hilbert's idea is then to interpret the constants implicitly defined by his axioms on this field, and then to show, by calculation within this field, that the axioms hold for this interpretation. A point, for instance, is interpreted as a couple of elements of Ω, and a straight line as an appropriate equivalence class $Cl(<p, q, r>)$ of triples of these elements. Under this interpretation, a point $<p, q>$ is then taken to be a point of the straight line $Cl(<r, s, t>)$ if and only if $rp + sq + t = 0$.

In this way, the consistency of Hilbert's geometrical axioms is proved only under the hypothesis that the arithmetic of the field Ω is consistent in turn. Hilbert takes this for granted. But Frege is not arguing against this. He rather rebuts that such a proof cannot rule out that a deeper analysis of the concept of point, straight line and plane will in the end detect contradictions that an argument like Hilbert's does not reveal.

The proof shows that these axioms admit of an interpretation: that there is a domain of objects, assumed as existent, satisfying the relational structure they describe. But it does not establish that this structure is satisfied by the very points, straight lines and planes of Euclidean geometry. Here lies the difference between Hilbert and Frege. The former thinks that Euclidean geometry is nothing but the theory of this structure; the latter thinks that it cannot but be the theory of these very objects, and that only by showing that these (and not other) objects satisfy the axioms of the structure it is possible to show that the axioms are consistent (Blanchette, 1996, p. 330).

The difference between Frege and Hilbert is then parallel to that between those who believe that mathematics is about objects with some particular intrinsic nature (making them, for instance, be natural numbers or euclidean straight lines); and those who think instead that mathematics deals only with relational structures that are possibly instantiated by some domains of objects (that are, as such, independent of them). The former are platonists like Frege both as regards arithmetic and as regards geometry; for Frege, both arithmetic and geometry are about particular objects; the difference is that while the objects of arithmetic are logical and can then be described by analytic truths and known through conceptual analysis, those of geometry are spatial and can then be described only by synthetic (though *a priori*) truths and known only through intuition. The latter can also be considered as platonists, provided that platonism be intended as the view that mathematics is about abstract objects conceived as mere positions in structures. This is for example the view endorsed by Stewart Shapiro (see § 5.3).

To clarify this point, with respect to Hilbert, notice that according to him it is perfectly in order to speak of the truth of the axioms and of the existence of the items they implicitly define. Here is what he writes in a letter to Frege (29 December 1899, see Frege, WB, p. 51; PMC, pp. 39–40):

> if the arbitrarily given axioms do not contradict one another with all their consequences, then they are true and the things defined by the axioms exist. This is for me the criterion of truth and existence.

Clearly, Frege could not accept this way of talking of truth and existence. Some might spot here something close to Carnap's notion of internal existence, but the philosophical differences between Hilbert and Carnap counsel to be cautions on this point.

2.4.3 Hilbert, the Formalist Programme and the Rise of Metamathematics

As we have just seen, Hilbert's consistency proof requires that the arithmetic of Ω be consistent. But since Ω is a subset of the field of reals, its consistency follows trivially from the consistency of the theory of real numbers. But it is not hard to show that both the arithmetic of Ω and the theory of real numbers are consistent if the arithmetic of naturals is. The foundation of Euclidean geometry advanced by Hilbert is thus in good standing only if the arithmetic of naturals is consistent. But how can we know that it is? Had Frege's programme succeeded, it would have offered, philosophical implications aside, a solid argument for this. Its failure leaves the problem open. Nor is the problem solved by Russell's or Carnap's versions of logicism. One might take Hilbert's *Grundlagen* as suggesting a suitable alternative solution, after all: Peano axioms could be conceived as Hilbert conceives of his own axioms (what, *mutatis mutandis*, is just what Peano himself did) – namely as implicit definitions of the structure of natural numbers – and they could then be proved consistent by giving an interpretation of them on a domain of independently available objects. The most obvious way of doing this is to use set-theoretic models such as Zermelo's and von Neumann's. But in so doing, one merely shifts the problem. Worst, one shifts it in the wrong direction, for the axioms of ZFC are much less perspicuous than Peano's.

A completely different strategy is suggested by Hilbert himself in a series of essays published since 1922, many of which written together with Paul Bernays, who was his assistant in Göttingen and the author of several essays devoted to the same topic.[40] As a matter of fact, the basic conception underlying this new strategy results from an idea already present in the 1899 treatise. The consistency proof offered there does not only require that the arithmetic of Ω be consistent, but also that a third theory be available, concerning the relations between the axioms of geometry and the arithmetic of Ω, within which the proof is conducted. Hilbert states this proof in an informal way, and this hides the problem of fixing this theory in an appropriate way. Still, the mere fact of reasoning mathematically about the relations between two mathematical theories (his axiomatic geometry and the arithmetic of Ω) shows the possibility of doing mathematics about mathematics: whatever this third theory might be, it should be a metamathematical one. Hilbert's new suggestion is to look for some sort of metamathematics that could be suitable for conducting a consistency proof for arithmetic, and mathematics in general, not depending on the supposed consistency of any another mathematical theory.

At the core of this programme is a method of mathematical proof suitable for delivering its conclusions with total certainty. According to Hilbert, this is possible if the proof deals with immediately accessible objects that are given to us in a way that makes perfectly manifest those features of theirs that are relevant for the proof itself. This is the case for a finitary version of arithmetic, which Hilbert sets himself to develop. Properly speaking, its objects are not defined; they are rather displayed, since they directly consist in appropriate configurations of signs. Thus, proofs do not take, in it, the form of deductions either within an axiom system providing an implicit definition of these objects or within some other sort of formal (or informal) system. They rather rely on the immediate intuition of these configurations of signs. Here is how Hilbert suggests these configurations of signs are to be fixed (1922, p. 163; English translation in Ewald, 1996, p. 1122).

> The sign 1 is a number. A sign that begins with 1 and ends with 1, and such that in between + always follows 1 and 1 always follows +, is likewise a number: for example, the number-signs 1 + 1 [and] 1 + 1 + 1.

Notice that Hilbert's number-signs do not stand for natural numbers; within this version of arithmetic, they rather are the natural numbers. They are not, then, symbols for, or expressions of, something else: they are, as such, what this version of arithmetic is about. Beside these, Hilbert also makes use of other sorts of signs that "mean [*bedeuten*] something and serve for communication" (*ibid.*). Among them, some, like '2' and '3', are "abbreviations" for determinate number-signs ('2' is an abbreviation of 1 + 1, and '3' an abbreviation of 1 + 1 + 1, for example); others, like '$\mathfrak{a}$', '$\mathfrak{b}$' and '$\mathfrak{c}$', stand for indeterminate number-signs; and finally '=' and '>', are used "for the communication of assertions" (*ibid.*).[41] Hence, by writing '$\mathfrak{a} + \mathfrak{b} = \mathfrak{b} + \mathfrak{a}$' Hilbert is intending to communicate, for example, that "the number-sign $\mathfrak{a} + \mathfrak{b}$ is the same as [the number-sign] $\mathfrak{b} + \mathfrak{a}$" (which means that 1 + . . . 1 + 1 + _ _ _ 1 is the same sign as 1 + _ _ _ 1 + 1 + . . . 1 whatever number of times 1+ is repeated in place of '. . .' and '_ _ _', respectively), whereas by writing '$\mathfrak{b} > \mathfrak{a}$' he is intending to communicate that "the number-sign $\mathfrak{b}$ extends beyond the number sign $\mathfrak{a}$" (*ibid.*).

Through this minimal language it is possible to draw proofs. Here is one Hilbert presents by way of example. Take any two distinct numbers $\mathfrak{a}$ and $\mathfrak{b}$. One of them will be a part of the other, just like 1 + 1 is a part of 1 + 1 + 1. Suppose that $\mathfrak{a}$ is a part of $\mathfrak{b}$, what might be communicated by writing '$β > α$'. There will then be a number c such that $β$ is just $α + c$, what might be communicated by writing '$β = α + c$'. To assert that $α + α + c = α + c + α$ is then the same as asserting that $α + β = β + α$. By eliding '$α$' in the former formula, one gets then that $α + β = β + α$ if and only if $α + c = c + α$. But c is, by hypothesis, contained in $β$, as well as $α$ is. Thus, if we go on this way, in a finite number of steps we will arrive to the conclusion that $α + β = β + α$ if and only if $δ + δ = δ + δ$, for some number $δ$. Since it is clear that $δ + δ = δ + δ$, whatever number $δ$ might be, this proves that $α + β = β + α$, whatever numbers $α$ and $β$ might be.

This proof makes no use of axioms, and relies on deductive rules that are apparently as obvious as the conclusions reached in its various steps. This obviousness is due to the fact that the proof concerns a precise content: signs like 1 + 1 or 1 + 1 + 1. It is then certain since, as Hilbert says, it deals with "extralogical concrete objects that are intuitively present as immediate experience prior to all thought" (Hilbert, 1926, p. 171; van Heijenoort, 1967, p. 376): objects that are not merely identified as the items that satisfy some given conditions, but are rather directly displayed and given as such. The Kantian flavour of this quotation should be evident, and it is still more evident when one realizes that the numbers 1, 1 + 1, 1 + 1 + 1, etc. cannot but be types, and cannot be tokens, since each of them has different occurrences, which are just occurrences of the same (number-)sign. The adjective 'concrete' as referred to these objects should not be taken, then, in its usual meaning, according to which no type can be concrete, but rather in a different meaning, so as to concede that a type could be concrete when it bears an appropriate relation with its tokens.[42] Not only sensible intuition is at stake here, also pure intuition is.

This is even clearer when one realizes that the proof given above, and the corresponding theorem, concern an infinity of numbers. Nonetheless, their infinitary character is different form that of theorems like 'There are infinite primes'. The former theorem is a schema of theorems, each of which (for instance, '7 + 5 = 5 + 7') can be verified by finite computation, whereas this is not the case for the latter, since 'There are infinite primes' has no particular instance, and no finite computation provides us with a verification of it. An easy proof is available for the latter too, for instance along the lines Euclid indicates in proposition xx of the ninth book of the *Elements*. Let p be a prime number and P be the product of all prime numbers less than or equal to p. Take the number $P + 1$. Clearly, it cannot be divided by any of these prime numbers. Therefore, either it is prime in turn, or it has a prime divisor greater than p. Then whatever p is, there is another prime number greater than it.

This proof is very simple, but it appeals to properties of natural numbers that do not immediately stem from their being given as signs like $1 + 1$ or $1 + 1 + 1$, and that can then be fixed only through a set of axioms. The crucial point is just here: proofs like that of Hilbert, which do not follow from axioms but rather appeal to properties of natural numbers that immediately stem from their being given as signs, can only secure theorems of finitary arithmetic.

The difference between a (necessarily finitary) theory including only such proofs, like Hilbert's own finitary version of arithmetic, and an axiomatic (possibly infinitary) theory is crucial in his view. His basic point is that, since the former deals with objects that are, as said, "intuitively present as immediate experience prior to all thought", it is free from any risk of contradiction. His proposal is thus to complete finitary arithmetic by the introduction of "ideal elements", so as to get infinitary arithmetic (or mathematics, in general), and then to supply the latter with a proof of its consistency that works as a proof within his finitary version of arithmetic. Let us better explain this. In Hilbert's view, ideal elements are abstract items that are supposed to meet a number of conditions that characterize them. Typically, they are the items implicitly defined by an axiom system. These conditions, typically expressed by the axioms forming this system, are to be stated in a perfectly codified formal language. Hence, proving theorems about ideal elements consists in drawing formal deductions within this language, and this means nothing but writing appropriate sequences of strings of symbols belonging to it. These formal deductions are not to be confused, according to Hilbert, with genuine mathematical proofs. They are rather formal reformulation of genuine proofs using an appropriate system of signs. Hence, these signs and the strings they form are endowed with meanings that make them suitable for providing such a reformulation. Still, nothing prevents one from studying these signs and strings, and sequences of the latter, through a suitable metamathematical theory dealing with them just as Hilbert's finitary version of arithmetic deals with numbers-signs. Should this allow one to prove that no string of signs meaning a contradiction – what would be the case, for example, for '$1 = 0$' – could be formally deduced, this would also prove that the genuine mathematical theory whose proofs have been formally rephrased is consistent. And this proof would not be relative to the consistency of any other theory, since the only theory on whose consistency it would eventually depend would be the metamathematical theory in which it takes place, whose consistency would be actually warranted by its including only proofs warranted by their dealing with "extralogical concrete objects".

Hilbert's "formalist" programme in the foundation of mathematics is just the programme suggested by these considerations: to reformulate current mathematics through formal theories that, once conceived as purely deductive games, could be proved to be such that a string of signs meaning a contradiction could not be deduced in them, so as to prove the consistency of current mathematics itself.

Hilbert was under no illusion that this programme could be easily accomplished. He confines himself to offering just some examples, feeling confident in future success. His objective is clear, though (Shapiro, 2000a, pp. 163–4). Current mathematics, including ideal elements, or ideal mathematics for short, should satisfy the following conservativeness constraint with respect to finitary arithmetic (whose consistency is warranted by the possibility of stating it as a mere theory of number-signs): if T is an axiomatic formal version of an ideal mathematical theory, and τ is any statement in the language of finitary arithmetic, then it has to be possible to prove in T a reformulation τ^* of τ in the language of T only if τ is true (which means, if finitary arithmetic is stated as a mere theory of number-signs, that the properties that τ ascribe to the relevant number-signs are actually enjoyed by them). But it is natural to require that an ideal mathematical theory be such that every statement of finitistic mathematics could be rephrased in the language of a formal axiomatic version of it or of an appropriate extension of it. It is then natural to require that if τ is a truth of finitary arithmetic, there should be a theorem τ^* of an axiomatic formal version T of any ideal mathematical theory that is a reformulation of τ in the language of T. One can then reason as follows: imagine that T is not conservative with respect to finitary arithmetic, namely that there is a truth τ of finitary arithmetic such that the negation $\neg\tau^*$ of its reformulation τ^* in the language of T is a theorem of T. Hence, both τ^* and $\neg\tau^*$ are theorems of T, and thus T is inconsistent. But if T is inconsistent, then every statement in its language is a theorem of T, and then there certainly is a theorem of T that is a reformulation of a false statement in the language of finitary arithmetic. Thus T is non-conservative relative to finitary arithmetic. It follows that an axiomatic formal version T of an ideal mathematical theory is consistent if and only if it is conservative relatively to finitary arithmetic. The consistency proof comes down to the proof of conservativeness, then: one must prove that it is not possible to prove in T the reformulation of any false statement of the language of finitary arithmetic. Let us consider a truth of this arithmetic, '1 = 0' for instance. To prove that T is consistent, it is enough to prove (finitistically) that the reformulation of the negation of this truth, '1 = 0', is not a theorem of T, namely that there is no sequence α of formulae in the language of T, following from T's axioms, the last of which is the reformulation of '1 = 0'.

2.4.4 Gödel's Theorems

The hopes for a consistency proof for a large part of ideal mathematics, including ideal arithmetic, along the lines suggested by Hilbert were halted by an obstacle that many view as insurmountable: the two incompleteness theorems proved by Gödel (1931). Due to their importance, these theorems are discussed in many texts, at different levels of technicality and details. The reader might find helpful, for instance, Nagel and Newman (1958), Smullyan (1992), Potter (2002, ch. 10), Smith (2007), Berto (2009). Here we'll give a short summary, based on Shapiro, 2000a, pp. 165–8. Suppose that T is a formal theory sufficiently strong to contain a large part of ideal arithmetic (appropriately formulated). Suppose also that T is effective, namely that there is an algorithm establishing whether a given sequence of symbols of its language is a well-formed formula, and whether a given sequence of well-formed formulae is a proof.[43] It seems natural to claim that the role Hilbert assigns to his reformulations of actual mathematical theories could only be played by theories like this. Gödel's theorems prove two essential limitations for such theories.

The first of these theorems proves that there is a statement γ^* of the language of T that is a reformulation of a statement γ of the language of finitary arithmetic, such that if T is consistent, then neither γ^* nor its negation are among the theorems of T.[44] The proof of this theorem consists in the effective production of γ^*. It is thus independent of ideal mathematics. Now, if we interpret γ^* in a standard way, it asserts that γ^* itself is not provable in T. Hence, if T is consistent, γ^* is true under this interpretation, but it is not provable.

Gödel also shows that the proof of this first theorem can be carried out in T, and that under appropriate restrictions it is possible to prove in T a statement asserting that if T is consistent, then γ^*. Suppose that T is consistent, and that it is possible to prove within T that it is. It follows that in T it is possible to prove γ^*, contrary to what the first theorem establishes. Hence, if T is consistent, this cannot be proved in T. This is the second theorem. But we assumed that T contains a reformulation of finitary arithmetic (which is contained in any significant part of ideal arithmetic). Hence, if T is consistent, it is not possible to prove that it is by relying on finitary arithmetic, as Hilbert's programme would require.

Different ways out can be imagined for overcoming the obstacle posed by these theorems, and for carrying on with Hilbert's programme, once appropriately modified. Some of them are highly technical in character (Gentzen, 1936, 1938). One, discussed by Detlefsen (1986), is based instead on the idea that finitary arithmetic is intrinsically informal, and that no reformulation of it can be contained in a formal theory like T. Be that as it may, it is a fact that Hilbert's programme did not lead to any unanimously accepted proof of the consistency of arithmetic. Nevertheless, it had the effect of suggesting both a way of thinking of mathematics, and a way of proving mathematical theorems about mathematics (which is studied today by the so-called 'proof theory'), and these are still both extremely relevant in the current debate in the philosophy of mathematics.

2.4.5 The Intuitionistic Strategy

Both the logicist programme (be that in Frege's version or in Russell's and Carnap's) and the formalist programme aim at justifying some mathematical theories (or possibly all of them), considered in their entirety, without proving that, or asking whether, there are parts of these theories that are well-founded alongside others that are not (Hilbert's reasons for justifying finitary arithmetic are distinct from those justifying ideal mathematics; but the goal of founding arithmetic is unattainable for him, if a justification for the latter in its entirety is not available). A different view has been championed by Luitzen Egbertus Jan Brouwer in a series of essays published in the first half of the twentieth century.[45] According to Brouwer, many among the arguments that are commonly produced in mathematics are inevitably tied to the assumptions that there are mathematical objects whose existence does not in any way depend on the activity of a subject, and that the truth of statements about them is equally independent of the subjects' asserting them. If one rejects these assumptions, one should also reject these arguments and the results they allegedly prove. This makes a vast part of generally accepted mathematics unwarranted and unjustifiable.

Brouwer's ideas gave rise to a still popular philosophical movement, called 'intuitionism' due to the role that intuition plays in it. According to Brouwer, mathematics stems from a free mental activity, wholly governed by the pure intuition of time, this latter being understood as an internal, elementary and fundamental experience. There is an explicit reference to Kant here. Still, the aprioricity of space is rejected, and hence the foundational role of pure spatial intuition as intuition of the form of outer sense is denied. Geometry, or better, the various geometries, are thus for Brouwer, just like arithmetic and analysis, the results of mental constructions ultimately founded on the pure intuition of time and, more specifically, in the intuition of the "pure two-oneness". Here is how Brouwer himself introduces this point (1912, p. 80; English translation p. 80):

> However weak the position of intuitionism seemed to be after this period of mathematical development [in the second half of nineteenth century], it has recovered by abandoning Kant's apriority of space but adhering the more resolutely to the apriority of time. This neo-intuitionism considers the falling apart of moments of life into qualitatively different parts, to be reunited only while remaining separated by time, as the fundamental phenomenon of the human intellect, passing by abstracting from its emotional content into the fundamental phenomenon of mathematical thinking, the intuition of the bare two-oneness.

This intuition immediately produces finite natural numbers (taken as cardinals) and, through indefinite iteration, gives rise to the intuition of the linear continuum, and from here to all of mathematics, through appropriate mental constructions.

Brouwer's continuum, however, is not our usual (classical) real continuum. From the intuition of the pure two-oneness, mathematics is generated by successive mental constructions, and these are always governed by inner intuition; there is thus no place in it for the assumption of infinite totalities whose elements are not supposed to be generated step by step through an appropriate procedure. This view might be considered obscure, and partly it is. Brouwer, however, does not merely state his ideas in general; he also criticizes those mathematical arguments which he takes to violate his ideals, and offers alternative ones: this gave rise to a new mathematics (intuitionistic mathematics) whose certainty should be warranted, according to him, by its intuitive origins, namely by its being formed through a completely controlled mental activity. An example is given by Brouwer's real continuum, significantly different from the classical one. According to him, a real number is an infinite sequence of natural numbers, each of which is either computed by an algorithm or is arbitrarily chosen. In the latter case, however, the only information about a real number that can be used in a proof is those concerning an initial, actually obtained segment of the relevant sequence. Several results hold for this continuum that do not hold for the classical one, and vice versa. Intuitionist mathematics includes vast portions of classical mathematics, even parts of what to Hilbert would be ideal mathematics. Their justification is not made to rely, however, upon the consistency of axiomatic systems, but rather upon the introspective activity that is supposed to ensure that they can be constructed. Notice that we are talking about constructibility proper here, not of actual construction. As Poincaré (1900, 1902, 1906) did before him, Brouwer (who appeals to several of Poincaré's insights in shaping his own proposal both in his (1912) and in later writings) postulates a faculty (which Poincaré had called 'power of the mind (*esprit*)') that can warrant that some mental acts can be repeated indefinitely. Moreover, he postulates the faculty of acknowledging good constructive processes even when they have not been actually carried out (and presumably cannot be actually carried out, in fact).

It is clearly impossible to introduce here, not even briefly, intuitionistic mathematics. It is important to notice, however, that Brouwer's criticisms are mostly directed towards the acceptance of the law of excluded middle and all inferences related to it. For any statement *a*, this principle asserts that it is either the case that *a* or the case that not *a*. An immediate consequence of the law is the double negation elimination rule. This asserts that for every statement *a*, the negation of the negation of *a* is equivalent to *a*. The rule is used in some *reductio ad absurdum* proofs, where, in order to prove that *a*, one proves that *a*'s negation leads to contradiction. This often enables one to prove that there is an object of a given kind with a given property by proving that the statement that every object of the same kind has not that property leads to a contradiction.[46] Brouwer does not take all instances of the excluded middle, and the inferences relying on it, as fallacious: he accepts, for instance, that every natural number, however big, is either prime or non-prime. The reason for this is that, at least in principle, one of the two alternatives can be verified through effective calculation, which for Brouwer is equivalent to a construction. The same holds for existence assertions: in order to justify them it is necessary and sufficient to produce, or at least identify, an effective construction of the relevant objects. Since each natural number can be easily got by such a construction, the existence of natural numbers is thus for Brouwer certain and *a priori* ascertained. They are obviously abstract objects, and are perfectly described by the intuitionistically warranted theorems about them, these being thus true. But all this depends on the human activity: what exists and what is true in mathematics is all and only that to which this activity (at least in principle) allows us to access through a process that is warranted by intuition.

On these grounds, Brouwer rejects any identification of mathematics with linguistic systems governed by logical rules. Not only mathematics cannot be reduced to logic, it cannot even be stated through formal systems. Language is nothing but a means for communication, and within mathematics it should be used only to communicate the safe results of inner experiences, favouring their repetition in other subjects. Logic is then only one way of describing the inner process leading to these results, and by itself, can warrant no inference and, *a fortiori*, no mental construction.

Despite this principled denial of any substantial role of language and logic in mathematical production, Brouwer's ideas suggested to his student Arend Heyting (1930, 1956) the possibility of building a system of intuitionistic propositional and predicate logic, together with a suitable semantics that takes the relevant statements as assertions about the availability of proofs. The statement '$a \lor \neg a$' is not a theorem in this logic, of course; when interpreted by the corresponding semantics, it asserts that either a proof of 'a' or a proof of '$\neg a$' is available. Its truth depends then on the nature of a and, arguably, on the advancements of (intuitionistic) mathematics. Analogously, from a proof that the negation of a statement 'a' leads to contradiction follows (provided the system is consistent) that '$\neg a$' is not a theorem, but not that 'a' is, since this provides no proof of this latter statement. Hence '$\neg\neg a \Rightarrow a$' is no more a theorem in this logic.

Subsequent developments of intuitionism are mainly due to Heyting's system and its further developments. Beside technical contributions, there is today a renaissance of Brouwer's philosophical views on mathematics, its ontology, and its epistemology. In this context, these ideas are often connected to Edmund Husserl phenomenological interpretation of mathematics.[47] These themes are certainly cognate with what is discussed in our book; but unfortunately space forces us to leave them out of our discussion.

2.5 Gödel's Platonism and the Rise of Mathematical Intuition

Intuitionism survived the threats that logicism and formalism presented because it denied that logic (however understood) could justify any part of mathematics and even provide the objects of arithmetic, as the logicists maintained, or that it could help in formulating these theories in such a way as to make a consistency proof possible, as suggested by Hilbert. Having deprived logic of any foundational role, Brouwer comes back to Kant, looking at pure intuition for the ultimate warrant of mathematics. Something similar is done by Gödel. The notion of intuition, however, is treated differently in Gödel and in Brouwer. Brouwer amended Kantianism by rejecting the mode of pure intuition that was for Kant directed to the outer world; Gödel believes that we must go beyond Kant by acknowledging a form of non-sensible intuition suitable for providing access to abstract objects and/or truths which we could not access either through our inner sense (as suggested by Brouwer), or by merely granting the repeatability of some mental acts (as suggested by Poincaré). Mathematical intuition is thus pure according to Gödel, but it is not constitutive of either objects or truths.

Gödel's interest in philosophy (not only philosophy of mathematics and logic) lasted for all his life, and is witnessed by several notes, part of which have been published posthumously in the third volume of his works (CW3, see also UPE), and part of which are held at Princeton University, where he taught from 1940. These notes are inspiring studies by many scholars, and seem to point to strong connections with Kant's ideas and, towards the end of Gödel's life, with Husserl's thought (see Tieszen 1992, 2005; Føllesdal, 1995; Atten and Kennedy, 2003). Some (like Cassou-Noguès, 2005) have seen in these documents a progressive weakening of the platonist theses that were suggested in the two main philosophical works published during Gödel's life (1944, 1947–64). The impact of these essays on the subsequent debate has been, however, so huge that many of those who believe Frege's platonism went bankrupt see in Gödel the example of the archetypical platonist.

The first of these essays is explicitly devoted to a discussion of Russell's logicism, and more specifically of the effort made by Russell and Whitehead in *Principia Mathematica* to abide by the Vicious Circle Principle, steering clear of impredicative definitions. We have already seen, discussing Carnap's views, that a similar effort appears unjustified when one thinks of mathematics as a mere linguistic framework. But it is equally unjustified if one believes that mathematics deals with already given objects that our definition merely describe. This is just Gödel's view, who maintains that the Vicious Circle Principle is plausible only if a "constructivist . . . standpoint toward the objects of logic and mathematics" is endorsed, namely whether one assumes that "the entities involved [in mathematical definitions] are constructed by ourselves", whereas "there is nothing in the least absurd in the existence of totalities containing members, which can be described (i.e., uniquely characterized) only by reference to this totality", if "it is a question of objects that exist independently of our constructions" (1944, p. 456; Gödel is here referring to Ramsey, 1926). If this last view is endorsed, impredicative definitions can at best point to a limitation in our epistemic powers.

The constructive point of view Gödel is opposing here is not Brouwer's, even though the label 'constructivism' is often reserved to the latter. As an example of the relevant kind of constructivism, Gödel indicates the no-class theory of Russell and Whitehead. For, according to this view, mathematics only deals with logical constructions based on immediately given data, though the problems Russell and Whitehead encountered in their foundational programme suggest a "more conservative course", according to which classes and concepts are "objectively existing entities" (*ibid.*, p. 468). Gödel better expresses this conception as follows (*ibid.*, pp. 456–7):

> Classes and concepts may, however, also be conceived as real objects, namely classes as 'pluralities of things' or as structures consisting of a plurality of things and concepts as the properties and relations of things existing independently of our definitions and constructions. It seems to me that the assumption of such objects is quite as legitimate as the assumption of physical bodies and there is quite as much reason to believe in their existence. They are in the same sense necessary to obtain a satisfactory system of mathematics as physical bodies are necessary for a satisfactory theory of our sense perceptions.

Gödel (*ibid.*, p. 449) takes the parallel between mathematics and physics, which he often stresses, to have been advanced by Russell himself in "one of his earlier writings" (probably Russell, 1903, *Preface*, or Russell, 1906, pp. 172–3; see also Russell, 1907 – which, however, Gödel could hardly have known – and, on the same issue, Irvine, 1989, pp. 309–16 and Mancosu, 2001, pp. 104–05) and to be supported by the developments of mathematics:

> He compares the axioms of logic and mathematics with the laws of nature and logical evidence with sense perception, so that the axioms need not necessarily be evident in themselves, but rather their justification lies (exactly as in physics) in the fact that they make it possible for these 'sense perceptions' to be deduced; which of course would nor exclude that they have a kind of intrinsic plausibility similar to that in physics. I think that (provided 'evidence' is understood in a sufficiently strict sense) this view has been largely justified by subsequent developments, and is to be expected it will be still more so in future.

But, in fact, Gödel revises Russell's view in a non-logicist sense, not by appealing to the evidence of logic, but rather to that of arithmetic and settheory (*ibid.*):

> It has turned out that (under the assumption that modern mathematics is consistent) the solution of certain arithmetical problems requires the use of assumptions essentially transcending arithmetic, i.e. the domain of the kind of indisputable evidence that may be most fittingly compared with sense perception. Furthermore it seems likely that for deciding certain questions of abstract set theory and even for certain related questions of real numbers new axioms based on some hitherto unknown idea will be necessary.

As it is presented in his essay on Russell, Gödel's platonism then consists in the claim that sets exist. This is justified, for him, by the development of set theory itself: far from making less plausible the claim that sets exist as independent objects, the need of stronger axioms, required in order to solve problems raised within the theory, even more reassures us of their existence and reveals their hidden properties. Rather than looking for a justification of mathematics down at the basis of our epistemic powers, he looks for it high in its more extreme limits, in the assumptions of set theory.

This conception is confirmed, and made still more evident, in the second essay, devoted to the discussion of the Continuum Hypothesis (henceforth, CH). There are two versions of this essay. When Gödel writes the first (Gödel, 1947), he has already proved (1939a, 1939b, 1940) that if ZF is consistent, then also is ZFC + CH (the system obtained by adding CH to the axioms of ZFC).[48] Hence, if ZFC is consistent, the negation of CH cannot be proved in it. In his essay, Gödel conjectures that if ZFC is consistent, not even CH itself can be proved in it, and thus that CH is independent of ZFC. This will be later proved by Cohen (1963). More precisely, Cohen will prove that if ZF is consistent, then so also is ZFC + ¬CH (the system obtained by adding the negation of CH to the axioms of ZFC). Gödel revises his essay when Cohen is still working on this, but sends out his manuscript (that will be published one year after the publication of Cohen's result; Gödel, 1964) shortly before coming to know of Cohen's proof, for which he accounts in a postscript. The main difference between the two versions, however, lies not in this postscript, but in another addition, a supplement where Gödel informs his readers of other results that had meanwhile be proved concerning CH, and criticizes the views of Alfred Errera (1953), according to whom the independence of CH from ZFC (later proved) would have made the question of its truth wholly meaningless, similar to what happened to Euclid's fifth postulate (which, as well known, is independent from the remaining postulates of Euclidean geometry, and whose negation results in two different axioms respectively included in two systems of non-Euclidean geometry). It is in replying to this objection, that Gödel appeals to the notion of mathematical intuition.

Already in its 1947 version, Gödel's essay is a manifesto for a research programme that is still very lively (see note 36): the search for suitable axioms to be added to those of ZFC so as to make CH decidable or, more specifically, its negation provable. Gödel's final statement is indicative (1947, p. 524; 1964, p. 264; the quotation is from the latter):

> I believe that adding up all that has been said one has good reason for suspecting that the role of the continuum problem in set theory will be to lead to the discovery of new axioms which will make it possible to disprove Cantor's conjecture.

The use of the term 'discovery' is a clear symptom of Gödel's philosophical views: a decision concerning CH is a decision concerning its truth or falsity, and this can be reached by deepening our analysis of the meaning of the non-logical constants occurring in set theory, beginning with the term 'set'. The aim of this analysis is not only that of elucidating this meaning and making our use of these constants more perspicuous; it will also help to reveal the real properties of sets. The passages where Gödel presents this idea are subject to significant changes between the two versions, but the main point seems to stand; (*i*) the inference from the non-decidability of CH in ZFC to the meaninglessness of CH relies on a constructivist view (in the sense specified above) of mathematical objects; (*ii*) this conception must be replaced by another, where it is acknowledged that "the set-theoretical concepts and theorems describe some well determined reality, in which Cantor's conjecture must be either true or false" (1947, p. 520; 1964, p. 476; the quotation is from the latter); (*iii*) hence, the non-decidability of CH in ZFC only indicates that "these axioms do not contain a complete description of that reality" (*ibid.*); (*iv*) this complete description can be attained through the addition of new axioms.

Gödel sees this as an aim within our reach, since the axioms that have already been accepted are based on a "concept of set" that itself "suggests" new axioms. In other terms, Gödel seems to think that the system formed by the former axioms can be extended in a "natural" way, "without arbitrariness", by adding new axioms to them. Leaving Gödel's examples aside, his point is that the new axioms can be suggested by considerations of two sorts: considerations relative to their capacity of solving open problems that are not solvable by the already accepted axioms, and to the comparison between the consequences these latter have when taken in isolation, as opposed to when they are taken together with the new ones; considerations relative to the "fruitfulness" of the new axioms, namely to their delivering a significant simplification of proofs also obtainable through the already accepted axioms.

In both cases, only technical work within set theory or about it (through Hilbert's metamathematical method) can reveal the suitable properties of sets still unknown. In other terms, the appropriate hints are visible only to the trained mathematician. This is exactly the point that Gödel seems willing to clarify in his 1964 supplement, through the notion of mathematical intuition. He observes that the theories respectively obtained by adding CH and its negation to the axioms of ZFC are not symmetric: "the latter (but not the former) has a model which can be defined and proved to be a model in the original (unextended) system" (Gödel 1964, p. 483). Considering models makes the situation similar to that we would have if we were willing to decide on the truth of Euclid's fifth element by taking geometry not as an abstract mathematical system, but rather as a theory "referring to the behaviour of rigid bodies, rays of light, etc." (*ibid.*) The difference is just that, in this latter case, our guidance would derive from our inquiry on physical reality, whereas in set theory we are guided by the consideration of another kind of reality. Mathematical intuition is what allows us, or better allows the trained mathematicians, to gain access to this reality (*ibid.*, pp. 483–4):

> Despite their remoteness from sense experience, we do have something like a perception also of the objects of set theory, as is seen from the fact that the axioms force themselves upon us as being true. I don't see any reason why we should have less confidence in this kind of perception, i.e., in mathematical intuition, than in sense perception, which induces us to build up physical theories and to expect that future sense perceptions will agree with them and, moreover, to believe that a question not decidable now has meaning and may be decided in the future.

But how does this intuition work? To explain this, Gödel employs once more a simile with physics (*ibid.*, p. 484):

> It should be noted that mathematical intuition need not be conceived of as a faculty giving an immediate knowledge of the objects concerned. Rather it seems that, as in the case of physical experience, we form our ideas also of those objects on the basis of something else which is immediately given. Only this something else here is not, or not primarily, the sensations. That something besides the sensations actually is immediately given follows (independently of mathematics) from the fact that even our ideas referring to physical objects contain constituents qualitatively different from sensations or mere combinations of sensations, e.g., the idea of object itself, whereas, on the other hand, by our thinking we cannot create any qualitatively new elements, but only reproduce and combine those that are given. Evidently the 'given' underlying mathematics is closely related to the abstract elements contained in our empirical ideas. It by no means follows, however, that the data of this second kind, because they cannot be associated with actions of certain things upon our sense organs, are something purely subjective, as Kant asserted. Rather they, too, may represent an aspect of objective reality, but, as opposed to the sensations, their presence in us may be due to another kind of relationship between ourselves and reality.

Though Gödel mentions Kant in order to oppose the latter's views, he certainly does not mean to reject all of Kant's philosophy. For even though he is speaking of ideas and thought (when he explains what we cannot do "by our thinking"), Gödel seems to have in mind something very close to Kant's concepts and understanding, respectively. We cannot have direct access to objects, both abstract and concrete or physical; in order to do this, we must use certain fundamental ideas (concepts) that thought (understanding) cannot provide. For this can surely combine ideas that are already given, but cannot create new ones. As far as physical objects are concerned, these ideas cannot even come from sensations, since this, in so far as it is sensation of objects, is already informed by those ideas. Similarly, in the case of mathematics, these cannot come from intuition either, if this is understood as a faculty providing an immediate relation between us and mathematical objects: mathematical intuition must contain something more. This something does not provide for other objects, but rather for appropriate and fundamental concepts: ideas in Gödel's but not in Plato's sense. Gödel's intuition differs then from Plato's *nous*, but seems rather close to Proclus', at least if one concedes that *nous* for Proclus generates *logoi* (see §§ 1.2 and 1.4). But what are these fundamental ideas (or concepts) that thought combines but cannot create? In the preceding quotation, Gödel says clearly that in the case of sensation, namely of our relation to physical objects, they include the very idea of object; but the quotation does not make clear which fundamental ideas are required in order to access mathematical objects.

To understand this, we must consider a footnote that Gödel adds to his text, after the words 'empirical ideas' (where he explicitly speaks of the "concept of set", making clear how the term 'idea', as used by him, is to be understood; *ibid.*):

> Note that there is a close relationship between the concept of set . . . and the categories of pure understanding in Kant's sense. Namely, the function of both is 'synthesis', i.e., the generating of unities out of manifolds (e.g., in Kant, of the idea of one object out of its various aspects).

What we need if we are to have access to mathematical objects is then the concept of set, which cannot come from our intuition of those objects, for this is already informed by that concept (or idea). It thus seems that mathematical intuition plays a twofold role: on the one side, it allows us to access to mathematical objects, though in a non-immediate way, that is, only through the use of a fundamental concept that, on the other side, it acquires from a reality (that Gödel calls 'objective', but apparently conceives as formed by concepts) that is prior to that of mathematical objects themselves, and similar to that of the categories of understanding, in a Kantian sense. From the fact that a single fundamental concept is evoked, and that it is the concept of set, it seems to follow that mathematical objects are just sets. If there is mathematical intuition of sets as objects, there is thus, by necessity, also mathematical intuition of the concept of set.

But not even this is enough. Gödel introduces a new ingredient, making his notion of mathematical intuition even more involute (*ibid.*, pp. 484–85):

> However, the question of the objective existence of the objects of mathematical intuition (which, incidentally, is an exact replica of the question of the objective existence of the outer world) is not decisive for the problem under discussion here. The mere psychological fact of the existence of an intuition which is sufficiently clear to

produce the axioms of set theory and an open series of extensions of them suffices to give meaning to the question of the truth or falsity of propositions like Cantor's continuum hypothesis.

Mathematical intuition seems thus to be conceived by Gödel, at the same time, as intuition of objects, namely sets, as intuition of concepts, or at least of the fundamental concept of set, and as propositional intuition, the intuition that sets are so-and-so, that some statements about them are true.

Though he is not willing to renounce the intuition of sets, and thus takes it that he can no more renounce the intuition of the concept of set, Gödel seems to suggest that only the propositional intuition that sets are so-and-so is crucial to mathematical practice. But if it is on this practice that one has to rely, then the elaboration of this involute philosophical apparatus related to intuition appears as a useless appendage of a specific inquiry on how we come to identify truths about sets, and, among them, those that are to work as axioms. This is a question Gödel had already dealt with in the previous parts of his essay included in its 1947 version, but to which he comes back again at the end of his 1964 supplement (*ibid.*, p. 485). In so doing, he also comes back to a criterion he had already briefly discussed earlier in the essay, that of the fruitfulness of axioms, and adds that it might not only concern the use of axioms in mathematics, but also their use in physics. This provides, Gödel claims, a "criterion of the truth of mathematical axioms" that is distinct from intuition and is "only probable"; a criterion that "though it may become decisive in the future, cannot yet be applied to the specifically set-theoretical axioms" (*ibid.*), in particular to those axioms that might be added to ZFC in order to make CH decidable.

This "criterion" has been often likened to the indispensability argument, which we will discuss in Chapters 6 and 7. Before this, we will consider, in Chapters 3 to 5, some reactions that the views presented so far, Frege's and Gödel's in particular, have elicited and still elicit among contemporary philosophers of mathematics.

3
Benacerraf's Arguments[1]

One of the main strands in the contemporary debate on Plato's problem has originated from two essays by Paul Benacerraf (1965; 1973). Before entering this debate, which will occupy us in Chapters 3–5, let us sum up some relevant aspects of Plato's problem that have emerged so far.

One way of formulating Plato's problem is to ask an ontological question: are there mathematical objects? In order to properly understand this question, and to answer it, at least three points should be addressed: (*i*) what does it mean to assert (or to deny) the existence of abstract objects? (*ii*) under what conditions can be asserted of something that it is an abstract object? and (*iii*) what makes it possible to conceive of an object as mathematical?

Question (*iii*) is only seemingly simple. As we noticed in the Introduction (pp. 3–4), if one wishes to answer it through a general characterization of what makes an object a mathematical object, the question becomes so complex that one would better avoid it. A way of answering it is to narrow its scope by focusing on paradigmatic examples. One obvious option is to deal only with arithmetic and set theory. The two often go together, especially if one concedes, as Russell did in his *Principles of Mathematics* (1903), that if arithmetic is about objects, then these are sets. The choice of confining oneself to the case of arithmetic, its possible extensions, and set theory, is shared by most of those who take part in the debate we will survey.

Given this restriction in scope, Plato's problem arises in the form of questions such as: do arithmetic and set theory speak of, or describe, objects? Are there objects such as natural numbers and sets (i.e. objects that are identical to natural numbers and sets)? Are natural numbers and sets objects?

A common way of answering question (*ii*) in the preceding paragraph is the following (cf. Introduction, pp. 4–6): *x* is an object if and only if it falls under a sortal concept. Whatever theory of concepts is adopted, it is natural to concede that some objects fall under a concept *P* if and only if there are *P*s. If this is conceded in order to answer question (*ii*), then it should also be conceded that nothing can be an object if it does not exist, so that the claim that *x* is an object entails the claim that *x* exists. This also answers question (*i*): to assert that there are objects – be they abstract or concrete – is tantamount to assert that there are objects falling under appropriate sortal concepts. This answer does not bring us very far, but it establishes a useful terminological convention. It also suggests that in order to argue in favour of platonism we must define appropriate sortal concepts, or, at least, we must say how they should each time be defined, and show that some objects fall under them. A way of showing this is to single out some objects independently of the use of those concepts, and then to show that those objects fall under those concepts, i.e. satisfy the conditions given in the definitions of the latter. This is exactly what is done by those who believe that natural numbers are identical to certain sets, and exist because these sets exist. Supporters of this strategy in the case of arithmetic must first define an appropriate sortal concept under which natural numbers are supposed to fall, say by appropriate axioms such as Peano's axioms, and must then show that certain sets meet the conditions in the given definitions, and thus fall under that concept. But he/she will also have to concede that those sets exist. The problem of the existence of natural numbers thus becomes that of the existence of (certain) sets.

The essential problem then is: what does it mean to define a sortal concept? Let us suppose that a condition of identity for what falls under a concept *P* has been fixed, i.e. that it has been established by stipulation that if *x* and *y* fall under *P*, then *x* is identical to *y* if and only if a given condition is fulfilled. Let us also suppose that if *x* and *y* fall under *P*, it is possible to ascertain whether this condition is or is not fulfilled. Can we claim that the sortal concept *P* has been defined?

To avoid being deceived by our familiarity with the concept involved, let us suppose that *P* is an entirely new concept, the concept of flagton for instance, and that all that we are able to say about this concept is that it is associated with a certain identity condition for anything that supposedly falls under it (which means that all we can say about flagtons is that *x* is the same flagton as *y* if only if *x* bears a certain equivalence relation to *y*). Have we with this defined the concept of flagton? One might answer in the negative on the following lines: in order to define a sortal concept it is necessary, but not sufficient, that an identity condition for anything that supposedly falls under it is fixed; one must also fix an application condition, namely a condition that any single x has to meet in order to fall under this concept (so that we would not have defined the concept of flagton if we were not able to say what makes x is a flagton, that is, if we were not able to fix a condition that x has to meet in order to be a flagton). But is this true? Couldn't one argue that, in some cases at least, the identity condition is sufficient? And could not the case of numbers be one of these cases?

As we saw in § 2.1.3, Frege rejected this possibility in Grundlagen. He noticed that an attribution of a natural number to something consists in making an assertion about a concept, namely the assertion that such a number belongs to a certain concept; from this he inferred that in order to define the natural numbers it is necessary to establish what this assertion means, thus ruling out the possibility that natural numbers be defined through the mere stipulation of an identity condition like that laid down by Hume's Principle: the number that belongs to the concept F is the same as the number that belongs to concept G if and only if F and G are equinumerous. A similar definition would be purely contextual: it would not determine what natural numbers are, but would only fix a truth-condition for statements of identity in which the terms employed for denoting natural numbers occur (statements of the form '$\alpha = \beta$', in which 'α' and 'β' are names for natural numbers).

If Frege rejected that natural numbers could be defined through the stipulation of Hume's Principle and opposed any attempt to define any other mathematical object through contextual definitions, the reason was that he thought that definitions of this kind run into the Caesar problem (see § 2.1.3), and that the latter could be overcome only by an explicit definition fixing a condition of application hence allowing us to say what natural numbers, or any other mathematical objects are. This is just the point with which Benacerraf's first argument is concerned.

3.1 What Natural Numbers Could Not Be (According to Benacerraf)

Frege's strategy for founding arithmetic and thus answering Plato's problem for that part of mathematics did not give the desired results. However, this is no reason to conclude that Plato's problem, as far as arithmetic is concerned, cannot be solved through a strategy structurally similar to Frege's, namely by establishing that every natural number n satisfies an identity statement of the form '$n = s$', where 's' is the name of an object that is identified independently of the consideration of the concept of natural number. Benacerraf (1965) argues against this possibility. Actually, he aims to do more than this: he wants to show that natural numbers "could not be objects at all" (*ibid.*, p. 79).

In § 2.3 we noticed that it is possible to identify natural numbers with particular sets generated one after the other, according to a recursive clause that establishes an ordering relation and allows us to obtain a model of Peano's axioms. We also said that this, by itself, does not solve either Plato's problem, or the problem of the foundations of arithmetic. However, if one wants to follow a strategy similar to Frege's, the most obvious candidates to be identified with natural numbers are sets, for example the set that form Zermelo's or von Neumann's progressions.

The first part of Benacerraf's argument (1965, Part I and II) is meant to show that natural numbers are not these sets, and that, in fact, they are not sets at all.

He tells a story about two young boys, Ernie and Johnny (these being the nicknames, respectively, of 'Ernst' – Zermelo's name – and 'John' – von Neumann's name), who have been trained to identify natural numbers, respectively, with von Neuman's and Zermelo's sets (Benacerraf deliberately attributes von Neumann's views to Ernie, and Zermelo's views to Johnny). For Ernie and Johnny, Peano's axioms are theorems about these sets. On these sets, they can easily define (explicitly) the usual arithmetical operations (addition, multiplication, exponentiation), and use these same sets to count objects,[2] proving that if k is the name of one of these sets, then every set of objects has k members if and only if it can be put into one-one correspondence with the set of sets including all and only those among these sets that are smaller than k.

Moreover, the progressions formed by Zermelo's and von Neumann's sets, that Johnny and Ernie, respectively, take as their numbers, are also ordered by a decidable ordering relation (i.e. a relation < such that for any two distinct elements of these progressions, x and y, it is possible to determine whether $x < y$ or $y < x$: notice that all progressions are by definition totally ordered, but it is possible that their order is not decidable; in the case of Zermelo's and von Neumann's, it is).

Ernie's and Johnny's numbers thus have all arithmetical features that are usually assigned to natural numbers, and they fulfil all conditions that natural numbers must fulfil.[3] A feature of sets, however, is that they can be members of other sets. Therefore, if natural numbers are sets, it makes perfect sense to ask whether one of them is a member of another one. Let us take two non-consecutive natural numbers, and let us ask whether the first belongs to the second. If natural numbers were

Zermelo's sets, the answer would be: no. If they were Von Neumann's sets, the answer would be: yes. Take an example. For Johnny, the numbers 2 and 4 are, respectively, the sets

$$\{\{\emptyset\}\} \text{ and } \{\{\{\{\emptyset\}\}\}\}$$

the former of which does not belong to the second (whose unique member is the set $\{\{\{\emptyset\}\}\}$). For Ernie, on the contrary, they are, respectively, the sets

$$\{\emptyset, \{\emptyset\}\} \text{ and } \{\emptyset, \{\emptyset\}, \{\emptyset, \{\emptyset\}\}, \{\emptyset,\{\emptyset\}, \{\emptyset, \{\emptyset\}\}\}\}$$

the first of which obviously belongs to the second.

But if natural numbers were certain particular sets, the aforementioned question should have a unique correct answer (in the affirmative or in the negative). Which of these answers is the correct one, then? Benacerraf's main point is that, insofar as Ernie's and Johnny's sets fulfil all conditions that natural numbers must fulfil, there is no way to reach a verdict other than to appeal to arguments extraneous to arithmetic and its applications. It follows that if natural numbers are certain particular sets, arithmetic and its applications do not give us any ground for deciding which sets they are.

At this point, Benacerraf adds two more premises to his argument: (*i*) "if the numbers constitute one particular set of sets, and not another, then there must be arguments to indicate which" (1965, p. 58); (*ii*) these arguments can come only from arithmetic and/or its applications (this latter assumption is mainly left implicit in Benacerraf, 1965). Since neither arithmetic nor its applications can supply these arguments, it follows that natural numbers are not sets.

The second part of Benacerraf's argument (1965, Section III.C) goes even beyond this conclusion, arguing that natural numbers are not objects at all. Benacerraf notices that what makes it possible for Zermelo's and von Neumann's sets to fulfil all conditions that natural numbers must fulfil is not their intrinsic nature, but rather the fact that they form a progression (of the appropriate kind).[4] It follows that it is possible to take other objects in place of them: if these other objects form a progression (of the appropriate kind), they will fulfil the relevant conditions too. Therefore, neither are there reasons for claiming that natural numbers are certain sets rather than others, nor are there reasons for claiming that they are certain objects rather than others. It follows that not only they are not sets: they are not even objects of any sort.

But what is arithmetic about, then? The following quotations suggest the answer (*ibid.*):

> So what matters, really, is not any conditions on the *objects* . . . but rather a condition on the relation under which they form a progression . . . in giving the properties . . . of numbers you merely characterize an *abstract structure*, and . . . the elements of the structure have no properties other than those relating to them to other elements of the same structure. If we identify an abstract structure with a system of relations . . . we get arithmetic elaborating the properties of the 'less-than' relation, or of all systems of objects . . . exhibiting that abstract structure . . . Arithmetic is therefore the science that elaborates the abstract structure that all progressions have in common merely in virtue of being progressions.

The answer is thus that arithmetic is about the "abstract structure" of progression. Benacerraf makes then an opposition between the singularity of a structure and the plurality of the objects of arithmetic: if any object pertains to the latter at all, it is just one, and this is the structure of progression. This is made clear in Benacerraf's positive proposal (*ibid.*):

> there is a modified form of . . . [formalism] . . . denying that number words are names, which constitutes a plausible and tempting extension of the view I have been arguing . . . On this view the sequence of number words is just that – a sequence of words or expressions with certain properties. There are not two kinds of things, numbers and number words, but just one, the words themselves.

Benacerraf devotes just a few lines to presenting this idea, and concludes by saying that in order to sustain it in a proper way, a whole book would be needed. Such a book has not yet been written. Benacerraf himself, however, comes back on his own argument about twenty years later (1996a, §. 3), and finds four shortcomings in it.

First, it is doubtful that only arithmetic and/or its applications can supply arguments for determining which objects the natural numbers are. From the fact that arithmetic and/or its applications cannot supply any, it does not follow that there aren't any. Metaphysical (or other) considerations relative to the concept of number could perhaps give us some.

Second, were these arguments lacking at all, it would not follow that natural numbers are not objects: they could be objects even though no argument could allow us to decide which objects they are.

Third, the argument would not help against those who claimed, applying Occam's razor, that natural numbers are sets (or some other objects) since sets (or these other objects) suffice for doing what we want to do with natural numbers, to the effect that there is no need of assuming the existence of any other entity. One could easily acknowledge that different progressions could play the role of natural numbers: were one willing to fix in an univocal way the reference of terms denoting the natural numbers, it would be enough to choose one or other among these structures, and this would not commit one to any ontological assumption. This might even be superfluous, since it could be claimed, as Resnik (1997, p. 239) suggests, that there is no fact of the matter underlying the question whether natural numbers are one or another progression.

Fourth, the second part of the argument rules out with no reason the possibility of objects whose intrinsic nature depends only on their standing in certain relations with other objects: the possibility that natural numbers are sets being ruled out, it could be claimed that they are not even objects of any other particular nature, that could be singled out independently of the relations among the elements of a progression. This would not imply, one could argue, that natural numbers are not objects at all: they could be objects characterized only by their bearing such relations to each others.

Although, according to Benacerraf, these objections do not refute his argument, they point to possible strategies for a platonist response. We will see that all these strategies have been adopted (this being what prompted Benacerraf to self-criticism). In § 5.1 we will discuss the neo-logicist proposal, embodying the first two objections. The indispensability argument could, with some provisos, point to a way of adopting the strategy suggested by the third objection. Lastly, non-eliminative structuralism, which we will discuss in §§ 5.3 and 5.4, is straightforwardly inspired by the fourth objection.

More generally, this objection suggests that in order to argue that a certain portion M of mathematics is not about objects, it is not enough to argue against the possibility of solving Plato's problem for M by establishing that for any relevant individual constant 't' occurring in M an identity of the form '$t = r$' holds (where 'r' is the name if an object that can be singled out without appealing to the concepts of M). As a consequence, an argument such as Benacerraf's could only hold against some versions of platonism (Fregean in spirit), namely those who take the one just described as the proper (if not the unique) way of solving Plato's problem. Other versions of platonism could maintain, and indeed do maintain, that mathematical objects, or some of them, are *sui generis* objects, namely objects that are distinct from any other independently given objects (Parsons, 2008, p. 101).

There is also another simple way of rebutting Benacerraf's argument: one could argue that even if correct, it would only rule out that each name of a natural number refers to a unique object, but would leave the possibility open that such a name refers somehow to a plurality of objects. The following are at least two ways of elaborating this idea.

According to Nicholas White (1974, p. 112), "we can, in our arithmetical discourse, replace our numerical singular terms, such as 'three', by corresponding atomic general terms, such as 'is a three' or 'threes'". These general terms must be relativized to progressions, that is they must occur in phrases like 'x is a three in Σ', where Σ is a progression. Natural numbers would thus be "objects which occupy positions in progressions, and of which such predicates as '$\exists(\Sigma)(x$ is a three in $\Sigma)$' are true" (*ibid.*, p. 114; notation has been slightly modified in order to accord with present conventions). White's proposal seems then that names of natural numbers should be interpreted as predicate constants designating properties that are possessed by several distinct objects. There would not be the number three, but several numbers three, and a statement in which the term 'three' occurs would not be about $\{\varnothing, \{\varnothing\}, \{\varnothing, \{\varnothing\}\}\}$, nor about $\{\{\{\varnothing\}\}\}$, nor about any other particular object, but rather it will be about every object that, like these, can be considered as a three in a progression. The surface grammatical form of a statement like 'Three is a prime number', therefore, would not mirror its semantic structure. The statement should rather be rephrased as: 'For any progression Π, Π's three is prime in Π'. When so regarded, the statement would then be true.

On another score, Field (1974, pp. 209 and 220–2) has suggested that some singular terms partially refer to several distinct objects, each of which consists in a part of the term's reference. Names of natural numbers would be singular terms of this kind, and would thus partially refer to each object functioning as an appropriate element in a progression.

The ensuing two accounts of arithmetic are not subject to Benacerraf's argument, but may be countered with other objections. For example, both acknowledge the existence of more than one progression. In order to support them, one would have to justify, rather than the existence of the natural numbers, conceived of as single objects, the existence of several progressions. Moreover, the benefit of simplicity that common versions of arithmetic possess would be lost, since the recourse to paraphrases of arithmetical statements would be required.

3.2 Benacerraf's Dilemma

Let us come back to Benacerraf's first two self-criticisms. We noticed that neo-logicism embeds the first two of them. However, there is a much less demanding way to go through these criticisms. This consists in arguing – as Wright himself (1983, p. 127) has argued, independently of his own adoption of neo-logicism – that Benacerraf's argument is just an "argument for the inscrutability of reference of numerical singular terms with respect to the domain of classes". What it shows is that the use we make of those terms is compatible with "various interpretations", but it raises no problem that is peculiar to those particular terms. On the contrary, the argument just shows that those terms behave, as far as their reference is concerned, like any other singular term of ordinary language, at least if we accept Quine's views on reference and translation (see Quine 1960, especially ch. 2): if these terms have a reference, our ordinary use of them cannot help to determine what it is. It could also be claimed that arithmetic is not the source of our knowledge about natural numbers, but rather its outcome. It is not that natural numbers are what arithmetic says they are. Rather, arithmetic is what it is because natural numbers are what they are. This amounts to supporting a version of arithmetical platonism. If good reasons for accepting this view were available, Benacerraf's argument would prove helpless against it.

Both these rejoinders agree that Benacerraf's argument does not undermine platonism, but rather it makes it stronger, at least in the following sense: if natural numbers are objects existing independently of us, they exist independently of arithmetic, which is, for sure, an intellectual product of ours. It is thus not surprising that arithmetic does not tell us everything about natural numbers. But the two rejoinders are not without consequences. Frege's platonism is, as it were, a sober platonism: it claims that natural numbers exist because they are identical with objects whose existence is secured by independent reasons, namely extensions of concepts. Even if one rejected Frege's reasons for the existence of extensions of concepts, one could well conceive of reasons for accepting the existence of other objects that could be identical with the natural numbers. If we adopt one of the two rejoinders, however, this sober form of platonism might give way to an arrogant platonism, claiming for the existence of natural numbers independently of any attempt on our part to determine their nature, or even of our ever being in a position to do that.

Even though arrogant platonism is not an inescapable result of those rejoinders (neo-logicism is, for example, a way of avoiding it while endorsing them), many would see it as their natural outcome. The argument in the second essay by Benacerraf is devoted to resisting this result. It is not just an argument against platonism: it is rather an argument purporting to show that two different requirements concerning the truth of mathematical statements and mathematical knowledge, though independently plausible, cannot both be satisfied at the same time.

The first requirement is that "any theory of mathematical truth be in conformity with a general theory of truth . . . which certifies that the property of sentences that the account calls 'truth' is indeed truth" (1973, p. 666).[5] One can obviously raise doubts on the implicit assumption that such a general theory of truth is available at all. However, Benacerraf makes clear that his point is that a semantics structurally similar to that of the statements of ordinary language should be applied to mathematical statements, and that a distinction between truth and justification should be drawn in addition to this. He offers an arithmetical example. Let us take the statements 'There are at least three perfect numbers greater than 17' and 'There are at least three large cities older than New York'. The first requirement is satisfied if the semantic analysis of the latter is taken to be analogous to that of the former and, in particular, if both statements are taken as true when there are objects with the relevant properties, independently of our justification for this.

The second requirement is that the explanation of mathematical truth be compatible with that of mathematical knowledge: "since our knowledge is of truths, or can be so construed, an account of mathematical truth, to be acceptable, must be consistent with the possibility of having mathematical knowledge" (*ibid.*, p. 667). This requirement might appear by itself rather weak, at least for those who take knowledge to be justified true belief. But Benacerraf endorses a conception of knowledge that has additional bearing on this requirement: a subject knows that p if and only if his/her belief that p is "causally related in a suitable way with what is the case because p is true" (*ibid.*, p. 672). He does not state precise conditions on the nature of this connection; what matters is that a subject knows that p only if there is something that makes p true and the subjects is related to this something in a way that makes it is possible for her/him to determine if the truth conditions of *p* are satisfied.

Let us consider a platonist view satisfying the first requirement. If such a view is adopted, it is "difficult to see how mathematical knowledge is possible" (*ibid.*, p. 673), since it is difficult to account for the connection that, according to the second requirement, must hold between the subject's beliefs and what makes the corresponding statements true. More precisely, if we stick to the case of arithmetic, if natural numbers are abstract objects existing independently of us, "then the connection between the truth conditions for the statements of number theory and any relevant events connected with the people who are supposed to have mathematical knowledge cannot be made out" (*ibid.*). The second requirement would thus not be met. On the other hand, if we concede – according to a conception that Benacerraf calls "combinatorial" – that mathematical statements are true if and only if they are proven, or, in any case, that their truth depends on "certain . . . syntactic facts about them" (*ibid.*, p. 665), then the second requirement would be met, but the first wouldn't, since we would not have any explanation of why this should properly count as a condition for truth.

Benacerraf's point, in short, is that a good semantics for mathematics goes together with a bad epistemology, and a good epistemology goes together with a bad semantics. This is what has become known as Benacerraf's dilemma. Benacerraf makes two examples by way of illustration. The first is Gödel's platonism, that violates the second requirement to the extent that it does not explain adequately in which sense "axioms force themselves upon us as being true" (see § 2.5). The second examples traces back to Quine (1936), who presents a "combinatorial" conception of mathematical truths and fails in connecting the conditions for the truth of mathematical statements with their truth in a proper sense.

The strength of Benacerraf's second argument can be differently appreciated. It might be argued that Benacerraf gives no reasons for the adoption of his requirements. However, the argument has the merit of pointing to the apparent incompatibility between a semantic and an epistemological requirement, and this is so even if no appeal is made to questionable assumptions concerning the nature of mathematical knowledge, if not for the fact that it should be knowledge of human subjects induced by appropriate states of affairs. Indeed, as Field (1989b, p. 233) has remarked, Benacerraf's dilemma can be stated in such a way that it "does not depend on any assumption about necessary and sufficient conditions for knowledge" but instead "depends on the idea that we should view with suspicion any claim to know facts about a certain domain if we believe it impossible in principle to explain the reliability of our beliefs about that domain". More generally, Benacerraf does not much insist on the nature of mathematical knowledge, but rather on the possibility, for the subjects of this knowledge, to gain access to what that is supposed to be knowledge of. The problem is then that it seems implausible to assert at the same time that mathematical objects are abstract, that they exist independently of us, and that they are accessible to us. This is the so-called problem of access.[6]

3.3 A Map of Responses to Benacerraf's Dilemma: Contemporary Solutions to Plato's Problem

The problem of access has accompanied platonism since its origins. But Benacerraf's dilemma does not raise a problem only for platonists. As Bob Hale and Crispin Wright remarked (2002, p. 103), "the general issue of reconciling semantics and epistemology for mathematics is not just a challenge for would-be platonists – it faces all philosophical positions which allow that pure mathematics presents . . . a substantial proper part of human knowledge". It is thus natural that a vast part of the debate on Plato's problem of the last forty years has focused, more or less explicitly, on the attempted responses to the dilemma. These attempts have been accompanied as far as platonists are concerned, by the attempts to rebut Benacerraf (1965)'s argument.

Hale and Wright (2002, pp. 103–4) have suggested a classification of responses to the dilemma in "conservative" and non-conservative responses. The former are those that "[take] the dilemma head-on", conceding that: (*i*) mathematical statements must be taken at face value, according to their grammatical structure; (*ii*) pure mathematics, conceived of in accordance with condition (*i*), "represents, at least for the greater part, a body of *a priori* knowledge". At least one of these conditions is rejected by non-conservative responses.

Among the latter, Hale and Wright list Field's nominalism, Geoffrey Hellman's eliminative (modal) structuralism, and the conceptions stemming from the indispensability argument. Among the former, they distinguish between those – called 'intuitional' – according to which the problem of access can be solved only through the appeal to "a special faculty – traditionally 'intuition' – which enables an awareness of systems of abstract . . . objects and of their characteristic properties"; and those – called 'intellectual' – that acknowledge that "access to the objects of pure mathematics is afforded by our general abilities of reason and understanding". The only intuitional response they consider is the view of Charles Parsons. Among intellectual responses, they consider Shapiro's *ante rem* structuralism and their own neo-logicism.

All conservative responses find places, though with relevant differences, in a platonist framework. We will deal with them in Chapter 4, and to those mentioned by Hale and Wright we will add Edward Zalta's theory of abstract objects (that Zalta views as a kind of logicism). Non-conservative responses are more closely tied to a nominalist and/or empiricist tradition. In Chapter 5 we will consider, beyond those mentioned by Hale and Wright, also Maddy's empiricist and cognitivist platonism, and Yablo's fictionalism. We will leave the indispensability argument for a separate treatment in Chapters 6 and 7.

4
Non-conservative Responses to Benacerraf's Dilemma

In the present chapter we discuss some of the major responses to Benacerraf's dilemma. In particular, we will discuss some of those that in § 3.3 have been classified, following Hale and Wright (2000), as non-conservative. We will consider four of these: Field's nominalism; fictionalism, both in Field's and in Yablo's version; Hellman's eliminative modal structuralism; and the version of platonism advanced by Maddy on cognitive grounds.

4.1 Field's Nominalism: Mathematics Without Truth and Science Without Numbers

The first non-conservative response to Benacerraf's dilemma that we discuss has been proposed by Field (1980, 1982). Field advances a nominalist view, denying the existence of abstract objects that mathematical theorems are allegedly supposed to describe, but he accepts (Field, 1982, pp. 45–6) clause (*i*) among those defining conservative responses (see § 3.3 above). He agrees that the appropriate conception of the truth of mathematical statements makes it depend on the existence, independently of any cognitive activity on our part, of certain abstract objects having certain properties. This amounts to the satisfaction of Benacerraf's semantical constraint. As a consequence, mathematical theorems are not true according to Field, or are at best vacuously so. On the assumption that it is possible to know p only if p is true, only the vacuously true mathematical theorems might represent a body of knowledge. However, it seems safe to deny that the truth of these theorems is in any sense relevant to the acceptance of clause (*ii*). Field must therefore reject this clause.[1]

In his two above-mentioned works, Field does not put forward any positive argument in favour of his nominalism. Rather, he intends to oppose all those arguments that, in order to conclude that mathematical theorems are true, and that mathematical objects exist, appeal to the crucial role that mathematics plays in the sciences, in physics above all. This is the basic idea underlying the indispensability argument. According to Field, all these arguments rely on an implicit, often unquestioned, assumption: that the truth of the theorems of mathematics is a necessary condition for the successful applicability of mathematics to science. This is what he denies. More generally, he denies that the virtues of mathematics depend on the truth of its theorems, that these theorems must be true in order for mathematics to be "good" (1982, p. 46; 1985, p. 239).

According to Field, however, those arguments are the only non-circular arguments that have been advanced for platonism and/or semantic realism concerning mathematics. Rejecting them amounts to opposing these views and to defending nominalism. To claim that mathematics is a body of truths, it is not enough to note that it is "logically derived from an apparently consistent body of axioms"; one should also explain why we "regard the axioms as truths, rather than as fictions that for a variety of reasons mathematicians have become interested in" (1980, pp. vii–viii). If we try to explain this without taking the applications of mathematics into account, we end up appealing to an "unjustifiable dogma" (*ibid.*, p. 5). If the indispensability argument is blocked, then, "there is no reason to regard any part of mathematics as true" (ibid., p. viii) nor to concede that the abstract objects it speaks of exist. This does not amount to a positive argument for nominalism, nor even to a direct argument against platonism. But it gives, according to Field, good reasons to reject the latter (1989a, p. 45):

> We also can't have direct evidence against the hypothesis that there are little green people living inside electrons and that are in principle undiscoverable by human beings; but it seems to me undue epistemological caution to maintain agnosticism rather than flat-out disbelief about such an idle hypothesis.

We will see in Chapter 7 how Field suggests blocking the indispensability argument; here we will focus on the positive part of Field's view, that is on the reasons he offers in support of the thesis that mathematics can be "good" without being true, these being, *ipso facto*, grounds for a response to Benacerraf's dilemma.

It goes without saying that Field does not want to deprive mathematicians of the right to determine what counts as good mathematics, and to pass it out to philosophers. He just wonders (1982, pp. 46–9) whether there is, beyond consistency, any other general condition – that is a condition independent of the specific content of a given mathematical theory – that a given mathematical theory must fulfil in order to be acceptable (that is, in order to be considered as good). That there are no such conditions conflicts with the generally accepted requirement that mathematics could be useful, that is, that it could be applied in science (notice: the requirement is not that mathematics is useful, but that it can be useful, that is, that it is applicable). It seems obvious that consistency alone does not suffice to satisfy this requirement. There are, however, different ways for arguing that it isn't. According to Field, a mathematical theory can be applied to the study of a given phenomenon only if it is not pure, that is if it contains, alongside theorems concerning "mathematical entities" (the usual theorems speaking of numbers, sets or other allegedly abstract objects) also "bridge-laws" (1980, pp. 8–11) associating these entities with the entities involved in the phenomenon under study. This is what happens, for example, in the case of set theory with (non-mathematical) *urelement*, a version of set theory in which sets can have objects that are not sets as their members. In particular, granted that according to Field physics deals with concrete objects, these bridge-laws are intended to associate mathematical entities with concrete entities. Field calls 'nominalist' a language containing only terms denoting concrete objects and whose variables are supposed to range only over these. We can thus say that in order to be applicable to science, mathematical theories must be not pure, but nominalistically impure (this label is not due to Field, we introduce it for the sake of simplicity).

Now, the crucial point is that a theory of this sort, consistent as it might be, can have false nominalistic consequences: its axioms (including, as we have seen, bridge-laws) can have nominalistic consequences that are false. Field takes it as obvious that a similar theory is unacceptable. And he believes that the same holds for a nominalistically impure mathematical theory that, even if consistent, has nominalistic consequences that could be false. He takes it that only those (consistent) nominalistically impure mathematical theories whose nominalistic consequences, if any, are necessarily true are acceptable (or else these mathematical theories would have a nominalistic, and thus empirical, content, and this would be unacceptable).[2] Field seems to take this condition as obvious, and takes it as the nominalist equivalent of the condition that is often expressed in platonistic terms by requiring that the theorems of a mathematical theory be necessarily true – or, as he briefly says, the former condition is "more or less equivalent" to the latter (1982, p. 51). This is the general condition whose satisfaction Field envisages as necessary in order for a nominalistically impure mathematical theory to be accepted.

An important clarification is in order. It is clearly possible, in a nominalistic language, to formulate statements entailing that mathematical objects, or some of them, do not exist. This is the case, for example, with a statement that explicitly denies the existence of such objects, like 'There are no non-concrete objects', or with a statement that imposes restrictions on the cardinality of the domain of objects, like 'There are no (concrete or abstract) objects that cannot be named differently from other existent objects (in a countable alphabet)' (which implies that there are no non-denumerable domains of objects). The negation of these statements is not necessarily true. However the negation of the former follows from any mathematical theory (both pure and nominalistically impure), at least in so far as it is understood in accordance with its surface grammatical structure, as Field requires, to the effect that its truth entails the existence of abstract objects. Analogously, the negation of the latter statement follows from many commonly accepted pure mathematical theories, like analysis (the theory of real numbers), and then, *a fortiori*, from all the nominalistically impure mathematical theories enclosing these pure theories. Hence, if one required, for a mathematical theory (either pure or nominalistically impure) to be acceptable, that the negations of statements like these were not among its consequences – by advancing that they are nominalistic statements not necessarily true – it would follow that no mathematical theory (both pure and nominalistically impure) would be acceptable (in the former case), or that many commonly accepted mathematical theories (both pure and nominalistically impure) would not be acceptable (in the latter case). To avoid such an implausible conclusion, one should then supplement the aforementioned argument with a clause of separation between nominalistic and mathematical statements that would rule out cases like these and grant that the nominalistic consequences of a nominalistically impure mathematical theory that are required to be necessarily true for this theory to be acceptable be "agnostic" with respect to the existence of mathematical objects (Field, 1980, pp. 11–12; Shapiro, 1983, p. 523, note 4).[3] For the sake of exposition, we will assume in what follows that this clause is always met.

A mathematical theory that has no nominalistic consequences vacuously fulfils the aforementioned necessary condition of acceptability supplemented with such a clause. This is generally taken to be the case for pure mathematical theories. But someone who admitted the possibility of pure mathematical theories with nominalistic consequence[4] should presumably require, for Field, that also these theories fulfil this very same necessary condition. Since Field takes it that mathematical theories can only be either pure or nominalistically impure, he can lay down the necessary condition of acceptability for a mathematical theory *tout court* as follows: "for an interesting[5] mathematical theory to be good, it need only be *consistent with every internally consistent theory about the physical world*" (1982, p. 49), that is, it must be such that if N is any consistent body of nominalistic statements and M the relevant mathematical theory, then M + N is consistent.[6] He then calls 'conservative' a mathematical theory that is consistent with every internally consistent theory about the physical world. It follows that a mathematical theory is acceptable only if it is conservative.

Once this condition is laid down, everything Field says about mathematics concerns conservative mathematical theories. This is straightforward: in the light of this condition, if an alleged mathematical theory were not conservative, it could not be acceptable, and thus it could not presumably be an actual mathematical theory, that is a theory actually accepted by a community of mathematicians. We can thus say that according to Field all mathematical theories are conservative: in short, mathematics is conservative. Moreover, mathematics is also good in so far as it is conservative, and it is on this that its utility hinges. Conservativeness is the condition that must replace truth if the virtues of mathematics are to be accounted for.

There are two sides to this thesis: it asserts that mathematics is good, that it is acceptable, in so far as it is conservative; and also that its conservativeness explains its applicability. The first part of the thesis is an immediate consequence of the way in which conservativeness is defined, and of the assumption that conservativeness is the only necessary condition for acceptability. The second part must be independently justified. Let us see how Field proceeds.

Suppose that N is a consistent body of nominalistic statements, and M a (consistent) mathematical theory, possibly nominalistically impure. According to Field's condition, then, to claim that mathematics is conservative, in the sense just specified, is the same as claiming that, whatever N and M might be, a nominalistic statement φ is a consequence of N + M only if it is a consequence of N alone (what is usually expressed by saying that N + M is a conservative extension of N). In other words, mathematics is conservative if and only if the addition of any mathematical theory M to any body of nominalistic statements N does not inflate the body of N's consequences.

The preceding equivalence can be justified as follows. Suppose that φ is a consequence of N + M but not of N alone. Then N+¬φ would not be consistent and M would not be conservative. Vice versa, suppose that N is a consistent body of nominalistic statements and M + N is inconsistent. Then every statement would be a consequence of M + N, whereas not every statement would be a consequence of N. There would be a nominalistic statement φ that would be a consequence of M + N but not of N.[7]

Joseph Melia (2006, p. 203) suggests that Field's point here is that "mathematics must be conservative as otherwise it would be impossible to make sense of the popular view that mathematics must be true in all possible worlds or *a priori* true" (the phrase 'true in all possible words or *a priori* true' is Field's). This reading seems misleading, however, and the same can be said of many formulations by Field and by his commentators. According to Field, mathematics is not true in all possible words, or *a priori* true, at all. Rather, as we said above, its only statements that are true are those that are vacuously true. The issue here is due to something we have tried to make clear above: pure mathematics is obviously conservative, or should at least be obviously taken to be so. The crucial point about the conservativeness of mathematical theories concerns then nominalistically impure mathematical theories. These can have nominalistic consequences that, according to Field, must necessarily be true, true in all possible worlds, and/or *a priori* true. When Field claims that the only mathematical statements to be true are those that are vacuously true, he is referring only to pure mathematics or to non-nominalistic statements of nominalistically impure mathematical theories.

Let us come back to the thesis that the conservativeness of mathematics explains its applicability. In order to properly understand this, we can recur to a metaphor. Imagine that we want to draw some consequences from a given argument on the basis of a number of premises formulated in Italian. Suppose that we are native English speakers, and thus have a limited mastery of Italian, even though we know it well enough. We could avail ourselves of some bridge-laws establishing relations between certain Italian words and certain English words, even if we intend to formulate our conclusions in Italian. These stipulations would help us in finding some of the consequences of our premises more easily. However, they would be appropriate only in so far as the consequences that they allow us to draw were already consequences of the premises in Italian alone. What Fields seems to tell us is that (good) mathematical theories always provide bridge-laws, which are appropriate in this sense. Imagine that N is a body of statements of a given physical theory, and that the latter, as any good physical theory should do according to Field, includes only nominalistic statements. We could then avail ourselves of any (good) mathematical theory M for then "it might be much easier to see" N's nominalistic consequences – thus "facilitating deductions" from N to this consequences (1982, p. 50) – with no risk of incurring in nominalistic statements that are consequences of N + M but not of N alone. If we suppose that physics can be conceived of as a body of nominalistic statements, we would not only explain why mathematics can usefully be applied to physics, but we would also ground our explanation on a feature that is peculiar to the former, its conservativeness.

Let us see why conservativeness should be peculiar to mathematics. On Field's account, the role of mathematics within physics cannot be likened to that of stipulations introducing "theoretical entities" such as (some) sub-atomic particles (1980, pp. 7–8). These stipulations are useful since "a wide range of phenomena" can be accounted for by their means that cannot be appropriately explained by any known alternative theory. They are thus "theoretically indispensable", whereas mathematics is wholly dispensable from physics. Moreover (1982, pp. 50–2), if we slightly modify the conservativeness condition for mathematics in order to apply it to physics – and so we define a physical theory T as conservative if and only if for any body P of statements concerning observables and for any statement ψ also concerning observables, ψ is a consequence of P + T only if it is a consequence of P alone – we must then conclude that a conservative physical theory is unacceptable (if it is not part of a larger physical theory that is non-conservative and acceptable), since it cannot be tested through its observable consequences.

But even if all this is conceded, before claiming that the conservativeness of mathematics explains its applicability at least to physics, if not to the whole of science, we must make sure that actual physical theories can be conceived of as bodies of nominalistic statements. So far, Field has argued that (good) mathematics is conservative over bodies of nominalistic statements. Suppose that everything that a physical theory tells us about the physical world could be expressed by such statements only. This theory could be thus conceived of as a body of nominalistic statements and Field's argument would apply and show that good mathematics is conservative over it. But suppose that, on the contrary, a physical theory told us something about the physical world that could not be expressed by nominalistic statements. This is wholly plausible: a statement such as 'Gravitational acceleration at poles is equal to 9,823 m/s^2' tells us something about the physical world, but it is not nominalistic. The issue is whether what it tells about the physical world could be expressed by nominalistic statements only (we will come back to this in §§ 6.3.2 and 7.1). Should it happen that a physical theory tells us something about the concrete world that could not be expressed by nominalistic statements only, it could not be conceived of as a body of nominalistic statements, and Field's argument could not be applied to show that (good) mathematics is conservative over it. Field must then show that things stand differently.

Mathematics is not only ubiquitous in actual physical theories; its language is also employed within the latter in formulating consequences about the physical world. In order to make his argument cogent, Field must make sure that this is a contingent state of affairs, due to the major convenience of mathematical language: he must make sure that both the basic assumptions of actual physical theories and their consequences about the physical world can be appropriately reformulated nominalistically. What must be shown is that mathematics is not only conservative over bodies of nominalistic statements, but also that it is eliminable from actual physical theories without loss of empirical content. In other words, that it is possible to do with physics what Field (as opposed to other nominalists) does not believe possible to do with mathematics: to reformulate the physical theories in an appropriate way without in any sense appealing to abstract objects and their properties and relations, but also without any relevant loss, that is, in case of physical theories, any loss as regards what these theories say about the physical world.

In order to show this, Field presents an example. He shows how it is possible to reformulate Newton's Gravitational Theory in a nominalistic language: a language containing no terms for real numbers (nor, thus, for natural numbers) and no variables that are supposed to range over them (and, thus, no function or predicate letter designating properties of, relations among, or functions defined above, those numbers). This paradigmatic example is meant to indicate how one should proceed in other cases (the paradigmatic character of this example has been challenged, for example, by Malament, 1982, pp. 532–4; see also Leng, 2010, pp. 60–70).

Field first shows how to eliminate terms that are supposed to denote natural numbers and variables that are supposed to range over them.[8] For this, it is sufficient to appeal to numerical existential quantifiers, such as '$\exists_i x[P(x)]$' (i = 0, 1, . . .), that is read as follows: 'There are exactly i xs that are Fs.[9] These can indeed be recursively defined in first order logic, for instance along Frege's suggestions in § 55 of *Grundlagen*, as follows:

$$\exists_0 x[F(x)] \Leftrightarrow \forall x[\neg F(x)]; \quad \exists_n x[F(x)] \Leftrightarrow \exists y[F(y) \wedge \exists_m x[F(x) \wedge x \neq y]]$$

where m is the predecessor of n.

Things are more complex for terms supposed to denote real numbers and variables supposed to range on them. Field first shows that "the use of real numbers in geometry can be accounted for by the conservativeness of mathematics, without assuming the truth of the theory of real numbers" (1980, p. xi). He refers, in particular, to Hilbert's axiomatization of Euclidean geometry, and notices, correctly, that it does not appeal to real numbers, these being rather used in order to give models on which axioms are interpreted in consistency and independency proofs (see § 2.4.2). Field also notices that, in dealing with these models, Hilbert has actually proved a "representation theorem" showing that a system in which only genuinely geometrical entities occur fulfils the structural conditions of the field of real numbers. According to Field, this could be taken to be a nominalistic system, involving only concrete objects to which properties and relations are assigned. These concrete objects are points and space–time regions. Hilbert (and, at least to a first degree, Descartes), showed that operations and functions defined over real numbers can be recast as operations and functions defined over points in space and segments joining them. The mechanical theory itself, with which Newton has formulated his theory of gravitation, does not appeal to real numbers (whose theory was not available at that time). Field has to show that the lack of any recourse to real numbers is no limit to the appropriate and rigorous development of Newton's theory, up to its modern version, to the effect that such a modern version can be actually recast by appealing only to points and space–time regions, and defining on their basis (via implicit definitions) every other relevant magnitude: velocity, acceleration, force, mass, density, temperature, and so on.

We can only sketch the basics of this reformulation here. Let us consider the case of temperature (Field, 1980, ch. 7). According to Field, in the modern presentation of Newton's theory, laws for temperature are given as laws relative to a real-valued function T defined on quadruples of real numbers ($T: \mathbb{R}^4 \to \mathbb{R}$). If we can define over space–time points two appropriate functions with values, respectively, among quadruples of real numbers and among real numbers ($\varphi: ST \to \mathbb{R}^4$ and $\psi: ST \to \mathbb{R}$), we can then replace the function T with an appropriate function defined on the basis of these latter. If these are, in their turn, defined through other functions defined on space–time points and with values on these points, we can use these latter functions to replace T without appealing to real numbers.

Such an argument explicitly appeals to real numbers and real-valued functions defined on real numbers. As Field himself remarks (*ibid.*, pp. 5–6), this argument, like others in his book, is a platonistic argument. This is not a problem for Field, since if it is possible to prove "*platonistically* that abstract entities are not needed for ordinary inferences about the physical world or for science" (italics in the original), then, *a fortiori*, every argument in favour of platonism based on the indispensability argument will be illegitimate. This reply would be safe if Field's aim were just to reject the indispensability argument. But we have seen that the eliminability of mathematics from physics is also an essential premise for establishing its conservativeness over physics. If Field must appeal to platonistic assumptions in order to support this premise, he can no more appeal to conservativeness in explaining the applicability of mathematics to physics without requiring that the former's theorems are true (Chihara, 1990, pp. 162–3). His response to Benacerraf's dilemma would be jeopardized. Chihara (*ibid.*) has more to say on this issue: if appeal to platonistic arguments were needed in the explanation of the applicability of mathematics, the explanation would not be viable to a nominalist, for it would make use of statements that a nominalist deems untrue. Not even Field, then, should believe in his own explanation, nor could he hope to convince other fellow nominalists of its soundness.

This being set aside, Field's argument has elicited several objections (some of them are collected and discussed in Chihara, 1990, pp. 153–73 and Appendix; in the Appendix Chihara also considers arguments advanced by Field, 1984, that he takes to be "the beginnings of a reply" to his own objection, see also MacBride (1999). A recent discussion of the objections to Field's programme is given by Leng, 2010, ch. 3). We will discuss just two among them.

The first objection is due to Shapiro (1983). Field's formulation of the conservativeness condition is ambiguous: it can refer both to semantic and to syntactic consequences. If we restrict our attention to one of the several equivalent formulations given by Field, the former reading states that, in order for a mathematical theory M to be conservative over a body of nominalistic statements N, a nominalistic statement φ can be derived from N + M only if it can be derived from N alone. The other reading, on the contrary, states that φ is true in every model of N + M only if it is true in every model of N. If N and M make use only of a first-order language, Gödel's (1930) completeness theorem guarantees that syntactic and semantic consequences are co-extensive (for it asserts that in first-order theories it is possible to derive a given well-formed formula from another if and only if the latter can be satisfied in every model of the former). But Field's reformulation of Newton's Gravitational Theory makes use of a second-order language, containing variables for space–time points and variables for space–time regions (that is, for sets of such points), even though Field himself suggests a way of using only a first-order language in which variables range only over regions (conceived of as primitive entities, rather than sets of points). Field (1980, note 30) explicitly opts for the semantic reading, and notices that this might require some modifications in his arguments if second-order languages are allowed for. Shapiro's point is that both if a second-order language is allowed and if a restriction to first-order languages is made, Field's argument runs into trouble (this point hinges on some fairly plausible conditions relative to N and M that will not be presented here).

Let us suppose that N makes use of a second-order language. Field's arguments on the conservativeness of mathematics would then be sound only if they concerned semantic consequences. However, semantic conservativeness is not the kind of conservativeness that is relevant to his argument, since it gives no reasons for believing that mathematics is useful in physics to the extent that it helps to simplify deductions. For if mathematics were only semantically conservative and φ were derivable from N + M it would be for sure a semantic consequence of N, but nothing would grant that it would also be derivable from N. If it were not, then adding M would allow us to derive a (semantic) consequence of N that cannot be derived from N alone. Shapiro supports his point with the case of a nominalistic statement that is not a theorem of Field's nominalistic physics but can be derived from the latter together with set theory. This and other examples (see, for instance, Burgess, Rosen, 1997, pp. 118–23) have originated a vast debate (a brief summary of which can be found in Shapiro, 2000a, pp. 235–7). An important issue is that the very fact of appealing to the notions of semantic consequence in a second-order language (where this notion has no syntactic counterpart that could ground an explanation of it) should be troubling for a nominalist, since the notion of truth in every model of a given (consistent) body of statements appears inevitably non-nominalistic (on this point, however, see Field, 1984; see also Leng, 2010, pp. 51–7).

Let us now suppose that N only makes use of a first-order language. Field would then have to show that his reformulation of Newton's theory could be recast in first order language and would then be appropriate. But Shapiro denies that this is possible, since he takes it to be impossible to prove in the thus-reformulated theory, that space–time points (defined as regions with no sub-regions) satisfy the same structure satisfied by the quadruples of real numbers. Shapiro's point here (1983, pp. 529–30) is that if we suppose that N only makes use of a first-order language, we have no proper explanation of how mathematics gets applied to Newton's Gravitational Theory.

Overall, what Shapiro shows is that "either the mathematical theory is not conservative in the philosophically relevant way or the mathematics is not applicable to the physical theory in the usual way".[10]

The second objection (for various versions of which, see among others Malament, 1982, pp. 531–2; Resnik, 1985, pp. 164–9; Petitot, 1995, pp. 155–6; Shapiro, 1997, pp. 75–7; Leng, 2010, pp. 57–8) is directed towards the thesis, which is essential if Field's argument is to be taken to oppose platonism, that points and regions of space–time are concrete objects, and Euclidean geometry, in Hilbert's axiomatization, is a theory of these concrete objects (what, according to Field, would make it a physical rather than mathematical theory). This was clearly not Hilbert's view, and just a few mathematicians, physicists and historians of science would be prone to accept it. But let us discuss the issue in more detail, in order to avoid arguments from authority.

Field concedes that on these points and regions it is possible to define operations, relations and functions whose essential features depend on those points forming a continuum, that is a non-denumerable set, something that, after all, is explicitly imposed by one of Hilbert's axioms. Field acknowledges that this is a strong assumption, but claims (1980, p. 31) that according to his version of nominalism "the nominalistic objection to using real numbers [is] . . . not on the grounds of their uncountability [that is, on the fact that they form a non-denumerable set] or of the structural assumptions . . . typically made about them [but] . . . rather to their abstractness: even postulating one real number would have been a violation of nominalism as I'm conceiving it". He then adds (*ibid.*, p. 33) that the structural equivalence between a system of concrete objects like space-time points on the one side and a systems based on real numbers ($\mathbb{R}^4$) on the other should come as no surprise, given that, historically, "the theory of real numbers . . . was developed precisely in order to deal with physical space and physical time".

From an historical point of view, this is highly questionable. Apart from this, however, the issue is not whether the nominalist is arguing against the non-denumerability of real numbers, or rather against their abstractness, but whether a system of concrete objects can be a continuum and have all the other related properties attributed by Field to the system of space–time points. As Shapiro (2000a, p. 230) has remarked, the issue is not just whether space–time should be conceived of as a physical reality or as a mere relational structure, thus taking a stance in the celebrated dispute between substantivalist and relationalist views.

Even is one agrees with the former, as Field (1980, pp. 34–6) does, and concedes that "space–time points . . . are entities that exist in their own right", one could argue that it is one thing to say that space–time points are physically real, and another one to say that they are concrete. The problem hinges on how Field distinguishes concrete from (alleged) abstract objects. Oddly enough, Field remains vague on this point, however. He notices (*ibid.*, pp. 31–2) that "space–time points are not abstract entities in any normal sense", since "the structure of physical space is an empirical matter" and so our knowledge of these points is not *a priori*. This is a rather weak argument, since it is possible to maintain that "the essential properties of some abstract entities [such as sentences and propositions] are known empirically" (Resnik, 1983, p. 516).

Apart from this, one could maintain that concrete objects should (as necessary conditions for their being so): (*i*) satisfy Leibniz's principle of indiscernibles (the object x is identical with the object y if and only if, for every monadic property F, x has F if and only if y has it); (*ii*) be capable of existing as such independently of a totality to which they supposedly belong (to the effect that concrete objects allow at best for definitional, but not for ontological, impredicativity: see § 2.2, 2.4.1, 2.5). But if space–time points form a continuum, they could hardly fulfil both these conditions: the difficulty with condition (*ii*) should be obvious; as regards condition (*i*), one could observe that if this condition were to obtain, it should be possible, at least supposing that our senses can develop indefinitely, to single out objects through perception and to distinguish any concrete object from any other one concrete object, whereas it does not seem possible to discern the points in a continuum on the basis of their monadic properties).

The discussion might be carried on by pointing to other ways, not necessarily incompatible with the ones just presented, of distinguishing between abstract and concrete objects. But let us concede, for the sake of the argument, that space–time points are indeed concrete objects (see also Field, 1982, pp. 59–63 for other arguments for this claim). It will still remain an open question if on their basis it is possible, with no appeal to any other abstract objects, for instance sets, to define all the apparatus that is required in order to obtain the geometry that is involved in Field's reconstruction of Newton's Gravitational Theory. Field obviously thinks that this is possible, and argues (1980, pp. 36–40) that space–time regions (and, among these, segments) can be conceived of as mereological sums of points, rather than sets of them.[11] Were this so, not only it would be possible to think of geometry, at least the Euclidean one (supposing that the structure of space–time is

Euclidean), as a theory of concrete objects; it would also be possible to define real numbers, and the whole apparatus of analysis, on the basis of these objects, or even to view real numbers as concrete objects. Field's complicated strategy would thus then be superfluous, since it would be possible to reply to Benacerraf's dilemma at least for a large part of mathematics, and to give a satisfying philosophy for this part of mathematics, by claiming that it deals with concrete objects.[12] One would still have to explain how (apparently *a posteriori*) knowledge of these objects could be obtained, and how access to them could be granted. We don't know whether, and how, this could be attained. But surely, should this turn out to be unfeasible, Field's nominalism, however formulated, would prove unable to answer to the problem of access, and thus to reply to Benacerraf's dilemma.

4.2 Mathematics as Fiction: Field and Yablo

The aim of Field's complex argument is to deny that the theorems of mathematics must be true in order to be successfully applied to physics. But how should we understand mathematical statements as such, and account for the natural and widespread opinion that when we assert the theorems of mathematics we make (non-vacuously) true assertions?

We saw that Field's argument does not rely on the attempt to reformulate mathematical statements in a nominalistic language. Field, on the contrary, argues several times that the correct interpretation of mathematical statements is one according to which they are about putative mathematical objects. Since these objects do not exist, those who take the statements of mathematics as (non-vacuously) true are simply mistaken. Field's view thus presents itself as a sort of error-theory, similar to that proposed for ethics by John Leslie Mackie (1977): if believing in p means that the statement p is believed true, and believing that p is true implies believing that there exist the objects whose existence is a necessary condition for the truth of p, it follows from Field's view that anyone who believes that $2 + 2 = 4$ or that 8 is the number of planets is in error.

Field argues, however, that it is possible to speak of the truth of mathematical statements in another sense (1980, p. 2; 1989a, pp. 1–14), according to which truth does not depend on the existence and properties of any abstract object. It is a matter of adopting, towards these statements, the position known as 'fictionalism'. There are several way of describing fictionalism (see, for example, Papineau, 1988; Yablo, 2001; Kalderon, 2005; Eklund, 2011; Balaguer, 2011; Leng, 2010).

Broadly speaking, we can say that fictionalism towards a given area of discourse is the thesis according to which the statements in this area must be seen as analogous to those of a fictional story. More specifically, the singular terms occurring in those statements must be interpreted in analogy with the names of the characters of a piece of fiction: there are not, in fact, objects such as, for example, natural numbers, but we can still speak of natural numbers insofar as we take them to be useful fictions for particular purposes. We can thus claim that mathematical statements are (non-vacuously) true if (and only if) we do not mean with this to claim that they are true in a proper sense, or true *simpliciter*, but rather that they are true-in-a-fiction. In the same way we speak of Sherlock Holmes and say that the statement 'Sherlock Holmes lives in Baker Street in London' is (non-vacuously) true, but do not think that Sherlock Holmes really exists, nor that it is literally true that he lives in Baker Street in London. The predicate 'true-in-the-story' would play, within a story, the same role that the predicate 'true' plays in ordinary (non-fictional) discourse: for example, if the statement 'There are prime numbers greater than 2', taken at face value, is true-in-the-story, then it is true-in-the-story that there are numbers, some of which are prime and greater than 2. Nothing grants, however, that a statement that is true-in-the-story is also true *simpliciter*. This can of course be the case, if, for example, the fiction contains the statement '221b of Baker Street is between Park Road and Melcombe Street'. But what makes the statement true-in-the-story is what is said in the story, whereas what makes it (non-vacuously) true *simpliciter* is the way things stand in reality. It is only in the story of arithmetic that there are certain mathematical objects, the natural numbers, having determinate properties that make the statements of arithmetic true-in-the-story.

A key problem for fictionalists is to make clear in which sense mathematical theories can be seen as analogous to fictional stories. In a novel, for example, the author enjoys a degree of freedom in establishing what has to count as true-in-the-story that seems to be lacking in mathematics. This is not an unsurmountable problem for fictionalists, granted that one is prepared to admit that a fictional story can be extremely codified and supplied with definite criteria for establishing when a statement must be counted as true-in-the-story, also on the basis of other statements that are held true-in-the-story.

Let us rest content, for the moment, with this sketchy description of fictionalism, and let us come back to Field. Fictionalism helps Field to amend some implausible aspects of his view. It helps him, in the first place, to at least partially explain in what sense the statements of mathematics can be held (non-vacuously) true by many of those who assert them. In the second place, it allows him to reply to a possible objection, that he would not be able to account for the objectivity of mathematics. Take the two following statements: '3 + 2 = 5' and '3 + 2 = 7'. The former is usually held true, the latter is not. For Field, however, neither statement is true *simpliciter*, or, at best, they are both vacuously true (if interpreted as universal generalizations such as 'For every x, y, z, if $x = 3$, $y = 2$ and $z = 5$, then ...'). Thanks to his fictionalism, Field (1998) can claim, however, that the former is usually held (non-vacuously) true, contrary to the latter, since it is, contrary to the latter, true-in-the-story of arithmetic (in the same sense in which 'Sherlock Holmes lives in London' is true-in-the-story *The Hound of the Baskervilles*, contrary to 'Sherlock Holmes lives in Manchester'), insofar it follows from the axioms currently accepted for arithmetic.

Mark Balaguer (2009, pp. 136–7) has pointed to a limit of this view. In § 2.5 we mentioned the issue of the independence of the Continuum Hypothesis (CH) from the set theory ZFC, that is to the fact that the axiom systems ZFC + CH and ZFC + ¬CH are both consistent. Let us now suppose that the mathematical community comes to identify and accept a new axiom α, that this axiom is seen as self-evident, and that it can be added to the axioms of ZFC so to obtain a new theory ZFC + α from which ¬CH follows. Many would react by claiming that ¬CH has always been true, even if we did not know that it was, and that what we have here is a genuine discovery. If recourse to truth-in-the-story must allow Field to account for the widespread conviction that the theorems of mathematics are true, according to Balaguer, Field should be in a position to account also for this reaction. But the notion of truth-in-the-story only concerns mathematical theories that are currently accepted, and these theories (specifically, ZFC) are consistent with both CH and ¬CH. All Field could say is that neither CH nor ¬CH are currently true-in-the-story of set theory, and that ¬CH would be true in a different story, that of ZFC + α. But he cannot account for the widespread intuition according to which it is possible to discover a truth that was not known before, even if it was a truth: to do this, he should claim that there is, as it were, an ideal story of set theory, that this ideal story can be different from all the theories of sets that are currently accepted, and that a statement can be true-in-the-(ideal)-story of set theory even if it is not possible to say whether it is so only on the basis of the currently accepted theories. It is unclear, however, why should not be open to Field to reply that the intuition Balaguer speaks of, however widespread, is of any significance to a platonist, but not to a nominalist.

Field's is only one among several fictionalist views. In order to offer some guidance through the various proposals on the market, it will be helpful to introduce the distinction, that John Burgess and Gideon Rosen have suggested, between revolutionary and hermeneutic nominalism.[13] Skipping some details, we can say that revolutionary nominalists maintain that: (*i*) mathematical statements (be they pure or impure), taken literally, say what the platonist takes them to say; (*ii*) since there are no mathematical objects, these statements, taken literally, are never true (if not vacuously); (*iii*) the usual mathematical and scientific theories that include parts of mathematics should be replaced by appropriate nominalistic reformulations. Hermeneutic nominalists, and, in particular, those Burgess and Rosen (2005, p. 517) call "content-hermeneutic" nominalists, agree with clauses (*i*) and (*ii*). However, they believe that when mathematical statements (whether pure or impure) are taken literally, their real content does not get properly expressed. Revolutionary nominalists maintain that the surface form of mathematical statements suggests that they have the content that also platonists believe them to have. Hermeneutic nominalists agree, but they take the surface form of those statements to be misleading: when they are correctly understood, their real content proves to be wholly acceptable to a nominalist. Therefore, hermeneutic nominalists do not agree with revolutionary nominalists on clause (*iii*): they do not aim at offering nominalistic reformulations of mathematical statements, but rather at showing that these statements already possess a nominalistically acceptable content. Their task is thus to show how this content can be expressed in nominalistic terms or, if it cannot be, in some way nonetheless acceptable to a nominalist.[14]

Another version of hermeneutic fictionalism is the one Burgess and Rosen (*ibid.*, p. 516) call "attitude-hermeneutic" nominalism, referring with this to the propositional attitude[15] of mathematicians and scientists who accept mathematical statements, as they are usually formulated, without explicit reservation, and rely on them for the purposes of their activities. The basic claim of this view is that this behaviour does not in any way indicate that these mathematicians and scientists believe what those statements assert, or, in other words, that they hold them true, taken literally.

This second version of hermeneutic nominalism is the most controversial, given that it relies on a factual claim concerning what mathematicians and scientists actually believe that can hardly be confirmed or falsified (it is at least doubtful that a psychological or sociological inquiry could establish, with sufficient generality, the actual attitude of mathematicians and scientists). What is relevant, however, is that also this version of hermeneutic nominalism, as opposed to the revolutionary one, does not ask for a replacement of the statements of the usual mathematics with appropriate nominalistic reformulations.

Jason Stanley (2001) shows how the distinction between revolutionary and hermeneutic can be applied to the case of fictionalism. Revolutionary fictionalists adopt a prescriptive view: they concede that mathematical statements, taken literally, speak of abstract objects, but contend that this speaking of abstract objects should be viewed in analogy with a fictional story's speaking of its characters. Hermeneutic fictionalists adopt a descriptive view: their intention is to show that the actual understanding of mathematical statements that is at play in the scientific and mathematical practice is such that these statements are understood in the way in which the statements in a fictional story are.

It is not easy to locate Field's view according to these distinctions. On the one side, he wants to show how we can reformulate nominalistically our scientific theories, and this would suggest that he is listed among the revolutionary fictionalists (as Burgess and Rosen do). But, on the other side, he offers no nominalistic reformulation of mathematical theories (not even those that are used in scientific theories), and rather proposes to eliminate them (at least from science). Moreover, Field never suggests that current scientific theories should be replaced in practice with their nominalistic versions.[16] As far as his fictionalism is concerned, however, it is safe to claim that Field is alien to hermeneutic views. Even if he thinks that it is an error to believe literally in a mathematical statement, Field (1989a, p. 8) also maintains that it is implausible to attribute, in general, any specific view (whether platonist or fictionalist) to ordinary speakers and to mathematicians and scientists, for these do not generally have "anything like a consistent view" of what belief in mathematical statements implies from an ontological point of view.

An openly hermeneutic version of fictionalism is due to Yablo (1998, 2001, 2002, 2005).[17] Yablo has, more than others (but see note 23 below), tried to explain to what extent mathematical theories can be likened to fictions, or to figurative language (especially to a metaphorical one) in general. His position merges aspects of both content-hermeneutic and attitude-hermeneutic fictionalism. According to Yablo: (*i*) we should distinguish, in a mathematical statement, its "literal content" (platonist) from its "real content" (nominalist); (*ii*) the former is true-in-the-story of mathematics, the latter is true *simpliciter*; (*iii*) when both ordinary speakers and scientists assert a mathematical statement (unless explicitly assuming a platonist reading), they assert its real content. Those who assert a mathematical statement do so, according to Yablo, in a "fictionalist spirit" (2001, p. 74): they do not take it as literally true, but rather make "as if" there existed mathematical objects; they rather assert its real content. The literal content of a mathematical statement is then, at best, "quasi-asserted", and this requires distinguishing assertions from quasi-assertions.

According to Yablo, Field cannot account for this distinction. His view would be a sort of "meta-fictionalism", according to which, when we assert a mathematical statement φ, say '3 + 5 = 8', what we really assert is that φ satisfies some suitable correctness condition C (in Field's case, that φ follows from accepted mathematical theories). Yablo finds three problems with this. A modal problem: while '3 + 5 = 8' seems to be necessarily true if true at all, it is not necessarily true that it follows from our accepted theories (we could have no theories, or have different ones). Second, a problem of concern: if we assert 'The number of starving people is large', we surely do not want to say that the number of starving people is large according to our current mathematical theories; our concern is not with these theories. Third, a phenomenological problem: when we assert φ, we do not take ourselves as saying anything about a mathematical theory as such.

Yablo then suggests, as a way out, a form of "object fictionalism". According to this, while the literal content of a statement φ belonging to fiction $\mathcal{F}$ (namely, what we quasi-assert when we assert φ) is what is true-in-$\mathcal{F}$, its real content (what we really assert) is that some real circumstance, call it '$k_\phi^{\mathcal{F}}$', obtains, this being the real circumstance that makes φ true on the basis of what is established by $\mathcal{F}$, namely what makes φ true-in-$\mathcal{F}$.

Let us see an example. In the fiction of arithmetic it is established that there are entities which we can appeal to in order to say how many objects of some sort – for instance, Martian moons – there are in reality. This fiction thus says that there is an entity, the cardinal number 2, such that it can be used to express that the Martian moons are as many as they actually are, since this is, according to the fiction, the number of Martian moons. The real circumstance which makes the statement 'The number of Martian moons is 2' true-in-the-story of (applied) arithmetic that 2 (rather than other natural numbers) is the number of Martian moons is that Martian moons are as many as they actually are. The real content of this statement is that this circumstance obtains, and this can be described by means of numerical quantifiers (see § 4.1). Its literal content is rather that 2 is the number of Martian moons.

Therefore, both the metafictionalist and the object fictionalist think that, when we assert a statement φ belonging to fiction $\mathcal{F}$, we quasi-assert what is true-in-$\mathcal{F}$, (namely, φ's literal content); but the former thinks that we really assert that φ is true-in-$\mathcal{F}$, while the latter thinks we really assert that the real circumstance $k_\phi^\mathcal{F}$ holds.

This is easily applied to impure mathematical statements like 'The number of Martian moons is 2'. But suppose φ is a statement of pure mathematics, like '3 + 5 = 8'. To apply this view to it, it must be explained what, according to the object fictionalist, we really assert when we assert φ, namely what is the real circumstance making '3 + 5 = 8' true-in-the-story of arithmetic. One option is to say that this is the circumstance that the axioms of arithmetic are such that from them follows that the sum of numbers 3 and 5 is the number 8, namely that arithmetic is actually such that things are this way.

Yablo chooses a second option (2001, p. 78; 2002, p. 178). Suppose we have introduced the fiction of arithmetic so that when in reality we have three Fs and five Gs we can use the fiction to say that the number of Fs is 3 and the number of Gs is 5. Assuming that Fs and Gs are mutually distinct, it follows that the number of the objects that are either F or G is 8. We can express the real circumstance making this statement true-in-the-story of arithmetic with the following statement: 'If there are three Fs, and there are five Gs, and no F is a G and *vice versa*, then there are eight (F or G)'. Call this statement 'ψ'. We can express ψ in a first-order predicate language by using numerical quantifiers as '$(\exists_3 xFx \wedge \exists_5 yGy \wedge \forall x\neg(Fx \wedge Gx)) \Rightarrow \exists_8 z\,(Fz \vee Gz)$'. Yablo takes ψ to give the real content of '3 + 5 = 8'. Since ψ is a truth of first-order logic (pending further qualification about quantified arithmetical statements; see Yablo, 2002, p. 179), he concludes that "the real content of any arithmetical truth is a logical truth" (*ibid.*).

This raises the following issue. If ψ is a statement in a proper sense, so that 'F' and 'G' are predicate constants, it expresses an implication which only holds for the objects falling under those particular predicates. But this is not surely what we mean by '3 + 5 = 8'. A possible solution is to say that the real content of '3 + 5 = 8' is not given by ψ, but rather by a doubly universally quantified second-order statement quantifying over predicate variables (namely, the previous statement prefixed by '$\forall F \forall G$'). This would then require an interpretation of second-order logic immune to Quine's objections (see Chapter 2, note 15).[18] Another option is to read ψ as a schema, and take 'F' and 'G' as schematic letters: it would then express an implication holding for every substitution of these schematic letters with appropriate predicate letters. This raises a new problem, however: ψ is intended to express the real content asserted when we assert '3 + 5 = 8', but it is not clear how we can be said to assert schemas of statements.[19]

We have said that Yablo's proposal is motivated by his ascribing three problems to Field's. Does his solution to these problems give reasons for favouring his account?

As regards the solution of the problem of concern, it seems so. What he says about the real content of 'The number of Martian moons is 2' suggests, indeed, that the real content of 'The number of starving people is large' is that starving people are as many as we consider to be a lot, which has strictly nothing to do with arithmetic, indeed.

Things are different with the solution to the modal problem. On the one side, one could dissolve the problem by denying that the theorems of arithmetic are necessarily true if true at all. On the other side, also Field has argued (1984), though on different grounds, that a large part of mathematical knowledge is just logical knowledge. At best, Yablo's solution of the modal problem could lend support to his own account only if it were easily extensible to all theorems of arithmetic and to any mathematical theory.

But can this solution, and more generally Yablo's view on what makes statements like 'The number of Martian moons is 2' and '3 + 5 = 8' true-in-arithmetic, actually be applied to other sorts of mathematical statements?[20] Take a statement like 'There are infinite prime numbers'. While it seems easy to say, according to Field's view, that what makes this statement true-in-arithmetic is that it follows from the axioms of arithmetic that there are infinite prime numbers, it seems hard to find a real condition, external to arithmetic, that makes this statement true-in-arithmetic.[21] This suggests that Yablo's solution of the phenomenological problem is appropriate only in some cases, but not in others.

The issue goes far beyond this problem, and rather concerns Yablo's view as a whole. Assume that the real content of a statement can be expressed by another statement (literally understood). We should then be able to define a function RC from statements to statements, mapping a statement φ to the statement $RC(\varphi)$ which, literally understood, expresses φ's real content (given a mathematical fiction $\mathcal{M}$, $RC(\varphi)$ is the statement expressing $k_\varphi^\mathcal{M}$; cf. Yablo, 2001, pp. 77–8). But how are we to define $RC(\varphi)$ and find its values for a given mathematical statements? In the examples above, this seems to be feasible. Other cases can be harder, though. Imagine, says Yablo (2005, pp. 103–8), that we had Goodmanian ancestors (speaking only a nominalist mereological language, such as that codified by the "calculus of individuals" by Goodman, 1951). These ancestors might in time have come to feel the need of simpler languages to cope with their practical needs. So they might have gradually introduced more and more suitable fictions allowing them to talk progressively of new kinds of numbers and sets. We might think of our current situation as one in which we have forgotten our Goodmanian origins, and are thus unconsciously using languages appealing to all those fictions. The values of $RC(\varphi)$ should then be retrieved by going back in this chain of fictions, up to the statements expressing (when literally taken) the real content of our current statements. This picture, however, is of little help for the exact determination of those values.

Yablo in fact denies that it is possible to determine the real content of every mathematical statement (we shall come back on this in § 7.4.2). But this may not save his view from objections, since problems emerge already for simple statements, which should certainly have a determined real content, like 'There is exactly one prime even number'. A fictionalist could not accept its real content to be that prime even numbers are as many as they actually are. Thus it cannot be treated similarly to 'The number of Martian moons is 2'. But we cannot even treat it as Yablo suggests '3 + 5 = 8' should be treated. We might still treat it differently from how Field would, by saying for example, that its real content is that the axioms of arithmetic are such that from them it actually follows that there is exactly one prime even number. More generally, for any statement of a mathematical theory $\mathcal{M}$ we should take $k_\varphi^{\mathcal{M}}$ to be an actual state of affairs, and we might see it as the historical circumstance that $\mathcal{M}$ is as it is, which would not be, strictly speaking, a circumstance about $\mathcal{M}$ itself, but rather one about the way in which mathematicians have elaborated their theories. But is this really satisfactory?

A further problem is given by statements like 'The number of natural numbers is 0'. Taken as a statement of arithmetic, its real content is that the circumstance that makes it true-in-arithmetic holds. But this statement is not true-in-arithmetic, and this circumstance does not actually hold, then. So its real content should be false. However, a nominalist wants it to be true. In order to escape this predicament, we can take the statement as belonging to applied arithmetic, like 'The number of Martian moons is 2': its real content would then be that the real circumstance that numbers are as many as they in fact are holds, namely that there are none.

To explain the difficulties raised by these examples, Yablo distinguishes two roles that terms apparently denoting mathematical objects can have: the role of "representational aids" and the role of "things represented". The former allows us to use mathematical fictions to speak of real facts; the latter allows us to talk of objects within the mathematical fiction. In 'The number of natural numbers is 0', 'natural numbers' is used to denote things represented, the natural numbers, and to say that there are in fact none. '0' is used as a representation aid: it is used with the meaning it has in the mathematical fiction in order to express the real content 'There are no natural numbers'.

This duality of roles echoes Carnap's (1950a) distinction between internal and external existence questions we discussed in § 2.4.1, whose spirit it is Yablo's aim to recover. As Carnap (1950a, p. 250, note 6) notices, this distinction is meaningful only if it is possible to distinguish between "questions of meaning" and "questions of fact", hence between analytic and synthetic statements. We mentioned Quine's (1936, 1951a, 1951b) objections to this. If Yablo is to rehabilitate Carnap's distinction, he has either to rehabilitate the analytic/synthetic distinction too, or to ground the former on new bases. Yablo (1998; 2001, pp. 86–91; 2005, pp. 96–100) follows this latter strategy, relying on discussion of what Kendall Walton (1990, 1993) calls 'games of make-believe'. Skipping over details, Walton acknowledges two roles statements might have in these games. They can be used to say something internal to the game ('This is Excalibur', as said while grabbing a toy sword), or to say something about those objects serving as "props" for the game ('Excalibur has broken', uttered to indicate that the toy sword has broken). According to Yablo, this distinction runs parallels to the distinction between the roles of terms as things represented and representational aids respectively.

Now, should this distinction be conceived as related to two contents a single statement might have depending on the role assigned to its terms, or rather to two different uses of language due to different propositional attitudes? From what has been said so far, Yablo would seem to opt for the first alternative (thus qualifying as a content-hermeneutic fictionalist). However, Yablo (2001, p. 90) suggests that two different propositional attitudes can be singled out (thus qualifying himself, to some extent, as an attitude-hermeneutic fictionalist). When we assert 'The number of Martian moons is 2' we are not deliberately and consciously pretending that there are numbers; our attitude is not belief, but rather what Yablo calls 'simulation'. Simulation resembles pretending, but it can be adopted unconsciously (similar to what happens with dead metaphors, whose figurative meaning is no more perceived). Provided we do not consciously give a platonist interpretation to our mathematical statements, our default attitude, says Yablo, is simulation.

Carnap's internal/external distinction is then reinterpreted in terms of the distinction between assertions made within and outside games of make-believe. Similarly to what Carnap takes to be the case with internal questions asked within linguistic frameworks, when we ask if a simulated mathematical statement is true, what we are asking is whether within the fiction there are the objects that are required to make it true-in-the-fiction, namely if its literal content is true-in-the-fiction. Sure, we can also ask whether its literal content is true *simpliciter*. For Carnap, this would consist in asking the question outside any linguistic framework, and thus to pose a meaningless question. For Yablo, the literal content of a fictional mathematical statement is rather false *simpliciter*.

The thesis that simulation is the attitude speakers sometimes adopt is an empirical thesis, and should be testable through psychological inquiries. This makes Yablo's proposal amenable to empirical objections, like the so-called 'autism-objection' raised by Stanley (2001, pp. 48–9). Some autistic subjects have impressive mathematical abilities, still we know they cannot engage in pretence (this is indeed a diagnostic test for autism: see Baron-Cohen, Leslie and Frith, 1985; Leslie, 1987, 1994). If Yablo is right, their mathematical skills are inexplicable (but see Liggins 2010a). A further problem with simulation, raised by Burgess and Rosen (2005, p. 526), following suggestions by Horwich (1991), is that positing the existence of a propositional attitude distinct from belief seems gratuitous if it cannot be explained how this attitude can be manifested by speakers in their use of language, which the very definition of simulation seems to rule out in this case.

Burgess (2004) has other objections to fictionalism in general (see also Burgess and Rosen, 2005). He notices that the analogy between mathematics and fiction is often vaguely presented. He takes both hermeneutic and revolutionary fictionalism as defective. The former reduces to an empirical thesis on speakers' propositional attitudes. As such, it is unsupported by empirical evidence. Revolutionary fictionalism is instead a revisionist thesis, and as such it is immodest and unacceptable: it should not be open to philosophers to question such a successful practice like mathematics on philosophical grounds. Leng (2005a) has replied that immodesty really lies in offering reformulations of actual theories that do not preserve what is relevant to scientific practice; but granting that this is avoided, revolutionary fictionalism is not immodest *per se*. Balaguer (2009) has also replied to Burgess. According to Balaguer, fictionalism is a purely philosophical thesis, and as such it is irrelevant to mathematical practice.[22] Moreover, he claims, appeal to fiction is just an analogy (see also Thomas, 2000, 2002), and making the analogy too stringent might be inopportune and raises irrelevant problems.

Clearly, it remains an open question whether fictionalism might capture significant aspects of mathematics.[23] We will come back on other relevant aspects of fictionalism in § 7.2.4.

4.3 Eliminative Structuralism and its Modal Version

As we remarked in § 3.1, in closing his first essay Benacerraf emphasizes the opposition between the singularity of a structure and the plurality of the objects of arithmetic. He supports this claim with the suggestion that names of natural numbers lack reference, being merely words that are used to order in a given way. This leaves unexplained what we mean when we use these words. Once the idea that we are referring to objects (be they abstract or concrete) is ruled out, the only options seem either to offer appropriate paraphrases for the statements involving these names, or to endorse fictionalism. Moreover, the opposition between a single structure and a plurality of objects seems to be perspicuous only locally, for instance as far as the sole arithmetic is involved. But what about group theory, for example? A structure is commonly conceived of as a domain of objects on which functions and relations satisfying certain conditions are defined. A group, then, is a structure: it is a couple $<G, *>$ where G is any domain and $*$ is a law of internal composition on G (a binary operation such that if x and y are elements of G, also $x*y$ is) satisfying three conditions: it is associative; it admits a neutral element (an element u of G such that for every x in G, $x*u = u*x = x$); it is such that every element in G has an inverse in G (for every x in G, there is a y in G such that $x*y = y*x = u$). Group theory studies groups as such, their different types and properties. According to this theory, there are different types of groups: some are commutative, some aren't; some are finite, some infinite; some are cyclical, some aren't; and so on. Even if we concede that each type of group occupies a place in a structure, and that group theory studies this structure, still the definition of group suggests that groups are abstract objects, and that they are abstract exactly in so far as they are structures. Therefore, group theory is both about a single structure and about various objects. And what should we say, again, of the theory of functions of real variables? Real numbers are the elements of a structure, and these functions are laws operating on them, that is they are, or so it seems, objects defined on the elements of a structure.

These and other considerations suggest that one should not oppose structures to objects, but conceive rather of the former as networks of relations among the latter. This leads to what Parsons has called "a structuralist conception of mathematical objects": "the view that reference to mathematical objects is always in the context of some background structure, and that the objects involved have no more by way of a 'nature' than is given by the basic relations of the structure" (2008, p. 40; see also 1990, p. 303; Parsons quotes Resnik, 1981, p. 530, for a different formulation). This idea can be further elaborated in at least two ways: one could maintain that being concerned with a structure dispenses with unwelcome ontological assumptions, or rather that it makes these assumptions permissible. Let us consider the case of real numbers. They form a structure called 'complete ordered field' (technical details are not relevant here). No philosopher would deny this well-established mathematical fact, but we can specify structuralist views according to how this fact is interpreted. All structuralists agree, according to Burgess (2008, p. 403), that it implies that the expression 'the real numbers' means the same as 'the arbitrary complete ordered field'. The difference between the two alternative conceptions of structuralism thus hinges on the reading of the adjective 'arbitrary'. It can be maintained that the expression 'the arbitrary F' does not denote anything by itself, whereas the statement 'the arbitrary F is a G' means the same as 'all Fs are Gs'. But it is also possible to maintain that 'the arbitrary F' denotes a specific object having the peculiar feature that it does not have any property that is not shared by all Fs. Structuralists endorsing the first claim believe that statements about real numbers are generalizations over all complete ordered fields (all systems of objects instantiating the structure of a complete ordered field). Structuralists endorsing the second claim believe that real numbers are *sui generis* objects, each of which has the feature of not possessing any property that is not shared by all objects occupying the corresponding place in some complete ordered field. We will discuss this second option in §§ 5.3 and 5.4. Here we will confine ourselves to the first, that Parsons traces back to Dedekind (1888). Parsons (1990; 2008 chs 2–3) contains a thorough discussion of this view, which is then rejected (for more contained presentations, see Shapiro (1997, pp. 85–90; 2000a, pp. 270–5; for other versions of structuralism, see Resnik, 1997; Chihara, 2004).

One first possible move for proponents of this view is to generalize White's suggestion concerning natural numbers (see § 3.1). A structure can be defined without defining any particular system of objects that instantiates it: it is enough that the conditions that the relevant functions and relations must satisfy are laid down. The definition of a group shows this very neatly: it does not establish the existence of the structure (at least if one accepts that a structure exists only if there is a system of objects instantiating it), but it suffices to establish that the objects that are supposed to form its domain must have some relational properties. The basic idea is thus that a mathematical statement cannot say more: such a statement should thus be reformulated as a statement universally quantifying over the systems of objects instantiating a given structure. For example, a theorem such as '3 is a prime number' should thus be reformulated as 'For any system of objects σ, if σ instantiates the structure of progression, then the object occupying the 3-place in σ is σ-prime', where the term '3-place' is defined for all progressions in general. Parsons takes this option to lead to a form of eliminative structuralism: a view that "begins with the basic idea of the structuralist view of mathematical objects and develops it into an analysis in which reference or commitment to such objects, or to mathematical objects of a specific kind such as natural numbers, is claimed to be eliminated" (2008, p. 51).

If, however, no system of objects σ instantiating the structure Σ existed, then any statement of the form 'For any system of objects σ, if σ instantiates the structure Σ, then . . .' would be vacuously true. Every mathematical statement would then be true, and so would be, for example, both the statement corresponding to '3 is a prime number' and that corresponding to '3 is an even number'. Eliminative structuralism qualifies then as a plausible option only if it can be granted that there are systems of objects instantiating the structures that mathematics is supposed to be about. Hence, by itself, it represents neither an alternative to platonism, nor a response to Benacerraf's dilemma.

One way of avoiding this difficulty is to adopt a modal version of eliminative structuralism, such as the 'one' Hellman (1989, 1990, 1994, 1996, 2005) has offered, elaborating on a suggestion made by Putnam (1967; see § 7.2.2). The basic idea is that statements universally quantifying over systems of objects instantiating a given structure are to be replaced by statements universally quantifying over possible structures. These latter statements can be non-vacuously true even if no actual structures exist. The statement '3 is a prime number' would thus be interpreted as 'For every possible system of objects σ, if σ instantiates the structure of progression, then the objects occupying the 3-place in σ is σ-prime'.

Hellman suggests a general strategy for obtaining a modal interpretation of any given mathematical theory. He lays down a set of instructions that would allow, for any mathematical theory, to get both the modal counterpart of any statement of the theory, and a set of conditions that must hold if the theory thus obtained, involving only modal statements, is to be considered an appropriate version of the original one. According to Hellman, interpreting a given mathematical theory in modal terms consists in specifying, for the relevant theory, a 'hypothetical component' and a 'categorical component' of the translation procedure. The hypothetical component specifies, for any given mathematical theory, a general "translation scheme" (Hellman, 1989, p. 15) showing how an appropriate conditional modal statement is obtained from any statement of the theory. The categorical component, on the other hand, provides axioms that must be added to the set of modal statements obtained. Details for both components vary according to the particular mathematical theory under consideration.[24]

Let us consider, for instance, second-order Peano arithmetic (Hellman, 1989, pp. 11–44; 1990, pp. 314–21). Let φ be one of its statements. Hellman suggests that φ should be replaced by an appropriate modal reformulation φ_{msi} (the subscript 'msi' stands for 'modal-structuralist interpretation'):

$$\Box \forall X \forall R [\&PA^2 \Rightarrow \varphi]^X \; (Suc/R)$$

where '&PA^2' denotes the conjunction of the second-order Peano axioms, 'Suc' is the constant designating the successor relation or successor function implicitly defined by those axioms (whether it is a relation or a function clearly depends on the chosen formulation of the Peano axioms), 'X' and 'R' are variables ranging, respectively, over classes of objects and over relations or functions of the same type of Suc (for first-order Peano arithmetic, see Hellman, 1989, p. 24). This statement is to be read as follows: necessarily, for every class X of objects, and for every relation or function R of the appropriate type, &PA^2 implies φ, provided that quantifiers in &PA^2 and φ are relativized to the elements of X and Suc is replaced by R in these statements. The relativization to the elements of X allows us to restrict &PA^2 and φ to particular domains of objects, while replacement of Suc with R is meant to avoid interpretations of the successor relation that are incompatible with the modal interpretation (see Hellman, 1989, pp. 21–3, for details). φ_{msi} thus expresses the fact that, when appropriately relativized, φ holds in any possible progression (Hellman, 2005, p. 553). If φ is a consequence of &PA^2, then, φ_{msi} is a truth of second-order modal logic (otherwise, $\neg \varphi_{msi}$ is), once it is granted that it is possible that there are progressions (and thus also if no actual progression exist).

The categorical component of the modal interpretation for second-order Peano arithmetic includes several assumptions.[25] One assumption ensures the possibility of the existence of progressions, so that a problem similar to the 'one' we have seen for non-modal eliminative structuralism cannot occur:

$$\Diamond \exists X \forall R \; [\&PA^2]^X \; (Suc/R)$$

Without this assumption, φ_{msi} could be vacuously true. In addition to this, the categorical component also includes (some details are introduced in Hellman's works after 1989): an axiom of infinity, asserting the possibility of the existence of an infinite number of concrete objects (atoms or mereological sums – the axiom is expressed in mereological terms and uses plural quantification) and a comprehension schema for mereological sums, establishing that there is the mereological sum of the individuals having any given property, once it is granted that there exist at least one object with that property. We will come back on the motivations for these two assumptions. For the time being, it is enough to notice that they, together with the first, ensure that modal structuralism need not require the existence of any object, but only the possible existence of structures.[26]

As a matter of fact, as long as arithmetic is concerned, the assumption of the possible existence of the relevant structures (namely progressions) is dispensable, since the existence of the structure of progression is derivable from the other two assumptions. Such a "modal existence assumption" is required, however, in cases where this situation does not occur (further axioms might also be required for theories different from arithmetic).

It is already easy to see that recourse to modality is intended by Hellman to yield not only a mathematics without numbers – as the title of Hellman (1989) suggests – but also a structuralism without (actual) structures, as the label 'eliminative structuralism' indicates.

One first problem for this proposal comes from its use of second-order modal logic, even when the original theory is first-order: whatever the given theory is, the translation scheme appeals, as a matter of fact, to second-order logic (see Hellman, 1989, pp. 18–20; 1990, p. 322). Some might contest (as Hellman himself acknowledges) that this should not be allowed when giving a foundation for mathematics, since second-order logic already is a mathematical theory (see Chapter 2, note 15). Hellman has both a philosophical and a technical reply to these objections.

He points out (2005, pp. 25–6) that his modal structuralism is meant neither as a foundation of mathematics nor as a reduction of it to logic. It just shows what the adequate interpretation of mathematical theories is, and nothing prevents appeal to mathematical notions in this case. Hence, even if second-order logic were acknowledged to be a mathematical theory, the modal structuralist could appeal to it without circularity.

One might retort that if second-order logic actually were a mathematical theory, it should in its turn stand in need of an appropriate interpretation on the lines of Hellman's proposal, and this seems to engender some sort of circularity. This further objection is blocked by Hellman's second reply (suggested in Hellman, 1989, § 1.6, and elaborated in Hellman, 1996, pp. 104–14; see also Hellman, 2005, pp. 20–1). He denies that second-order logic should be regarded as a mathematical theory, and offers a mereological interpretation of it (see note 11 above): monadic predicate variables are supposed to vary over mereological sums of individuals. It is with this interpretation in view that Hellman introduces, in the categorical component of the modal interpretation of arithmetic, the above-mentioned axiom of infinity (granting the existence of an infinite number of objects, atoms or mereological sums) and the comprehension schema for mereological sums. In formulating the former of these axioms, he uses plural quantification as introduced by George Boolos (1975, 1984, 1985; see also Linnebo, 2008). He also appeals to Burgess, Hazen and Lewis (1991) for a way of reducing quantification on polyadic predicates to quantification over monadic predicates, so that second-order variables can be restricted to monadic predicate variables. This makes it possible not to assume the notions of relation and function as primitive, and to obtain the same expressive power of second-order logic thanks to the sole resources of mereology. As a consequence, second-order logic, so interpreted, could be taken as a non-mathematical theory. But, Hellman continues, were someone to insist that it should (maybe because of the infinitary axioms of the categorical components of the various mathematical theories, that could legitimately be thought of as parts of that logic), (s)he could still reply that it represents a primitive core of mathematics, not in need of any further interpretation, since modal structuralists might in the end accept that "some core mathematical content must be built into one's primitive notions if structuralism is to be articulated at all" (2005, p. 559).

Be that as it may, what is really relevant is that, according to Hellman (1989, § 1.6), this interpretation of second-order logic allows us to place a restriction on the modal existence assumption for arithmetic. If we concede that mereological atoms are concrete objects and that a mereological sum of concrete objects is a concrete object, we may restrict that assumption to the sole possibility of the existence of progressions of concrete objects. Hellman's modal-structuralist interpretation for arithmetic would thus be radically nominalist: (*i*) it would not postulate the existence of structures, but only their possibility; (*ii*) it would not postulate the possible existence of structures whose objects are abstracts, but only of structures whose elements are concrete (atoms or mereological sums); (*iii*) it would give a non-set-theoretic interpretation of second-order quantification over predicates and relations, by reducing polyadic to monadic quantification and interpreting the latter in mereological terms.

It is important (see Hellman, 1996, pp. 206–7) to distinguish this last point from the question whether a similar strategy can be applied as long as we go up in the order of quantification (that is, not only in the nominalistic interpretation of quantification over sets of natural numbers, but also over sets of reals, and sets of sets of reals, etc.), so as to allow for a nominalist (modal-structuralist) interpretation of theories with non-denumerable models, up to set theory. This is a complex issue (see Hellman, 1989, ch. 2; 1996). Here we can say that Hellman believes that the strategy could be iterated just by assuming the existence of a denumerable infinity of concrete objects (atoms or sums) up to the third order (that is, up to the theory obtained from second-order arithmetic by extending both its language with the introduction of third-order variables and its axioms with the addition of a comprehension axiom for these variables), and that by the assumption of the existence of continuum-many mereological atoms the strategy could also be applied to the fourth order. According to Hellman, these theories are powerful enough to account for most of the empirical applications of mathematics in empirical science, thus dispensing with set theory. It still remains that the acceptance of a domain of concrete objects with the cardinality of the continuum is deeply problematic for the reasons that were already discussed, on pages 123–5 above, with respect to Field's thesis that space–time points, which form a continuum, are concrete objects. More generally, this relates to an objection raised by Parsons (2008, § 3.17). Parsons is here discussing a suggestion made by Putnam (1967) on the possibility of a nominalistic interpretation of set theory – a suggestion which has inspired Hellman (1989, p. 71) – but the objection affects eliminative modal structuralism generally. Parsons argues that since a domain of concrete objects big enough to satisfy the standard model of set theory could not exist, the modal structuralist should acknowledge the possibility of the existence of a suitable domain of abstract objects. According to Parsons, this means that a modal structuralist who wishes to account for set theory could not be a nominalist.[27] However, as Burgess (2008, p. 405) has noted, this objection bears some weight only to an "archnominalist" who maintained that the notion of abstract object is unintelligible, and to whom even the mere possible existence of abstract objects would be unacceptable. But one could be nominalist and at the same time admit that the relevant notion is intelligible and that the existence of abstract objects is possible, though not actual.[28]

Hellman himself makes clear, however, that his eliminative modal structuralism permits, but does not mandate, a nominalist stance. If no such stance is assumed, modal structuralism can be seen as a form of "modal *neutralism*" (Hellman, 1989, p. 117, emphasis in the original), neutral with respect to the problem of establishing which is the nature of the possible relevant structures (if this second route is taken, the mereological language employed is to be interpreted in such a way that atoms and mereological sums might include also abstract objects). The inability of offering a nominalist interpretation of some mathematical theories does not constitute, by itself, a fatal objection to Hellman's proposal, whose main aim is not to defend a version of nominalism, but to show that, for the most part, actual mathematics, especially that applied to empirical sciences, can be given a modal-structuralist interpretation, independently of the issue of nominalism, and even if not all mathematics can in the end be given such an interpretation (Hellman, 2005, p. 559). It should be added, however, that the assumption of the possible existence of infinite objects, together with the comprehension axiom for mereological sums, makes the assumption that atoms and sums are concrete objects far less innocent that what Hellman seems to acknowledge explicitly.

The main difficulty for Hellman's proposal lies elsewhere, however, namely in his appeal to modality. Logical modality (that is, the notions of 'logically possible' and 'logically necessary') can be explained in set-theoretic terms: if '$\Diamond$' is interpreted as a logical possibility operator and 'φ' is a non-modal statement, then '$\Diamond\varphi$' is true if and only if 'φ' is satisfied by some set. Clearly, we are not dealing here with possible sets (what would engender an obvious circularity), but sets whose existence must be granted by the semantics that is adopted for modal statements. If this were the sort of modality Hellman appeals to, he should not only acknowledge the possibility of the existence of structures, but also the (actual) existence of appropriate sets. His view would then be no better off than classical conceptions of platonism. But there is more to it: as Shapiro (1997, p. 89; 2000a, p. 275) notices, if Hellman's proposal were to be applied to set theory, the assertion that there exists a given set should be interpreted as an assertion about every logically possible system of objects instantiating the structure imposed by that theory, and this would be circular. Eliminative modal structuralism should thus interpret the existence of sets differently from the existence of other mathematical objects, or else it should appeal to a kind of modality different from the logical one.

Parsons (2008, § 3.15) discusses various interpretations of modality: the logical, the physical, the metaphysical, and the one he explicitly calls 'mathematical'. We have already seen difficulties concerning the first of these, but the remaining ones are also hard to pin down. It is obvious that it is metaphysically possible for a man to jump, with no instrumental aid, from the Earth to the Moon, even though this is not physically possible. It should also be clear that what is constructible through a given procedure that is admitted within a given mathematical theory, for example a point such-and-such that is constructed by ruler and compass, is mathematically possible. These are just elementary examples of intuitions that we all have concerning physical, metaphysical and mathematical modality. But it is difficult to state what these are in general.[29] It is just important to notice that all of them raise difficult questions for modal structuralism, as Hellman himself (1996, p. 103) admits. His solution is thus radical: the notion of modality should be taken as primitive. He (2005, p. 556–8) claims that we should not hope for a formal characterization of modality; he nevertheless offers some hints at an informal explanation of the notion, so as to make it philosophically acceptable.

Be that as it may, the most pressing problem that the interpretation of modality raises for modal structuralism, as Shapiro (1997, pp. 228–9) points out, is not so much to say what is meant by 'possible' and 'necessary', but to offer, for the chosen interpretation, a plausible epistemology: it must be explained what it means to say that we know that something is possible and, still more difficult, how we know what it is to be possible. Nothing guarantees that this epistemological problem has any more chance of being solved than that afflicting platonism in its various forms. Hence, modal structuralism, as such, seems to offer no answer to Benacerraf's dilemma. At best, it allows a reformulation in modal terms of the sort of problems that the dilemma brings to light.

4.4 Maddy and the Cognitive Origins of Set Theory

The last non-conservative response to Benacerraf's dilemma that we discuss is the form of realism for which Maddy (1980, 1989, 1990a, 1990b) has argued along empiricist and cognitivist lines. The reason for leaving this proposal just before moving to conservative responses is that its non-conservative character could be questioned. Maddy indeed accepts clause (*i*) of conservative responses. However, she rejects clause (*ii*) since she rejects a component sub-clause in it. Though Hale and Wright do not make this explicit, clause (*ii*) can indeed be further divided into two sub-clauses: (*ii*.a), that mathematical statements represent, at least for the greater part, a body of knowledge, that is of truths;

and (*ii*.b), that these pieces of knowledge, or truths, are *a priori*. Maddy accepts (*ii*.a), but rejects (*ii*.b).

According to Maddy, our knowledge of the most elementary parts of set-theory is based on our perceptual resources, and thus is not a priori. This does not prevent Maddy from considering her view a form of platonism, a "Compromise Platonism", as she calls it (1989, p. 1136), which is given by a compromise between Gödel's platonism and the form of platonism suggested by the indispensability argument. Maddy acknowledges that "successful applications of mathematics give us reason to believe . . . that mathematics is a science, that much of it at least approximates truth", but she notices that "this isn't enough to give an adequate account of mathematical practice", namely "an account of the obviousness of elementary mathematics, which Gödel's intuition is designed to provide", nor "an account of other purely mathematical forms of evidence", like the evidence given by proof. It can be doubted that intuition plays this role in Gödel, but Maddy's point is clear: if, on the one side, the successful applications of mathematics suggest that the latter cannot be accounted for only in terms of our free stipulations, and thus makes us lean towards some form of platonism, on the other side platonism cannot be endorsed if we lack a direct account of our beliefs in the building blocks of the mathematical edifice. In giving this account, Maddy appeals to perception (*ibid.*, p. 1140):

> We perceive sets of physical objects much as we perceive the objects themselves. Both abilities develop gradually as the course of childhood experience interacts with evolutionarily conditioned brain structures. The neurophysiological changes that constitute this development also produce a range of extremely general beliefs about these sorts of things, about the space-occupying character of physical objects, for example, and the combinability of sets. These intuitive beliefs, as I call them, underlie the simplest and most fundamental assumptions of our physical and mathematical sciences, those assumptions that "force themselves upon as being true" [cited in Gödel, 1947–64; see. § 2.5].

According to Maddy we thus perceive directly some sets, and not only the objects which are their members. Maddy is not just claiming that, in appropriate circumstances, a plurality of physical objects may get as a whole into causal connection with our perceptual apparatus, but that by this interaction we also acquire beliefs about the set of these objects, which would thus be causally responsible for the formation of those beliefs. As a justification for this claim, Maddy (1990a, pp. 58–67) notices that many experimental results – some of which have been discussed in Dehaene (1997), even though this is a fast-growing research field – show that our most elementary numerical beliefs are perceptual in kind. The role played by set theory in mathematics would suggest that these beliefs be understood as being about sets. The basic idea could be expressed as follows: given that the best formulation of our mathematical theories involve set-theory, and given that these mathematical theories are involved in our best physical theories and, in particular, in our description of perceptual phenomena underlying our elementary numerical beliefs, this description requires that some elements of the world are conceived of as sets (even though the subject who perceives them might not consciously identify them as sets).

Inspired by the theories of Donald Hebb (1949, 1980), Maddy also describes a possible neuronal process that would lead, after repeated experiences with collections of medium-sized objects, to the production of structural transformations in our brain, ensuing in the creation of a real "set-detector" (1990a, p. 65) that would yield our concept of set. Another process would allow the formation of non-inferential beliefs about the perceived sets, and then beliefs about sets in general. These beliefs are supposed to deliver a justification for the more elementary axioms of set theory. Maddy then claims that there is a further kind of justification beyond this, not given by perceptual processes but rather by "theoretical sources" (*ibid.*, p. 107). These two kinds of justifications would together allow us to build the whole edifice of set-theory. Theoretical sources are meant to consist for the larger part in the appreciation of the consequences of set-theory, in particular those consequences that can be obtained when new axioms are added to the more elementary ones, so as to allow the theories thus obtained to play what, according to Maddy, is their main function: to offer an appropriate framework for the reformulation of mathematical theories. Maddy has repeatedly emphasized this point (see Maddy 1990b, 1992, 1997, 1998, 2005, 2007, 2011), also supporting her claim with historical data on the development of set theory.

Maddy, then, appeals to empirical tests on numerical cognition. The results of these tests are well established, but their proper interpretation is controversial. Often, claims concerning how we form our numerical beliefs are advanced on the sole ground that some behaviour on the part of the subjects can be described by the use of arithmetical concepts. But it is one thing how some phenomena can be described if one is willing or disposed to employ a given system of concepts, and another to know what the real, neuronal and physiological causes of those phenomena are.[30]

But, for the sake of the argument, let us concede that these experimental findings are to be interpreted in a way that supports Maddy's claims. Her argument appeals not only to these findings, but also to the role that set-theory has in mathematics, and to the role that mathematics has in scientific theories. And this argument, which closely resembles the indispensability argument and apparently appeals to an instance of inference to the best explanation (see §§ 6.3.2.3 and 7.4.2), raises at least two problems.[31]

The first of these problems has been pointed out by Parsons (2008, p. 210–11). Let us grant that it is legitimate to draw conclusions about "the organism's basic relation to the environment" from the role mathematics plays in our description of perceptual phenomena that allegedly ground our elementary numerical beliefs. It still remains that the mathematical theories that are involved in this description, and even those mathematical theories that are involved in the psychological and neurological theories that are needed to link those phenomena to these beliefs, can be formulated without appealing to set theory. Even if we concede, then, that these theories are indispensable to the description of the above-mentioned perceptual phenomena, it does not follow that their set-theoretical formulation is. It is thus "not plausible that the formulation in terms of set theory reflects the nature of things to the degree that Maddy's view presupposes".

The second problem goes as follows. Both the role set-theory has in mathematics and the distinction between sets in a proper sense and mere collection of objects, or other sorts of aggregates of these, depend on the theory as a whole. Hence, Maddy's argument cannot ground the claim that we perceive the objects that a theory playing a crucial role in mathematics is about. For either this theory is set-theory proper, and then it is not properly about some objects we perceive (since we merely perceive, at best, some collections or aggregates, whereas set-theory is about sets), or this theory is properly about some objects we perceive, and then it is not set-theory proper. At best, what Maddy's argument can support is the following: that we perceive collections or aggregates; that this suggested to some mathematicians the idea of a more complex notion that might have proved useful in the foundation of mathematics; that this has led these mathematicians and their heirs to clarify this notion by developing set-theory and then to obtain the envisaged foundation of mathematics; and that they were led to go as far from the notions of collection and aggregate as their need for a theory suitable for founding mathematics required. What we perceive, according to Maddy's account, would then at best be fairly distant precursors of sets conceived as the inhabitants of the set-theoretical realm of mathematics, the difference between the two being that between the objects falling under a system of concepts suitable for an appropriate account of the perceptual experiences Maddy is concerned with, and a proper mathematical theory.[32]

If this is correct, Maddy can at best draw our attention to some cognitive invariants that, according to a platonist interpretation of set-theory, correspond to some elementary properties of sets (conceived of as the objects of set-theory). But this is by far not enough for delivering a plausible epistemology for mathematics and an appropriate response to Benacerraf's dilemma. If this epistemology is meant to combine with a platonist position, it must explain our relation to the relevant objects, namely sets in a proper mathematical sense. If this explanation hinges on the fact that set-theory evolves through a sort of internal necessity, thanks to intra-mathematical kinds of evidence, we would face the following choice: either we have good arguments to the effect that, after all, the objects of set-theory are indeed the objects we perceive (even if our perception is not able to reveal all their relevant properties) – an argument that Maddy does not offer, and that it seems hard to offer – or we are in fact replying to Benacerraf's dilemma by claiming that the basis of our mathematical knowledge, or of a large part of it, lies in our own theoretical constructions, namely in a system of stipulations providing direct access to the abstract objects they define. This latter option might still be compatible with a platonist view, but this compatibility is due to philosophical arguments that Maddy does not provide. In §§ 5.2–5.4 we will review several ways of providing one.

Similar difficulties affect any attempt to come up with an empiricist philosophy of mathematics claiming that the basic beliefs grounding our mathematical theories have empirical origins. The problem is not with the claim that even our most abstract theories can be traced back to an empirical basis. This point, however hard to be properly accounted for, should sound plausible to anyone acquainted with the historical evolution of mathematics (see Kitcher, 1983). Rather, the real issue is that the shift from our empirically grounded beliefs to the principles underpinning our mathematical theories, despite its crucial and obvious importance in the edification of mathematics, does not seem to be empirically justifiable.

5
Conservative Responses to Benacerraf's Dilemma

In this chapter we continue our survey of responses to Benacerraf's dilemma and discuss those that in § 3.3, following Hale and Wright, we have classified as conservative. Again, we will present four of them: Hale's and Wright's neo-logicism; Linsky's and Zalta's "Object Theory"; Shapiro's *ante rem* non-eliminative structuralism; and Parsons' version of non-eliminative structuralism.

5.1 Neo-logicism: A Revised Version of Frege's Programme

Neo-logicism represents a conservative and intellectual response to Benacerraf's dilemma. It agrees with clauses (*i*) and (*ii*) given in § 3.3, and it is supposed to solve the problem of access by appealing to our intellectual resources. Unlike the conservative and intellectual responses that will be reviewed in the next two sections, neo-logicism deals only with arithmetic (we will give a sketch of its possible extensions to analysis). This view has been advanced originally by Wright (1983), and then refined in the light of the vast ensuing debate.[1]

As the name itself suggests, neo-logicism can be located within a general project for revising Frege's original programme. Underlying this project is an interpretation of Frege's context principle (see § 2.1.3), already proposed by Dummett (1956, pp. 493–4; 1973, pp. 195 and 496–7), according to which this principle suggests the following claim (Hale and Wright, 2002, p. 115; equivalent formulations can be found elsewhere in the works of Hale and Wright):

> if, under a suitable explanation of the truth-conditions of an appropriate range of . . . statements [in which terms which purport to refers to numbers or other abstract objects occur], suitable such

statements are . . . true, then those of their ingredient terms which purport reference to numbers or other abstract objects will in fact refer . . . to such objects.

Here (as elsewhere), Hale and Wright employ the verb 'to refer' and its derivatives in a technical sense, according to the use that is customary in works inspired by Frege. According to this use (see § 2.1.2), claiming that a proper name or a singular term refer is tantamount to claiming that there exists an object denoted by that proper name or term. The expression 'purports to refer to' thus evokes occurrences of such a term that are prior, logically speaking at least, to the acknowledgement or the assumption that such an object exists (even though it is acknowledged that the term is apt to uniquely identify the object to which it would refer if it happened to refer).

When this methodological maxim is adopted, claiming that if statements of a given body of statements A are true, then the terms $t_1, t_2, \ldots$, occurring in them and purporting to refer to certain abstract objects do indeed refer to these objects, is logically equivalent to claiming that the statements in A are true only if there exist abstract objects that the terms $t_1, t_2, \ldots$, denote. Let us suppose that A contains only the statement '2 + 3 = 5'. The claim is thus that this statement is true only if the numbers 2, 3, 5 exist. This requirement is equivalent to the requirement that the statement 'Snow White is Nessy's mother' is true only if Snow White and Nessy exist; the existence of 2, 3, and 5 on the one side, and of Snow White and Nessy on the other side, are necessary conditions, respectively, for the truth of '2 + 3 = 5' and 'Snow White is Nessy's mother'. When appropriately generalized, this claim is fairly natural, and is involved in all those arguments – including the indispensability argument – whose conclusion is that mathematical objects, or some of them, exist, since mathematical theorems, or some of them, could not be true if they didn't. Hale's and Wright's point is different, though. They rather propose to reverse the order of the argument: it is not because natural numbers exist, and are so-and-so, that arithmetical statements are true (and could not be true if natural numbers didn't exist); rather, it is because some appropriate theorems of arithmetic are true that terms purporting to refer to natural numbers have a reference, or in other words, that natural numbers exist and are so-and-so. This is what the laborious logical argument by Frege presented in § 2.1.3 seems to suggest. Let us simply call this thesis 'Frege's thesis': simplifying, this is the thesis that natural numbers exist and have certain properties because some statements are true.

By itself, the thesis will not bring us very far. Frege's own argument involves additional ingredients, whose role is to show that the theorems of arithmetic are true in so far as they are consequences of the laws of logic. If we want to ground our philosophy of arithmetic on Frege's thesis, we then need to replace Frege's additional ingredients with new ones whose role is to warrant the truth of the relevant statements without causing the problems that led to the abandonment Frege's original argument (see § 2.1.5). This is what neo-logicists do.

Before reviewing their views, let us pause on Frege's thesis. In stating it we tried to single out, in what seems to us the clearest possible way, the essential trait of the neo-logicist reading of the context principle. However, this thesis is never explicitly endorsed either by Frege or by the neo-logicists. More precisely, neo-logicists claim that once it has been established by appropriate syntactic criteria that some expressions are singular terms, and once it is further established by appropriate non-ontological criteria that some appropriate statements, in which those terms occur, are true, no doubts can be raised against the claim that these expressions have a reference, and that the objects to which they refer exist (see, for instance, Wright, 1983, pp. 7–8 and 13–14; 1998b, pp. 350–1; Hale and Wright, 2001, p. 8). It has become customary to refer to this thesis as the 'syntactic priority thesis':[2] in answering to questions about the existence of some objects, priority is conceded to the syntactic structure of statements of an appropriate kind, whose truth has been ascertained independently of ontological considerations.

The syntactic priority thesis is more specific than Frege's thesis, for it involves the (alleged, at least) identification of syntactic criteria establishing whether some given expressions are singular terms and the identification of the particular kind of statements, in which these expressions occur, whose truth is relevant to establishing whether the latter refer. It is also more general, in so far as it is not restricted to arithmetic. More importantly, it is not concerned with some kind of causal priority – that might be suggested by Frege's thesis – but rather with a purely epistemic sort of priority, relative only to the order of arguments that allow us to establish that certain objects exist.

This suggests two possible, very different readings of Frege's thesis. On one reading, priority is understood in causal terms: Frege's thesis is taken to assert that the existence of natural numbers depends on the truth of certain statements. The most natural way of articulating this reading is to claim that the notion of existence, at least in so far as abstract objects are concerned, is not primitive, but is rather defined after the notion of truth. In other words, one should claim that the statement 'There are prime numbers' is, by definition, the abbreviation of a longer statement like 'Statements such-and-such, in which terms such-and-such occur, are true'. A different reading of Frege's thesis is rather inspired by the syntactic priority thesis. According to this reading, Frege's thesis does not assert that the existence of natural numbers depends on the truth of the relevant statements, but rather that it depends on what makes these statements true. One might then accept this interpretation of the thesis, and claim at the same time that natural numbers exist independently of the formulation of the relevant statements, even though it is only through their formulation and through the ascertainment of their truth that we might come to know that those objects exist and to have epistemic access to them.

It is not immediate to see how neo-logicists interpret Frege's thesis and the syntactic priority thesis. They never explicitly suggest that the notion of existence is not a primitive one, and on some occasions (see for instance Hale 1994a, pp. 137–8) they draw attention to the difficulties engendered by the supposition that there is more than one sense of existence. Moreover, they believe that natural numbers are not "creations of the human mind, somehow brought into being by our stipulation" (Hale and Wright, 2001, p. 16). However, they also add (*ibid.*, pp. 16–17) that "the (sortal) concept under which they fall is . . . introduced" by us, through the acknowledgment of a general principle stating necessary and sufficient conditions for the truth of the relevant statements. The introduction of this concept, they claim (*ibid.*, p. 16), should allow us to "reconceptualize" a state of affairs whose holding is by itself independent of the acknowledgement of the general principle. The truth of those statements would thus result from the principle and the state of affairs taken together, and natural numbers would just be the referents of the singular terms occurring in those statements, and would exist in so far as those statements are true.

Let us make this point clearer by anticipating something we will later discuss in more detail. The general principle to which neo-logicists appeal in order to introduce the (sortal) concept of natural number is HP (see § 2.1.3), as formulated within the logical setting that we will present below (as we shall make clear later, HP implicitly defines cardinal numbers, among which natural ones are rather selected through an explicit definition). The principle is a biconditional: it states that such-and-such cardinal number is identical to such-and-such cardinal number (in Frege's formulation: the number belonging to the concept F is identical to the number belonging to the concept G) if and only if a certain state of affairs holds that can be described without appealing to cardinal numbers (in Frege's formulation: if and only if the concept F is equinumerous to the concept G). The left-hand side of this biconditional states then that the cardinal number such-and-such is identical to the cardinal number such-and-such; the right-hand side states that the relevant state of affairs holds (in order to avoid confusion, let us notice that according both to neo-logicists and to Frege, natural numbers are included among cardinal ones).

Now, according to neo-logicists (for a particularly clear exposition of this point see Wright, 1990, pp. 87–91), the connective 'if and only if' is here to be interpreted as implying that the two sides of the biconditional have the same truth-conditions: this last state of affairs is just the same as that which makes the cardinal number such-and-such identical to the cardinal number such-and-such. However, neo-logicists claim, such a single state of affair is described in different ways in the two sides of the biconditional, by two statements that have different "senses" or, in Frege's terminology, express two different thoughts, since the very same state of affairs is described by referring to numbers on the one side, and without referring to them on the other side. In this, neo-logicists are guided by Frege's claim that, "we carve up the content" of the right-hand side "in a way different from the original, and this yields us a new concept" (Grundlagen, § 64).

It follows that neo-logicists must concede that: (*i*) two statements with the same truth-conditions can have different senses, and express different thoughts; (*ii*) two statements with the same truth-conditions can be such that in one of them singular terms occur that refer to objects such that no terms referring to them occur in the other one. It is fairly common to accept (*i*), and this should cause no concern. Accepting (*ii*) is way more troublesome. Given (*i*), a way of making (*ii*) innocuous would be to concede that (*iii*) if two classes of statements *A* and *B* are such that in the statements of *A* there occur terms referring to objects (let us call them the *a*s) such that in the statements in *B* no term referring to them occurs, and if every statement in *A* has the same truth-conditions of a statement in *B*, then the *a*s exist only in so far as the statements in *A* are true, and these statements are true only in so far as the statements in *B* are true in their turn.

Claiming this would be tantamount to claiming that the *a*s exist only in so far as the statements in *B* are true. But this means that the sense in which *a*s are said to exist is explained by appeal to what makes the statements in *B* true. As far as HP is concerned, this means that cardinal (and thus natural) numbers established only in so far as the statements in which singular terms referring to them occur are true, and these statements are true in their turn only in so far as the statements providing some appropriate instances of the right-hand side of the principle are true (namely only in so far as some appropriate concepts are equinumerous to some other appropriate concepts). It would follow that the sense in which cardinal (and thus natural) numbers are said to exist would be explained by appealing to what makes these latter statements true (that is, to what makes the appropriate concepts equinumerous). If we suppose, as it seems plausible to do, that the sense in which other objects are said to exist is not explained in a similar manner, it would follow that there are at least two senses in which objects can be said to exist. And we have seen that neo-logicists refrain from accepting this. More generally, they reject (*iii*).

The only open route for neo-logicists seems thus to concede that, in making a claim to the effect that the state of affairs making the appropriate instances of the right-hand side of the principle true holds, we *ipso facto* attribute some properties and/or relations to cardinal numbers. However, since that state of affairs can hold independently of the acceptance of HP, and since the concept of cardinal number is introduced by this principle (and since natural numbers are among the cardinal numbers), if this is to be conceded it must also be conceded that cardinal numbers (and natural numbers among them) exist – in the sole sense in which objects can be said to exist – wholly independently of the truth of any statement in which the corresponding singular terms occur. Neo-logicists must then reject the first, causal reading of Frege's thesis, and must accept the second. The syntactic priority thesis is thus, in their view, not an ontological thesis, but rather an epistemic one.

If we accept what has been said so far, however, aren't we faced again with the problem of access, and thus with Benacerraf's dilemma? Neo-logicists do not think so, and claim that it is indeed thanks to the syntactic priority thesis, in its epistemic reading, that the problem can be addressed: HP affords us an epistemic access to natural numbers. More precisely, what affords us an epistemic access to natural numbers is the fact that HP allows us to establish that certain statements in which terms purporting to refer to naturals occur are true, where their truth is an "independently constituted" fact (Hale and Wright, 2002, p. 121). We will present the neo-logicist proposal in more details in what follows, leaving it to the reader to evaluate how good this proposed solution to the problem of access may be.

To this purpose, let us come back to the syntactic priority thesis first. This requires that a syntactical distinction between singular terms and other expressions can be drawn. As Hale (1996a, p. 438) remarks, "semantically, a singular term is any expression the function of which is to convey a reference to a particular object." This criterion clashes, however, with Frege's thesis, in so far as it requires for its application that a verdict on the referential nature of a given expression be prior to the verdict concerning it being a singular term, whereas the syntactic priority thesis requires that a syntactic criterion establishing whether an expression is a singular term (a criterion that does not appeal to the referential nature of this expression) be at hand (see Hale, 1987, 1994b, 1996a). Hence, neo-logicists must make sure that such a criterion is available. At least, they must concede that purely syntactical reasons, or linguistic stipulations, are available that not only allow us to establish that terms such as 'zero', 'one', 'two', etc., or '0', '1', '2', etc. are singular terms, but also that the same holds for terms such as '3 + 5', 'the least perfect number', 'the number of the concept ⌐being distinct from itself¬', etc. And they must claim that whatever natural numbers are, the referents of those terms are natural numbers. The singular terms satisfying this condition will here be called 'natural-numerical terms'.

Furthermore, neo-logicists must determine which are the statements from whose truth the existence of natural numbers follows; that is the kind of statements to which the syntactic priority thesis can be applied. Their proposal is that these are identity statements such as '$a = b$', where 'a' and 'b' are natural-numerical terms.

The truth-conditions of these statements must then be spelled out. According to neo-logicists, these are fixed by HP, stated as a biconditional establishing identity conditions for numbers of (sortal) concepts, or cardinal numbers. One might then ask, with Frege (who speaks, with no relevant difference, of numbers belonging to concepts), what is the number of a concept. In other words, one might ask for an application condition for the sortal concept of cardinal number (see Chapter 3). This is the essential point where neo-logicists part company from Frege: unlike him, they take the condition expressed by HP to be sufficient for fixing this concept. The reason is not only, or not much, that they believe that HP, in particular in its right-hand side, "at least implicitly, impose constraints" (Hale and Wright, 2001, p. 368) on the application of the concept of cardinal number, and that these constraints suffice for a solution to Caesar's problem. The reason is also, and more importantly, that HP, when located in the appropriate logical framework, can alone ground the whole of arithmetic. Let us focus on this latter crucial point.

According to neo-logicists, HP is an implicit definition of cardinal numbers. These do not strictly coincide with natural numbers, being rather a proper subset of them. In other words, all natural numbers are cardinal numbers (or numbers of concepts), but there are concepts (among which is the very concept of natural number) whose number is not a natural number, and then cardinal numbers that are not natural ones. However, the fact that all natural numbers are cardinal numbers allows HP to deliver truth-conditions for identity statements between natural-numerical terms. This is different from giving a sufficient ground for the entire arithmetic. In order to obtain this, it is not enough to fix conditions of identity for natural numbers; one must also give a definition of a predicate corresponding to the property of being a natural number that is based on these conditions, must explain how any natural number could in principle be defined individually, must show that natural numbers exist, and must finally define on them a decidable ordering relation relative to which they form a progression, what will then allow to define the usual arithmetical operations (see § 3.1).

Now, here is what Wright (2000, pp. 319–20) claims:

> There is, *a priori*, no particular reason why a principle intended to incorporate an account of the nature of a particular kind of mathematical entity should also provide a sufficient axiomatic basis for the standard mathematical theory of that kind of entity. It's one thing to characterize what kind of entity we are concerned with, another thing to show that and why there are all the entities of that kind that we standardly take there to be, and that they compose a structure of the kind we intuitively understand them to do.

The crucial point is then that, as far as HP and arithmetic are concerned, these two things go together: if this principle is embedded in an appropriate system of second-order logic, it is possible to deduce from it, only thanks to the resources allowed by that system, the second-order Peano axioms: these axioms can be obtained as theorems of the theory whose axioms are those of the relevant system of second-order logic with the sole addition of HP.

We saw in § 2.1.3 that, in the *Grundlagen*, after having stated HP, Frege appeals to the notion of extension of a concept, in order to spell out what numbers of (or, in his language, belonging to) concepts are. He then bases on BLV a proof of HP, the definition of the natural numbers (that is, he both provides a definition of a predicate corresponding to the property of being a natural number, and explains how any natural number could be in principle defined individually), and the derivation of Peano's axioms. Wright (1983, pp. 154–69) notices that once HP has been proven from BLV, both this law and the related Law VI do not play any crucial role in the derivation of Peano's axioms (the point was already suggested in passing in Parsons, 1964). The proof can thus avoid recourse to the definition of natural numbers as extensions of concepts, and can be based only on their identification with the numbers belonging to the appropriate concepts. In other words, according to Wright, the very same argument given by Frege suggests that it is possible to avoid the use of the function $\dot{\varepsilon}[\varphi(\varepsilon)]$, and thus of laws V and VI, and still prove Peano's axioms by directly adding HP to the axioms of an appropriate system of second-order logic. Wright (*ibid.*) sketches the general lines for this proof, that has been later stated formally by Boolos (1987; 1990, Appendix), who has also shown (1987; 1998, pp. 151–4) that the system so obtained is consistent (see also Boolos and Heck, 1998; Burgess, 2005, ch. 3). The theorem stating that, in an appropriate system of second-order logic, Peano's axioms follow only from HP is known today as 'Frege's theorem' (for an overview, see Heck, 2011, ch. 1). Let us see it in some more detail.

Let L_2 be a system of second-order logic with the identity relation defined on objects, and including an unrestrictred comprehension axiom schema, namely:

$$\exists P \forall x [P(x) \Leftrightarrow \mathcal{A}(x)]$$

This schema ensures that, for any formula $\mathcal{A}(x)$ of L_2, possibly with a free variable x, there is a monadic predicate P defined by that formula. In other terms, it is sufficient to state in this language such a formula in order to get a monadic predicate which applies to any object if and only if this object satisfies the formula. Let us add to the axioms of L_2 the following new axiom:

$$\forall P \forall Q [(\#P = \#Q) \Leftrightarrow (P \approx Q)]$$

where '$P \approx Q$' abbreviates a formula of L_2 asserting that the objects that are P are in one-one correspondence with the objects that are Q, namely – if monadic predicates are interpreted as concepts – that the concepts P and Q are equinumerous, and '#' is a function symbol, introduced (that is, implicitly defined) by the axiom itself, that takes monadic predicates and gives singular terms (such that for any P and Q, '#P' and '#Q' are names of objects).

This axiom is nothing but HP when embedded in L_2. From a logical point of view, it is an abstraction principle, that is, an axiom of the following form:

$$\forall \alpha \forall \beta [(\$\alpha = \$\beta) \Leftrightarrow \alpha \equiv \beta]$$

where 'α' and 'β' are variables of the given language, '$\equiv$' is any equivalence relation defined on the entities over which these variables range,[3] and '$\$$' is a function constant introduced (that is, implicitly defined) by the axiom, which takes expressions of the same kind as 'α' and 'β' and gives singular terms. Not all abstraction principles are analogous (for a study of their different features, see Fine, 2002). Two differences among others are crucial. The first difference is that holding between abstraction principles in which the values of the function $\$$ are entities of the same type as those to which $\$$ applies, or possibly of a higher type, and those in which the values of $\$$ are entities of a lower type that those to which $\$$ applies. A well-known example of the first kind is the principle considered by Frege in §§ 64–67 of *Grundlagen*:

$$\forall u \forall v[(Dir.(u) = Dir.(v)) \Leftrightarrow (u \mathbin{//} v)]$$

This principle asserts that any couple of straight lines u and v have the same direction if and only if they are parallel. Directions are something that straight lines have, and not vice versa, and are thus of a higher type than the things that have them. HP is an instance of the second kind of abstraction principles. Given any two concepts F and G, #F and #G are objects, and objects fall under concepts and are then of a lower type. BLV is also an instance of the second kind of abstraction principle, given that extensions of concepts, according to Frege, are objects. The second crucial difference among abstraction principles can be shown by confronting BLV and HP: the latter is consistent with respect to the axiomatic system in which it is embedded, since its addition to L_2 leads to a consistent system; the former is inconsistent with respect to this system (at least if either this very principle or the comprehension axiom schema are not restricted in appropriate ways) since its addition to L_2, or to any other system of second-order logic equivalent to that employed by Frege, leads to an inconsistent system.

This is what allows neo-logicists to amend Frege's definition of natural numbers: once recourse to extensions of concepts is avoided, it is possible to escape the contradiction that is built into their definition. Given any two concepts F and G, the singular terms '#F' and '#G', introduced in L_2 by HP, can indeed be interpreted as names for the numbers of these concepts. By the comprehension schema it is then possible to define the number zero as follows:

$$0 =_{df} \#[x : x \neq x]$$

where $[x : x \neq x]$ is the concept defined by the formula '$x \neq x$', namely the concept under which all and only those objects that are not identical to themselves fall, and this is tantamount to saying that it is a concept under which no object falls. The comprehension schema ensures that this concept is available; HP ensures that '$\#[x : x \neq x]$' is a well-formed singular term. Moreover, from HP and the logical truth that $[x : x \neq x] \approx [x : x \neq x]$ it follows that $\#[x : x \neq x] = \#[x : x \neq x]$. Neo-logicists then conclude, in accordance with the syntactic priority thesis, that 0 exists. It is then possible (again from HP and the comprehension schema), to define 1 as follows:

$$1 =_{df} \#[x : x = 0]$$

where $[x : x = 0]$ is the concept defined by the formula '$x = 0$', namely the concept under which only the object 0 falls. Since the latter is just $\#[x : x \neq x]$, if the availability of that concept must be secured by the comprehension schema it is necessary that the schema is extended in such a way that the formula '$\mathcal{A}(x)$' is permitted to include the functional constant '$\#$'. Neo-logicists concede this extension, and thus appeal to HP and to the syntactic priority thesis in order to conclude that also 1 exists. Clearly, it is possible to go on in the same way and explicitly define as many natural numbers as one likes, one after the other.

Through this procedure, however, it is not possible to get to the set ℕ of all natural numbers. In order to get this, two routes are open. One can define an operation $suc(-)$, which can be applied recursively to 0 and to all the objects obtained from 0, and such that $suc(0) = 1$, $suc(1) = 2 =_{df} \#[x : x = 0 \vee x = 1]$, etc., and then claim that ℕ is defined recursively by this operation (see Chapter 2, note 27).

This is not the route chosen by neo-logicists. They do not intend to define ℕ recursively, but rather as a whole, following Frege's suggestions. Here is the basic idea. One begins by defining the relation ⌜(being) the successor of⌝, which we will denote with '$Suc(-,-)$', in such a way that for two objects m and n, $Suc(m,n)$ holds if and only if there is a concept P and an object y such that: $P(y)$, m is $\#P$, and n is $\#[x : P(x) \wedge x \neq y]$. It follows that for two objects m and n, $Suc(m,n)$ holds if and only if these objects are numbers of concepts, and that if 0, 1, 2, 3, ... are defined as above, then $Suc(0,1)$, $Suc(1,2)$, $Suc(2,3)$, etc. One then defines the ancestral relation of $Suc(-,-)$, that is a relation $Suc^*(-,-)$ that, intuitively, is such that given two objects m and n, $Suc^*(m,n)$ holds if and only if n is the successor of m, or the successor of the successor of m, etc., namely if and only if n follows m in the sequence in which any term bears the relation *Suc*(-,-) to its predecessor. It is now possible to stipulate that an object n is a natural number if and only if either it is 0 or it is such that *Suc**$(0,n)$. Once this definition is laid down, and the number 0 is defined as above, one can then define the predicate N as follows

$$\forall n[N(n) \Leftrightarrow (n = 0 \lor \mathit{Suc}^*(0,n))]$$

and take ℕ to be the set of all the objects enjoying N.

If one also admits a comprehension schema like the one above for diadic predicates (or two-place relations), it is then easy to show that, within the system obtained by the extensions of L_2 introduced so far, the natural numbers (that is, the objects enjoying N) form a progression, that is, they satisfy Peano's axioms, or, as it is common to say for brevity's sake, Peano's axioms are proved in this system. Once this is done, it is also possible to prove that the number of a concept F under which infinitely many numbers fall – for instance, the concept of natural number – is not a natural number. From this follows, as we said above, that there are concepts whose numbers are not natural numbers.

These proofs are sound beyond doubt. However, there are divergent opinions as regards the logical and epistemic status of the resources they involve.

The (philosophical) problem of drawing the boundaries of logic – and, in particular, of deciding whether second-order logic should be counted as logic or rather as a mathematical theory, and, more specifically yet, whether second-order quantification requires the existence of sets or any other abstract objects, such as concepts – is far from being settled. We will not pause on this issue here. The reader may refer to Shapiro and Weir (2000) for an allegation to the effect that the logical resources involved in the proofs above would not be "epistemically innocent".

We rather want to concentrate on the status of HP, which has engendered most discussion. According to neo-logicists, HP is an analytic principle, since its left-hand side just provides the sort of conceptual resources that are needed in order to reconceptualize the state of affairs that is described in its right-hand side (see for instance Wright, 1999, for a detailed defence of this claim). The notion of analyticity that is at stake here differs from Frege's (see § 2.1.3), and even more from the commonly accepted notion of logical truth (see Boolos, 1997). Nevertheless, neo-logicists maintain that the analyticity of HP entails that their definition of natural numbers, and the ensuing proof of Frege's theorem, constitute a suitable realization of Frege's programme, and lay the ground for a neo-Fregean version of arithmetical platonism.

A survey of the debate concerning the alleged analytical status of HP cannot be attempted here. It is however important to notice that, even if it were admitted that HP is, in some suitable sense, analytic, and that this suffices to claim that the definition of the natural numbers that it sustains is logicist and/or Fregean in spirit, in order to conclude that arithmetic is about the (abstract) objects 0, 1, 2, . . . as defined in accordance with HP, one should also claim that that definition and the proof of Frege's theorem yield not just a possible (that is, mathematically adequate) version of arithmetic, but rather yield arithmetic as such. To see this, just notice that several ways of defining the natural numbers in a mathematically adequate way, that is, in such a way that they satisfy Peano's axioms, are available. Neo-logicists are concerned with the edification of a particular formal theory, often called 'Frege arithmetic', that turns out to be mathematically adequate. This theory, however, is not only different from many others among the mathematically adequate ones; it is also not logically equivalent to many of them (see Burgess, 2005). Neo-logicists do not seem to see this as a problem, and rather take Frege arithmetic as revelatory of the true (and/or unique) nature of natural numbers.

This stance could be justified as follows. Natural numbers are logical objects; the theorems of arithmetic cannot thus but derive from principles that are purely logical, or at least in some appropriate sense analytic; HP is analytic in some such appropriate sense, and the proof of Frege's theorem, like any other derivation within arithmetic, involves only purely logical principles; ergo, this arithmetic is not only mathematically, but also philosophically legitimate, in so far as it complies with the nature of natural numbers as logical objects; not only, it is also the only (mathematically appropriate) version of arithmetic to be philosophically legitimate. To shed some light on this point, one might observe, following Hale and Wright (see Hale and Wright, 2000, pp. 315–18) that where an implicit definition of natural numbers given by the bare stipulation of Peano's axioms would be "arrogant", in that it would stipulate the existence of an "appropriately large range of objects", their definition through HP avoids this sort of arrogance, since the existence of the natural numbers as defined above, or in other terms (according to the syntactic priority thesis) the truth of statements in which natural numerical terms occur, does not follow from HP alone, but from it in conjunction with the "independently constituted and, in the best case, independently ascertainable truth" of other statements of the form '$\#F = \#G$', which are logical truths (see also Wright, 1997, pp. 109–10; Hale and Wright, 2001, p. 10).

However this latter remark is judged, the preceding argument avoids circularity only if the premise that natural numbers are logical objects is not justified by their being defined on the basis of HP, but, on the contrary, is independently endorsed, and taken to be what makes this definition legitimate. But if this is so, natural numbers should have the nature they have, and thus exist, independently of this principle, and thus the problem would be to understand how this nature could be given to us. We would thus be faced again with the problem of access and Benacerraf's dilemma.

The problem could be answered, instead, if the existence of natural numbers, and their nature of logical objects were taken to stem from HP. But this would be tantamount to claim that, when referred to natural numbers, the notion of existence would not be primitive, but should rather be defined on the basis of the notion of truth. If this were accepted, the fact that Frege arithmetic reveals the true (and/or unique) nature of natural numbers should be justified without appealing to any argument that itself appeals to the assumption that HP accords with the nature that natural numbers have by themselves. A way of doing this would be to appeal to the non-arrogant nature of HP. But in order to acknowledge that HP is not arrogant one should understand Frege's thesis, and then the syntactic priority thesis, in such a way that they justify the view that the existence of certain objects can be granted by logic alone. And whoever is not already convinced by logicism could see this as a *reductio ad absurdum* of the thesis that HP does not stipulate the existence of numbers of concepts (Boolos, 1987, p. 95). The preceding argument would thus seem either question-begging, or such to appeal, in its turn, to the assumption that numbers are logical objects. On the other hand, if HP were acknowledged an existential import, it would constitute an arrogant definition, just like Peano's axioms. The problem seems thus to show up again: how can one justify the claim that Frege arithmetic reveals the true (and/or) unique nature of natural numbers?

To have an example of this situation, let us come back to Caesar's problem. HP does not deliver, at least not explicitly, any condition of application of the concept of natural number, and does not tell us which objects the cardinal numbers, and thus the natural numbers, exactly are. But then, how can Caesar's problem be solved, and how can the first of Benacerraf's problem be addressed? We mentioned above that the neo-logicist solution to Caesar's problem relies on the idea that HP provides, at least implicitly, constraints on the application of the concept of natural number. More precisely, neo-logicists appeal to the following principle (Wright, 1983, pp. 116–17; Wright, 1998a, pp. 395–6; Hale and Wright, 2001, p. 369): F is a sortal concept under which natural numbers fall if and only if there are (or there could be) singular terms 'a' and 'b' apt to denote objects that fall under F and such that the truth-conditions of the statement '$a = b$' can be adequately explained in terms of the equinumerosity of appropriate concepts. According to neo-logicists, this principle is a non-problematic consequence of HP, and it not only implies that Caesar is not a number, but also that the sets forming Zermelo's progression aren't, since each of them is equinumerous with all the others (Wright, 1983, p. 123). However, Wright himself acknowledges that in order to determine what natural numbers are, it is not enough to rule out Zermelo's progression, nor all the objects that do not conform to the principle above. The neo-logicist reply to Caesar's problem does not then consist in showing which objects the natural numbers are, but just in pointing out that it is a consequence of HP that no object with some undesirable features can be a natural number.

This reply might however turn out to be ineffective. Zermelo's progression satisfies the axioms of arithmetic; why should not its elements be identified with the natural numbers? What reasons do we have for accepting that a principle preventing this possibility underlies the definition allegedly providing the true (and/or unique) nature of natural numbers? Whatever reasons these might be, they cannot be based on the structural properties of arithmetic. But then, on what could they be based, if they also cannot depend on some features of numbers revealed by an independent access to them? The only plausible answer available to neo-logicists seems to be that the conditions that natural numbers must satisfy are not only those established by the arithmetical practice and by the use of natural numbers in counting objects (these being accounted for by Peano's axioms), but also those suggested by the requirement that a (philosophically) legitimate version of a mathematical theory "must somehow build its applications, actual and potential, into its core . . . rather than merely "patch them on from the outside". (Wright, 2000, p. 324; the last quotation is from *Grundgesetze*, § 159).

This is the so-called 'application constraint', or 'Frege's constraint': a general principle that, as the remarks in § 2.1.4 suggest, Frege seems to consider as essential in the case of real numbers. According to Dummett (1991, pp. 272–3), this constraint is at work also in Frege's definition of the natural numbers. Neo-logicists adopt this constraint for natural numbers, and claim that their version of arithmetic fulfils it, to the extent that the application of natural numbers to counting collections of object is directly explained by their definition as numbers of concepts. Frege arithmetic would thus reveal the true (and/or unique) nature of natural numbers since, as William Demopoulos (2005, p. 155; *Forthcoming*, p. 28 in the manuscript) remarked, "the importance of Frege's theorem is that it vindicates the philosophical program of explaining pure arithmetical knowledge on the basis of its account of the application of numbers in ordinary judgements of cardinality", namely our ordinary judgements in reply to questions such as 'How many objects such-and-such are there'?

However this argument is judged, it shows, together with the aforementioned reply to Caesar's problem, why neo-logicism is representative of the strategy suggested by the first two among Benacerraf's retrospectively self-addressed criticisms (see § 3.1). But it also brings to light the obstacles the neo-logicist programme encounters in its extension to the case of real numbers. Wright (2000, pp. 228–9) has observed that the reasons for adopting the application constraint in the case of natural numbers do not extend to the case of real numbers. If for the former it is plausible to claim that "the flow of concept-formation" goes from empirical applications to theory, for the latter the flow seems to go in the opposite direction, since "the classical mathematics of continuity is made to inform a nonempirical reconceptualization of the parameters of potential variation in the empirical domains to which it is applied". In other terms, Wright notices that when we assign to an empirical magnitude, for example, to the side of a plot of land, a length expressed by a real number, we are already appealing to distinctions that become meaningful only within our theory of real numbers, since there is no empirical way of distinguishing a length measuring say 2 or $\sqrt{2}$ kilometres from a length measuring r kilometres, where r is a real number distinct from 2 and $\sqrt{2}$ but arbitrarily close to them. The problem is thus that while in the right-hand side of HP is involved an equivalence relation between concepts that is easy to define and understand without appealing to natural numbers or other arithmetical notions, it is hard to imagine an abstraction principle introducing measures of magnitudes, the right-hand side of which involves an appropriate equivalence relation between magnitudes that can be defined and understood without appealing to analysis at all.

It is not very difficult to define the set of real numbers by means of suitable abstraction principles, both by obtaining it, as proposed by Shapiro (2000b), as an extension of that of the naturals, or, as proposed by Hale (2000), following Frege's suggestions in the second volume of *Grundgesetze*, obtaining it as the set of ratios between magnitudes. But the problem remains of offering arguments in support of the claim that the definitions thus obtained, that seem to square nicely with that offered by neo-logicists for natural numbers, can reveal the true (and/or unique) nature of real numbers. In a conference held in Paris in March 2009, Hale has proposed a justification for the application constraint that would apply, with no relevant differences, both to naturals and to reals. The basic idea would be that the full possession of the concepts of natural number and real number requires an understanding of the use of these numbers respectively in counting objects and in measuring magnitudes. But even if this could suffice to justify with no circularity the application constraint in both cases, we would still lack an explanation of how, in the case of reals, this use could be understood with no appeal to an equivalence relation between magnitudes based on resources pertaining to analysis. Hale makes appeal to an equivalence relation directly defined on magnitudes, understood as elements of a sufficiently rich structure. But this not only brings the problem back to the definition of magnitudes, but also suggests that the theory of reals is, as a matter of fact, based on a prior mathematical theory of magnitudes, and not on our pre-mathematical judgements. If this is so, the application constraint cannot assign any philosophical priority to a given definition of the reals as opposed to other, equally appropriate (from a mathematical point of view), definitions.

5.2 Linsky, Zalta and 'Object Theory': Mathematics and Logic (or Metaphysics) of Abstract Objects

The crucial problem for neo-logicists, as regards both naturals and reals, is to justify the claim that the definition of these numbers, based on HP or some other abstraction principle respectively, can reveal their true (and/or unique) nature. This problem affects any philosophical view purporting to account for a part of mathematics through a definition of the relevant entities which not only is mathematically appropriate, but is also taken to be the right one. If this obstacle is to be overcome, a new strategy is required: the sought-for account should no more be based upon one definition as opposed to the many available ones that are mathematically adequate to the same extent. This is what Bernard Linsky and Edward Zalta have explicitly suggested. Their aim is to provide for an appropriate characterization of the notion of mathematical object to be used as the basis for a platonist account of mathematics, where this is supposed to apply to actual mathematical theories as they are, rather than as they may be according to any suggested recasting of them in new terms (Linsky and Zalta, 1995, 2006; Zalta, 1999, 2000). According to Zalta, this characterization lays the ground for a neo-logicist programme, although of a substantially different form from the one Hale and Wright propose.

This is how Zalta (2000, p. 219) puts it (see also Linsky and Zalta, 2006, pp. 60–1):

> We defend a philosophical thesis which may preserve some of the spirit of logicism. Our thesis is that mathematical objects just are (reducible to) the abstract objects systematized by a certain axiomatic, mathematics-free metaphysical theory. This thesis appears to be a version of mathematical platonism, for if correct, it would make a certain simple and intuitive philosophical position about mathematics much more rigorous, namely, that mathematics describes a realm of abstract objects. Nevertheless, there are two ways in which the present view might constitute a kind of neo-logicism. The first is that the comprehension principle for abstract objects that forms part of the metaphysical theory can be reformulated as a principle that 'looks and sounds' like an analytic, if not logical, truth . . . The second is that the abstract objects systematized by the metaphysical theory are, in some sense, logical objects.

The underlying basic idea is that the notion of abstract object can be dealt with, *mutatis mutandis*, as the notion of set, namely explained through a formal axiomatic theory: Object Theory, as Zalta calls it (the theory was first presented in Zalta, 1983, and then refined and elaborated in several papers).[4] In contrast with set-theory, this is not a mathematical theory, however, but a metaphysical one. It employs the usual language of higher-order modal predicate logic without identity, enriched by a new form of predication beside the traditional one. This is called 'encoding', and is formally expressed by formulas like 'aF' (to be read: 'a encodes F), where 'a' is a term for objects or an individual variable, and 'F' is a predicate constant or a monadic predicate variable. In its more precise formulation, Object Theory is a typed theory, though for our purposes we can well confine ourselves to an untyped version restricted to the second order.[5]

It is not necessary that we present here the logical axioms for the usual constants of predicate logic, the modal operators, and the two forms of predication; it is enough to point out that one among these axioms, namely the following:

$$\Diamond[aF] \Rightarrow \Box[aF]$$

(where 'a' and 'F' are schematic letters for objects and monadic properties respectively) ensures that for any a and F, a can encode F only if it encodes it necessarily. The theory also features a primitive predicate constant '$E!$', through which other two predicate constants are defined by the following stipulations:

$$O!a =_{df} \Diamond E!a \qquad A!a =_{df} \neg \Diamond E!a$$

Informally, '$E!$' designate the property of being a concrete object. In turn, '$O!$' and '$A!$' designate, respectively, the property of being an ordinary object and that of being an abstract object. Since, for any object x, $\Diamond E!x \vee \neg \Diamond E!x$, objects are classified into two complementary classes: ordinary objects that can be concrete, and abstract objects that cannot be concrete. Informally speaking, in a system of modal logic, objects belong to possible worlds. An ordinary object is thus an object that is concrete in some possible world, whereas there is no possible world in which an abstract object is concrete. This, however, is not enough for explaining what concrete, ordinary and abstract objects are. This explanation is given by Object Theory as a whole. The theory includes proper axioms that, together with the aforementioned definitions, implicitly define the constants '$E!$', '$O!$', and '$A!$'.

One among these axioms, namely the following:

$$O!a \Rightarrow \Box[\neg \exists P[aP]]$$

ensures that there is no possible world in which ordinary objects encode a monadic property whatsoever. From this it follows that an object a encodes a property only if it is abstract. The theory also includes the following comprehension schema:

$$\exists x[A!x \wedge \forall P[xP \Leftrightarrow \mathcal{A}]]$$

where '$\mathcal{A}$' is any formula of the language of the theory in which 'x' does not occur freely. It ensures that, whatever formula is taken, and in particular whatever formula expressing a condition that some monadic property P is supposed to satisfy (that is, a formula with a free occurrence of 'P'), there is an abstract object encoding all and only the monadic properties satisfying it.

In addition to these axioms the theory includes three implicit definitions introducing the relation of identity between objects and between monadic properties:

$$a =_E b =_{df} [O!a \wedge O!b \wedge \Box \forall P[P(a) \Leftrightarrow P(b)]]$$
$$a = b =_{df} (a =_E b) \vee [A!a \wedge A!b \wedge \Box \forall P[aP \Leftrightarrow bP]]$$
$$F = G =_{df} \Box \forall x[xF \Leftrightarrow xG]$$

The first of these introduces the relation of concrete identity (or identity between ordinary objects), stipulating that an object *a* is concretely identical to an object *b* if *a* and *b* are ordinary objects that necessarily (that is, in every possible world) enjoy the same monadic properties. The second introduces the relation of identity between objects *tout court*, stipulating that an object *a* is identical to an object *b* if *a* and *b* either are concretely identical or are abstract objects that necessarily encode the same monadic properties. The last definition introduces the relation of identity between monadic properties, stipulating that a monadic property *F* is identical to a monadic property *G* if *F* and *G* are necessarily encoded by the same objects.

From the second of these definitions, the comprehension schema, and the axiom stating that only abstract objects encode monadic properties, it follows that for any appropriate formula '$\mathcal{A}$' there is one and only one abstract object encoding all and only the monadic properties satisfying the formula. In order to denote this object the usual description operator 'ι' can be used. This makes it possible to denote the object by the following term:

$$\iota x[A!x \wedge \forall P[xP \Leftrightarrow \mathcal{A}]]$$

An abstract object is thus conceived, by Linsky and Zalta, as the correlate of a sheaf of monadic properties; and for any sheaf of monadic properties, only one abstract object is its correlate. Zalta thus offers a way of formally implementing the old idea that abstract objects result from the reification of some properties or sheaf of properties: given any sheaf of monadic properties, there is an abstract object that is its reification. In Zalta's formal language, this is the object encoding all and only the properties in the sheaf. Let us imagine the language of Zalta's theory to include two monadic predicate constants '*Mst*' and '*LcNs*', designating the property of being a monster and living in Loch Ness. In this language it is then possible to write down the open (relatively to '*P*') formula '$\forall y[P(y) \Leftrightarrow [Mst(y) \wedge LcNs(y)]]$'. The above comprehension schema thus grants the existence of the abstract object:

$$\iota x[A!x \wedge \forall P[xP \Leftrightarrow \forall y[P(y) \Leftrightarrow [Mst(y) \wedge LcNs(y)]]]]$$

This is the object encoding all and only the properties that satisfy the formula '$\forall y[P(y) \Leftrightarrow [Mst(y) \wedge LcNs(y)]]$': the notorious Nessy.[6] The existence of this object, like that of any abstract object, is then a purely logical, metaphysical and linguistic fact, which is warranted by the metaphysics and the logic of abstract objects and by the grammar of any appropriate predicate added to this logic. In our example, it is enough to possess a language which allows talk of monsters and of Loch Ness, for the assumptions of Object Theory to ensure the existence of Nessy as an abstract object. This last specification is crucial: the existence of Nessy is a purely logical, metaphysical and linguistic fact only to the extent that Nessy is conceived of as an abstract object stemming from our imagination. Whether there is something concrete in our world that instantiates all the properties Nessy encodes is obviously something that no logic, metaphysics or language can decide.

This last point is emphasized, in Zalta's theory, by the distinction between usual predication and encoding. Given any sheaf of monadic properties, there is an abstract object encoding all and only the properties in that sheaf. This does not imply that there is also an object enjoying these properties. This object might exist in some possible worlds and not in others, might exist in no possible world, or might exist in all possible worlds. Which among these different possibilities holds depends on the way the possible worlds are made and on the relevant sheaf of properties. Just to give another example: even if Zalta's theory ensures the existence of an abstract object like the square circle – namely of the abstract object encoding the property of being a square circle – it is obvious from the very definitions of the properties of being a circle and being a square that in no possible world can an object instantiating both these properties (namely, a square circle) exist.

This conception of abstract objects underlies Linsky's and Zalta's views on mathematical objects. According to them, these are abstract objects. The problem is to establish which, among the abstract objects whose existence is entailed by Object Theory and by our linguistic resources, are mathematical. It will not do to simply claim that an abstract object is mathematical when it encodes a sheaf of mathematical properties. The issue is not just that it is hard in general to specify what a mathematical property comes to; it is rather that, if we proceed that way, then we should conclude that the square circle is a mathematical object, and this seems implausible. Linksy's and Zalta's idea is then that the burden is on the accepted mathematical theories (not on logic, nor on the metaphysics of abstract objects, nor on our linguistic resources or whatever other philosophical reason) to establish which abstract objects are mathematical, and thus which are the mathematical objects. In short, their point is that in order for an abstract object a to be mathematical, there must be a certain mathematical theory T in which a instantiates certain properties. If this is so, this is an object of that theory, namely the abstract object encoding all and only the properties a instantiates in the theory. Let us reflect on this idea.

The first step taken by Linsky and Zalta consists in extending the notion of encoding of a property to that of encoding of a proposition (we follow here the exposition in Linsky and Zalta, 1995, pp. 538–41, with minor modifications suggested by other presentations). This is achieved by treating propositions as 0-place properties. Le p be any proposition. Let us associate to it the 0-place property ⌜(being) such that p⌝. This should not be confused with the monadic property ⌜(being) an x such that $p(x)$⌝, where '$p(x)$' has a free occurrence of x. The property ⌜(being) such that p⌝ is rather the property every object has just in case the world to which it belongs is such that p. It is designated by '$[\lambda y\, p]$' (intuitively speaking, it is the reification of this proposition). This will be called a 'propositional' property. For brevity's sake, one can also say that it is "constructed out of p" (Zalta, 2000, p. 230). We will not dwell on the complex issue of establishing what a proposition is (for references on this, see McGrath, 2007). It will be enough to say that a proposition p is something of which one can say that it is the case that p, and that p is true or false.[7]

One could then wonder, what should we mean when we say that a proposition is mathematical? Linsky and Zalta give no answer to this question, since they maintain that it is not among the duties of their philosophical view to give any. That some propositions are mathematical just depends on the fact that the mathematical community accepts them as such. The philosopher just takes this as a fact. This fact might stand in need of an explanation, but this by itself does not commit the philosopher to the need of finding one. The philosopher should rather concentrate on different problems, for example (s)he can wonder by which criteria it can be established that a certain theory, which is taken to be a mathematical one, deals with some objects, and how these objects can be described and/or characterized. This is tantamount to characterizing the notion of mathematical objects in general relative to the notion of mathematical proposition, taken as primitive (supposing that theories are, or can at least be rendered as, systems of propositions). This is what Linsky and Zalta suggest. In order to do this, they do not even set out to explain what is a mathematical theory, but rather rest content with describing a mathematical theory in terms of certain propositions that are supposed to be mathematical. This description is based on the assumptions that a mathematical theory is an abstract object in its turn, and that it has an author. A mathematical theory is said to "assert" mathematical propositions: the basic idea is that a mathematical theory asserts a mathematical proposition if and only if this theory encodes the propositional property constructed out of this proposition.

Let us explain this point (see Zalta, 2000, pp. 230–2). That Linsky and Zalta have no intention to explain in general what makes a proposition or a theory mathematical is made manifest by their use of two primitive predicate constants: '*Math*(–)', which is supposed to designate the property of being a mathematical proposition; and '*Aut*(–,–)', which is supposed to designate the relation of being the author of something. Both the property and the relation are assumed as familiar. The relevant description is then given by the following explicit definition:

$$MathTh(a) =_{df} A!a \wedge \exists z[E!z \wedge Aut(z,a)] \wedge \forall P[aP \Leftrightarrow \exists p[Math(p) \wedge P = [\lambda y\ p]]]$$

which, in its turn, serves to introduce the constant '*MathTh*(–)', which is supposed to designate the property of being a mathematical theory (the condition '*A!a*' is only informally stated in Zalta, 2000; it is rather added as a conjunct of the definition in Zalta, PM, p. 62). A mathematical theory is thus described as an abstract object for which there exists a concrete object that is its author, and that encodes the propositional properties constructed out of certain mathematical propositions. These propositions are exactly those that are asserted by the theory. Linsky and Zalta also make use of a different terminology: they say that the relevant mathematical properties are true in that theory, and connect this way of putting things to the following explicit definition:

$$T \models p =_{df} T[\lambda y\ p]$$

The symbol '$T \models p$' is not to be taken with its usual meaning: it does not mean that p is a logical consequence of T (that p is true whenever T is), but simply provides another way of writing down what can be written through the symbol '$T[\lambda y\ p]$': it just indicates that T encodes the propositional property constructed out of p. Since Linsky and Zalta suggest that '$T \models p$' should be read as 'p is true in T', also the property of being true in a theory for a proposition must be understood as a property defined through the notion of encoding; to say that p is true in T is just a short way of saying that T encodes the propositional property constructed out of p.

It is then natural to make room for another form of expression (we try here to make explicit a point that seems to us implicit in what Linsky and Zalta say). Let $p_{[F(a)]}$ be the proposition that an object a has the property F, namely the proposition expressed by the statement '$F(a)$'. Let us suppose that $\text{T} \models p_{[F(a)]}$ holds for a given mathematical theory T, namely that $p_{[F(a)]}$ is true in T, or, what is equivalent, that T encodes the propositional property obtained by $p_{[F(a)]}$. It is then natural to express this fact by saying that a has F in T, or directly by writing '$\text{T} \models F(a)$'.

The last component in Linsky's and Zalta's view is a "rule of closure". They stipulate that the propositions following from a body of true propositions in a given mathematical theory are themselves true in that theory (in the aforementioned sense). Clearly, Linsky and Zalta's intent here is to appeal to the usual proof-theoretic notion of deductive consequence, to the effect that what they state is that any deductive consequence of any body of true propositions in a given mathematical theory (or, more precisely, of statements that express these propositions) is true in that theory (or, more precisely, expresses a proposition that is true in this theory). But one could also take the burden to be on the relevant theory to make clear what it means that a proposition follows from other propositions. Apart from this, it is important to notice that encoding is not in general closed under any kind of consequence relation: it is allowed that a certain object a might not encode the property G even though it is both the case that it encodes $F_1, \ldots, F_n$, and that from an object b's having $F_1, \ldots, F_n$, it follows (in some sense) that b has G. The rule of closure for mathematical theories must then be introduced through a separate stipulation that is independent from the rest of Object Theory.

We are now ready to see how Linsky and Zalta spell out the notion of mathematical object. Let us suppose that T is a mathematical theory. We can then say (see Zalta, 2000, p. 232) that an object a is an object of T if and only if there is some monadic property P such that a has P in T. This is expressed through the following explicit definition:

$$Ob.of(a,\text{T}) =_{df} \exists F[\text{T} \models F(a)]$$

Let now 'κ_T' be a term of the language of T for which it is possible in this language to formulate, for some monadic predicate letter 'F', the statement '$F(\kappa_\text{T})$', expressing the proposition that the object κ_T has the property F. Linsky and Zalta state that:

$$\kappa_\text{T} = \iota x[A!x \wedge \forall P[xP \Leftrightarrow \text{T} \models P(\kappa_\text{T})]]$$

Note that this is not a definition. This is made clear but the fact that the term 'κ_T' occurs on both sides of the identity sign. Linsky and

Zalta rather term it a "theoretical description" (Linsky and Zalta 1995, p. 540). All they suggest, as it has been anticipated, is that a mathematical object $\kappa_{_T}$ of a mathematical theory T is that abstract object which encodes all and only the properties it has in T. This can also be expressed as follows: a mathematical object $\kappa_{_T}$ of a given mathematical theory T is the abstract object that encodes all and only the monadic properties P that are such that T encodes the property constructed out of the proposition $p_{[P(\kappa_{_T})]}$.

From this definition it does not follow that there are mathematical objects, even less which they are. All this depends on the existence and features of mathematical theories. The definition only shows how it is possible to give a general account of those mathematical objects whose existence and specific features depend on the accepted mathematical theories. To understand this point, let us look at a negative example (which is not Linsky's and Zalta's).

Suppose we have a reasonable and appropriate understanding of the propriety of being a natural number, and let 'NN' designate this property in a suitable extension of the language of Object Theory. According to Object Theory, there is a unique abstract object

$$\iota x[A!x \wedge \forall P[xP \Leftrightarrow \forall y[P(y) \Leftrightarrow NN(y)]]],$$

which is the reification of this property, namely that encodes all and only the properties satisfying the formula '$\forall y[P(y) \Leftrightarrow NN(y)]$' of our suitable extended language (that is, the property NN itself and all the properties an object has if and only if it has NN). Let us call this object 'the Natural Number', and let us abbreviate this with 'v'. For it to be a mathematical object it would be necessary (and also sufficient) that there were a mathematical theory T such that

$$\forall P[vP \Leftrightarrow T \models NN(v)].$$

Since it is obvious that vNN, were there to be such a theory it would be the case that $T \models NN(v)$. Thus, in order for the Natural Number to be a mathematical object it would be necessary (and also sufficient) that there were a mathematical theory T such that $T \models NN(v)$. But, by the rule of closure, if F were a monadic property which an object would have in T if and only if it had NN, then it would also hold that $T \models F(v)$. Thus the Natural Number would be a mathematical object only if there were a mathematical theory in which it were a natural number. But is there any such theory?

174 *Plato's Problem*

Neither Object Theory as such, nor the general characterization of the notion of mathematical object suggested by Linsky and Zalta imply that there is any. In order to know whether there is one, one should rather look at mathematics and, possibly, at its history. And there is very little doubt that anyone would then conclude that there is not: the very term 'the Natural Number' is indeed at odds with the language of mathematicians, and it would be quite hard to find a (present or past) mathematician agreeing that the Natural Number is a natural number, whatever her/his understanding of the propriety of being a natural number may be.

It is thus not surprising that Linsky and Zalta do not use such a simple example to illustrate their characterization of the notion of mathematical object. They clearly purport to stay as close as possible to mathematical practice and history, and to make Object Theory's analysis conform to them. But, then, how does this characterization apply when this practice is taken into account? Another example (this time, a positive one, coming from Zalta's work: see Zalta, 2000) is suited for answering this question. It concerns second-order Peano arithmetic (which we shall call 'PA^2', for short; Zalta, 1999 also provides an account of natural numbers understood as numbers of sortal concepts in Frege's vein).

Let us extend the language of Object Theory by adding to it appropriate non-logical constants to be used to state, in this language, the axioms of PA^2 (which are just Peano axioms formulated in a second-order predicate language; see Chapter 2, note 23) plus the non-logical term 'PA^2_O' to be used to denote, in this same language, PA^2 itself. Let us then add to the axioms of Object Theory the new axiom '*MathTh*(PA^2_O)' plus as many axioms as are those of PA^2, each of which being of the form 'PA$^2 \models \mathcal{P}$' where '$\mathcal{P}$' is then replaced by a distinct axiom of PA^2 as rephrased in the language of Object Theory, once this is extended as mentioned above.[8] From '*MathTh*(PA^2_O)' and the definition of the predicate constant '*MathTh*(−)', it follows that

$$\text{PA}^2_O = \iota x[A!x \wedge \forall P[aP \Leftrightarrow \exists p[\text{PA}^2_O \models p \wedge P = [\lambda y\, p]]]]$$

(see Zalta, 2000, p. 259, note 22, for the proof), which asserts, within Object Theory, that PA^2 is the abstract object encoding all and only the propositional properties constructed out of the propositions which are true in PA^2 itself. Now, insofar as PA^2 is categorical (see Chapter 2, note 33), one can take the natural numbers to be just the objects that satisfy its theorems. But, for any such theorem φ, the rule of closure involved in Linsky's and Zalta's account of mathematical theories ensures that PA$^2_O \models \varphi_O$, where 'φ_O' is the reformulation of φ in the language of

Object Theory extended as indicated above. It follows that the natural numbers as defined in PA^2 are identified in Object Theory as the abstract objects each of which encodes all and only the propositional properties constructed out of the reformulation in this language, of the theorems of PA^2 concerning it (in fact, Zalta describes these objects in more details, but we cannot go deeper into this issue here; we thus refer the reader to his aforementioned work). Should we conclude from this that Object Theory, together with the general characterization of the notion of mathematical object associated with it, delivers an argument for arithmetical platonism? Clearly, one could concede that the foregoing considerations allow identifying or describing in an appropriate way the objects PA^2 is about only if one were ready to accept that PA^2 (or arithmetic in general) is about some objects. But being able both to provide a theory like PA^2, or any other categorical theory, and being ready to accept that it is about some objects, and even that presenting it is enough for ensuring that these objects exist, is still not the same as having an appropriate conception of their existence and a precise notion of a mathematical object. Now, a supporter of Object Theory could argue that this theory and the the general characterization of the notion of mathematical object associated with it do indeed provide for the sought-for conception and notion; and that this is enough to transform the mere acceptance of these ideas in a robust philosophical view, so much robust that merely advancing it would constitute an argument for platonism. In order to get this argument, however, one should show that Object Theory, and the general characterization of the notion of mathematical object associated with it, are appropriate for this task. But, to our knowledge, not much has been said by Linsky and Zalta in order to show that this is the case.

The following quotations well illustrate the situation (Zalta, 2000, p. 233; Linsky and Zalta, 1995, p. 547):

> There is no distinguished 'model-theoretic' perspective to tell us what are the 'objects of' a theory T. From a metaphysical point of view, the objects of a theory are the ones described by its *de re* claims, for these attribute properties to objects.

> Knowledge of particular abstract objects does not require any causal connection to them, but we know them on a one-to-one basis because *de re* knowledge of abstracta is by description. All one has to do to become so acquainted *de re* with an abstract object is to understand its descriptive, defining condition, for the properties that an

abstract object encodes are precisely those expressed by their defining condition. So our cognitive faculty for acquiring knowledge of abstracta is simply the one we use to understand the comprehension principle. We therefore have an answer to Benacerraf's worry.

The comprehension principle is here, clearly, the instances of the comprehension schema relative to the properties involved in the definition of the relevant objects. Linsky and Zalta thus maintain that understanding their definition of an abstract object is enough for becoming acquainted *de re* with it, namely to become acquainted with it as such and not as the object which satisfies some given conditions. If this is so, and mathematical objects are abstract, Benacerraf's dilemma simply dissolves. It has the trivial response that we know mathematical objects since we define them. And since this is *de re* knowledge, it is perfectly compatible with a traditional conception of mathematical truth: the theorems of mathematics are true because there are mathematical objects with the relevant properties. However, for Linsky and Zalta, this is a consequence of the fact that mathematical theories are abstract objects encoding appropriate propositional properties. In other words, it is a consequence of the way in which they describe mathematical theories and their objects. The issue is then to ponder whether theirs is an appropriate description. The answer cannot be just that that description is appropriate since it is coherent, and it also implements the natural and old idea that abstract objects are reifications of properties. This answer should also take into account actual mathematical theories, and should appeal to the fact that these are as Linsky and Zalta describe them to be. But, once it is conceded that the property of being a mathematical theory is defined, as we saw, on the basis of the two primitive constants '$Math(-)$' and '$Aut(-)$', in order to verify whether this is so, one must compare the conditions imposed on these constants and these very theories, and ponder whether the latter can be appropriately conceived as reifications of propositional properties. Linsky and Zalta do not offer anything like this. They even restrain from imposing to the constants '$Math(-)$' and '$Aut(-)$' any other condition apart from those involved in the definition of '$MathTh(-)$'. Hence, their (dis)solution of Benacerraf's dilemma reduces in the end to the claim that actual mathematical theories (the mathematical theories that mathematicians actually accept and develop) do possess by themselves the tools for contrasting this alleged quandary: all that philosophers can do, is providing them within a metaphysical framework for articulating their reply to the quandary. The question is then whether the proposed metaphysical framework is the good one. This is a question we cannot enter into here. The next two sections will make clear, at least, that it is not the only one that has been suggested for the same purpose.

5.3 A First Version of Non-eliminative Structuralism: *Ante Rem* Structuralism

The idea that mathematical objects are reifications of properties can be developed independently of the Object Theory. The basic thought is suggested by the thesis – advanced by Hilbert (see § 2.4.3) and implicitly acknowledged by many – that the consistency of a theory is a sufficient condition for the existence of the entities the theory speaks of. If a solution to Benacerraf's dilemma must be drawn from this thesis, however, one must make the status of these entities clear, must establish the criteria that a consistent theory must satisfy if it is to be accepted as a mathematical theory, and must establish in which sense we can say to know it, or say that it is true.

Frege's thesis, appropriately extended beyond arithmetic, suggests a way to deal with these requirements. According to it, if we want to show that some mathematical objects exist and that we have knowledge of them it is enough to identify some suitable statements and to fix their truth conditions and their relation with those objects, and to ensure that these conditions are met. According to neo-logicists, HP establishes the truth-conditions of the statements that are relevant to determining the existence of natural numbers, in such a way that their existence is not the result of a sheer stipulation, but rather hinges on facts independent of arithmetic. But a different route can be taken: one might acknowledge that the truth of the statements relevant for an application of Frege's thesis follows from their being the statements of a given mathematical theory, and that it is because of the way this theory is done that some mathematical objects exist and have certain properties: it is mathematics that explains mathematics, and its ontology depends only on mathematics itself and nothing else. This idea underlies what Parsons (1990, § 8; 2008, § 18) calls 'non-eliminative structuralism', as opposed to eliminative structuralism (see § 4.3). We will discuss two versions of it, Parsons' version and Shapiro's (1997) *ante rem* structuralism. In this section we begin with the latter (leaving the former for the next section), following the exposition in Shapiro (2000a, pp. 270–3 and 275–89).

The basic idea of non-eliminative structuralism is that focusing on structures is not enough to dispense with abstract objects. Rather, it helps to clarify the nature of mathematical objects and shows that they are objects in a genuine sense: (at least some) mathematical objects are just places in structures, and have no property except those depending on their being such places (they can still have properties that cannot be defined within the given structure: the number 100, for instance, has the property of being the number of cantos of the *Divine Comedy*). *Ante rem* structuralism concedes this but, as Parsons (2008, pp. 51–2) remarks, adds a further claim: structures exist independently of their being instantiated by particular systems of objects. Let us first consider the first thesis, as it is elaborated by Shapiro (see also Parsons, 1990, § 8; 2008, § 18 for a discussion and defence from possible objections).

According to Shapiro (1997, pp. 73–83), a structure is "the abstract form of a system" and a system is a "collection of objects with certain relations". More specifically, a structure is conceived as a network connecting empty places that are supposed to be occupied by objects standing in certain relations with each other. When this happens, the system formed by these objects instantiates the structure, and the latter is the form of the system. For the sake of brevity, we will say that a structure is instantiated by a system if the relations holding between the objects forming this system accord with those that are supposed to hold between the places of the structure, to the effect that these places can be filled by those objects; if a system instantiates a structure, this is the abstract form of the system.

Shapiro (1997, pp. 10–11, 82–3) distinguishes two different ways of thinking of structures and their places. We can treat places in terms of the objects occupying them, when, for example, we say that Von Neumann's three ($\{\varnothing, \{\varnothing\}, \{\varnothing,\{\varnothing\}\}\}$) has more elements than Zermelo's three ($\{\{\{\varnothing\}\}\}$), or that the president of the USA is a Democrat (in January 2012). Here, the terms 'three' and 'president of the USA' denotes offices. These are distinct from the objects occupying them, namely $\{\varnothing, \{\varnothing\}, \{\varnothing,\{\varnothing\}\}\}$, $\{\{\{\varnothing\}\}\}$, and Barack Obama: we attribute properties to these objects, but not to the places (it is $\{\varnothing, \{\varnothing\}, \{\varnothing,\{\varnothing\}\}\}$ that contains more elements than $\{\{\{\varnothing\}\}\}$, not the place-three in the structure that has more elements than itself; similarly, it is Barack Obama, and not the presidency of the USA that is a Democrat). This is the view adopted by eliminative structuralism. Shapiro calls it the 'places-are-offices' perspective. He does not see it as misguided, but simply notices that there are other uses of language suggesting a different view. We can say that the president of the USA is the commander-in-chief of the US Armed Forces. We are not thereby attributing any property to Barack Obama himself, nor to anyone else in particular, but we are rather talking about the role that Obama temporarily has, and others have had. In order to see this, it is enough to notice that the president of the USA has always been, and will likely continue to be in the future, the chief of the US Armed Forces, whereas Obama before 2009 was not, and will not be in future. When we say that the president of the USA is the commander-in-chief of the US Armed Forces we are treating the role of the presidency of the USA as an object: we denote it by a term with the same grammatical behaviour of other terms that, in other statements, we use to denote those who have that role. This is what Shapiro calls the 'places-are-objects' perspective.

Shapiro suggests that mathematical objects, at least those of modern pure mathematics (we will come back to this point) are places in structures that are treated according to the places-are-objects perspective. Mathematical theories are about structures, and, Shapiro submits, "mathematics is the science of structure" (1997, p. 5), though it is clear from the context of his book that he intends to refer to pure mathematics only, in particular modern pure mathematics. However, even though the theories of (modern pure) mathematics are about structures, their theorems, strictly speaking, are not; they are rather about structures' places, and describe the properties of these. Those theorems are thus to be read in accordance with their surface grammar: they are statements describing objects. This, in the end, is a way of formulating a platonist view.

Some clarifications are in order. First of all, the distinction between structures, their places, and the objects occupying them is a relative one: there can be mathematical theories concerning structures whose places are, can, or must be occupied by other structures. Group theory is a case in point. Mathematical theorems can speak of structures then, if structures occupy places in other structures. Second, that mathematical objects are places in structures does not mean that it is not possible, in some contexts, to speak of particular mathematical objects as such, rather than as places in structures, exactly in the way in which we speak of the sets $\{\emptyset, \{\emptyset\}, \{\emptyset,\{\emptyset\}\}\}$ or $\{\{\{\emptyset\}\}\}$ in a discussion of Benacerraf's dilemma for arithmetic. For example, nothing prevents the *ante rem* structuralist from telling that the natural-number structure (or structure of progression) is exemplified by appropriate sets or even that natural numbers are ultimately nothing but sets: in both cases, however, by speaking of sets, (s)he would refer to places in the structure fixed by set-theory, with the result that what (s)he would be telling, in fact, would respectively be that an appropriate system of places in the structure fixed by set-theory exemplifies the natural-numbers structure, and that natural numbers are ultimately nothing but these very places. Moreover, the *ante rem* structuralist can also argue that the natural-number structure is exemplified by a system formed only by some of its places, for example by those providing even numbers. Third, if places in a structure are objects, then the structure itself is a system of objects and, as a system, it instantiates itself. This happens, however, only in so far as places considered as objects (conceived according to the places-are-objects perspective) occupy places considered as offices (conceived according to the place-are-offices perspective).

Let us come back to our exposition. If (at least some) mathematical objects are places in structures, asking according to the places-are-objects perspective whether Julius Caesar is one of them is similar to asking whether Barack Obama is the president of the USA. This seems to be a category mistake. And the same mistake seems to occur if we ask whether three is {∅, {∅}, {∅,{∅}}} or {{{∅}}}. Still, if we insist in giving answers to these questions – maybe because we think that no mistake is present, or because we just don't pay attention to it – then, says Shapiro, it won't be difficult to answer them in the negative: Caesar is not a place of a structure, and thus is not a mathematical object; likewise, {∅, {∅}, {∅,{∅}}} and {{{∅}}} are not places of the structure of progression, and thus are not the number three, and Barack Obama is not the president of the USA.

Non-eliminative structuralism thus offers a simple response to Benacerraf's first argument. This response radically differs from the neo-logicist one, not only in its being general rather than confined to the sole arithmetic, but also, and above all, since it rules out any other cognate question (it does not merely reject the possibility that {{{∅}}} is a natural number, but also argues that Benacerraf's question is not well-posed, so that no other cognate question can be rightfully advanced with respect to other sets or other objects that are not places in the structure of progression), and does not depend on any condition imposed on (pure) mathematics from the outside. This is because, according to non-eliminative structuralism, saying of three that it is a place in the structure of progression is to say that it is a perfectly and wholly determinate object: being this place is the very essence of number three. Contrary to neo-logicists, non-eliminative structuralists tell us which objects the natural numbers are, and not only how they are like. The reason is not that non-eliminative structuralists have some specific explicit definition of natural numbers as opposed to the implicit definition afforded by HP. Since natural numbers are places in the structure of progression, they can be defined if and only if the structure of progression is, and this seems to be appropriately definable only by a system of axioms. Moreover, as Parsons (1990, pp. 334–5; 2008, p. 106) has noticed, places in a structure are incomplete in two ways: they are incomplete like the characters of a novel, since it is not determined whether they have or have not certain properties; but they are incomplete also in the sense that their distinctive properties are structural, and are thus not defined outside a structure (whereas the properties that a novel's characters have can be conveyed even outside the novel itself). Moreover, numbers of concepts can themselves be conceived as places in the structure defined by the axioms of Frege's arithmetic. As such, they are – according to non-eliminative structuralism – well determinate mathematical objects. The difference between the non-eliminative structuralist and the neo-logicist replies to the first of Benacerraf's problem then lies in the different conception that the former have of the notion of mathematical object. This is based on the idea that it is possible to have "genuine reference to objects if the 'objects' are impoverished in the way in which elements of mathematical structures appear to be" (Parsons, 2008, p. 107).

As far as Benacerraf's dilemma is concerned, endorsing non-eliminative structuralism is enough to deliver a partial response. Indeed, if (some) mathematical objects are places in structures, they have properties if and only if these structures have some appropriate features in turn. Hence, saying that the theorems of mathematics are true according to a good semantics (in Benacerraf's sense) can only mean that the relevant structures have the features that make its places have the properties that those theorems attribute to the objects they are about. Moreover, if (some) mathematical objects are places in structures, our access to them is via our access to these structures, and these are exactly what has to be made in this-and-that way if the theorems are to be true. Hence our mathematical knowledge is just knowledge of what makes mathematical theorems true. But this response is not yet complete, since it tells nothing about the existence of the relevant structures: if (some) mathematical objects are places in structures, they exist, have certain properties, and are known if and only if those structures exist, have certain other properties, and are known, in turn; but then, what is to ensure that these structures exist, have those properties, and are known? If these questions were not answered at all, non-eliminative structuralism would just be a way of moving Plato's problem a step backward, or rephrasing it in another language. In a sense, this is so: non-eliminative structuralism shifts Plato's problem on to something else. But this shift is benign, since once it is given its new form, the problem can be solved in various ways, that is, by giving different responses to the questions above. This is just what different versions of non-eliminative structuralism do, so as to complete the foregoing partial response to the dilemma. We consider here only the responses suggested by *ante rem* structuralism (in the next section 5.4, we shall consider different responses suggested by another version of non-eliminative structuralism, namely Parson's version).

For *ante rem* structuralists, the crucial point in the solution is just what explains and motivates the label '*ante rem*' and characterizes this non-eliminative version of structuralism among others: structures exist independently of the systems that instantiate them. Let us see what this means, and how it can be justified.

We can think of cases where a given structure is described as the abstract form of one or more systems, no other alternative description is available, and it is only through these systems that it is possible to grasp the structure. In these cases, there is a clear sense according to which these systems come prior to the structure: it is only by an access to the former that we can have access to the latter. But this priority is only epistemic, since nothing prevents one from arguing that the existence of the structure is independent from the existence of any system instantiating it. One could think of a situation where the latter have ceased to exist, or we have lost our capacity of accessing them: we would then be able to describe the structure only as the abstract form of systems that do not exist any more, and we could not have any access to it. However, it seems odd to claim that the structure itself, as an abstract form, would cease to exist.

But what does it mean for a structure to exist? If mathematics is about structures and nothing mathematically relevant has to be supposed to be there before them, it is natural to maintain that to say that a certain structure exists means nothing more than it is a place in another structure, which itself exists. But we could not regress *ad infinitum*: at some point we will need to explain in which sense a structure can be said to exist without appealing to other structures of which it is a place and to their existence.

Before considering Shapiro's suggested way out, a brief digression will be helpful. One could think that in order to reply to Benacerraf's dilemma, and to solve Plato's problem, a much weaker concession than that according to which structures exist independently of the systems instantiating them would suffice. We already observed that, according to Shapiro (1997, p. 143), "structuralism is (only) a perspicuous account of the bulk of contemporary mathematics", or even of its essential part.

If this is so, could one be disposed to acknowledge that pre-modern mathematics, or some specific part of modern mathematics, concern mathematical objects that are not places in structures? And that these objects give rise to systems whose existence ensures the existence of the structures that are their abstract form? Could one also be disposed to acknowledge that some structures exist as abstract forms of systems that are composed of non-mathematical objects? Again, could one be disposed to acknowledge that the existence of these structures ensures the existence of other structures? Finally, could one not claim that by an appropriate historical or logical chain it would be possible, structure by structure, to establish the existence of all the required mathematical objects? In the next section we will see that Parsons suggests a similar option. Shapiro goes a different way instead, though he does not exclude, in principle, the possibility of a (non-modern) mathematics that does not conform to *ante rem* structuralism.

Rather, Shapiro proposes a structure theory, whose role is both to explain the notion of structure and to lay down conditions for the existence of structures. In Shapiro (1997, pp. 93–5), he offers an informal version of it, modelled upon the usual informal expositions of set-theory. Here we will give a rough description of its eight axioms. The first (Axiom of Infinity) affirms the existence of a structure with an infinite number of places. The second, third, and fourth (Axioms of Subtraction, Subclass and Addition) affirm that if there is a structure, then there are others, obtained by the first by deleting or adding places, functions and relations. The fifth (Axiom of the Powerstructure) affirms that if there is a structure, than another structure exists that has as many places as there are subsets of the former one. The sixth (Axiom of Replacement) affirms that if there is a structure Σ and a function f mapping every place x of Σ with a place $f(x)$ of a structure Σ_x (notice that this structure varies with x, and is not thus necessarily the same for all the places in Σ), then there is a structure at least as large as the set-theoretic union of the places in structures (and this means that there is a function g such that for any place z in each Σ_x, there is a place $g(z)$ in this other structure). These axioms make sure that there exist structures with as many places as are required by the usual mathematical theories. The two remaining axioms give sufficient conditions for the existence of structures in general. The seventh (Axiom of Coherence) affirms that if κ is a coherent body of formulae of a second-order language, there is a structure satisfying κ. The eighth (Axiom of Reflection) affirms that if κ is a body of formulae in the language of the theory of structures, and κ holds, there is a structure satisfying the first seven axioms with the addition of κ.

A first thing worth saying about these axioms is that the set-theoretic terminology is dispensable: they can be expressed in a non-set-theoretical second-order language. A second thing is that the requirement in the seventh axiom, that κ is a body of formulae of a second-order language, is required to ensure categoricity (namely, the fact that κ defines a single structure, or that at least that all the structures defined by it are isomorphic). A third thing concerns the coherence condition. This cannot be equated with consistency (the condition that no contradiction is derivable from κ) for two reasons: one is that nothing guarantees completeness at second-order, so that it is possible to have consistent though not satisfiable theories; the other is that the consistency condition can be expressed only in modal terms (by denying that certain derivations are possible), or else by considering derivations as abstract objects and by denying the existence of some of them. In both cases, stating a consistency condition would call for resources, external to the theory of structures, that could not be available if the prior existence of abstract objects is not admitted – and this would be tantamount to renouncing the advantages of the theory of structures. Shapiro's idea is that the notion of coherence should be taken as intuitive and primitive, essentially informal. The thought is that mathematics takes care of itself, and is able to distinguish between coherent and non-coherent theories. This might strike one as odd, but it is far more plausible when we consider that an absolute proof of consistency for arithmetic and set theory, and thus for all mathematical theories that can be reduced to them, is absent, and still this doesn't affect the confidence that many of us have towards these theories.

The same problem can show up in a more general form, though. There are so many similarities between the theory of structures and set-theory that it is natural to wonder why we should adopt the former rather than the latter (if we see the latter as warranting the existence of a domain of primitive mathematical objects on which the existence of all other mathematical objects depends). The only answer available to an *ante rem* structuralist is that the notion of structure, as it is appealed to in mathematical practice – think of Bourbaki's mathematical encyclopaedia, entirely based on that notion – is endowed with an intrinsic explanatory power. But then mathematics not only justifies itself, but also explains itself. This seems to turn *ante rem* structuralism into a demise of the philosophy of mathematics: all its complex apparatus of philosophical resources lead to the claim that the best account of mathematics is given by mathematics itself. Although one might feel sympathetic with this conclusion, it must be noticed that a purely structuralist account of mathematics *à la* Bourbaki is threatened today by algorithmic conceptions of mathematics, stemming from informatics. Should we then suggest that a new form of philosophy of mathematics is needed, to fit with this new form of mathematics?

Be that as it may, it is a fact that the structure theory, by itself, is not enough to answer Benacerraf's dilemma. The main problem is not with truth, for nothing prevents one from defining the truth of a statement about the places of a structure in such a way that it depends on the properties of that structure, and let the notion of truth have the common features of the usual notion of truth. The problem is with knowledge: granted that structures are like the structure theory that (implicitly) defines them to be how can we have knowledge of them? According to Shapiro (1997, ch. 4), there are three ways this can be done.

Some small structures can be known through perceptual connection with some systems of concrete objects instantiating them: this is the process that cognitive psychologists call 'pattern recognition'.

A structure can then be known through a process of linguistic abstraction: both by extending, possibly indefinitely, the process that generates structures known in the first way just mentioned; and also by the definition of suitable equivalence classes among already given objects, in such a way that these classes form a new structure. Here is an example of the first possibility (Shapiro, 1997, pp. 118–19). Let us suppose we are confronted with systems of vertical strokes like |, | |, | | |, and that in them we recognize instantiations of some structures that we call 'one', 'two', 'three' (these would not be natural numbers yet, but just their perceptual predecessors). We can thus learn how to extend these systems by adding |, and then to acquire knowledge of the structure of progression as a structure instantiated by |, | |, | | |, . . . Let us now see an example of the second method. Let us think of the natural numbers as given, together with the operation of multiplication. We can form pairs of such numbers (the second of which is different from zero) and define upon them an equivalence relation by letting (n, m) and (p, q) be equivalent if and only if $nq = mp$. The equivalent classes relative to this relation will form a new structure, that of positive rational numbers.

Last, a structure can be known through its implicit definition, this being given, in accordance with the seventh axiom of structure theory, by a suitable system of axioms.

In all these cases, the basic idea is that we know structures because we are able to characterize, describe and obtain them in such a way that it is possible for us to determine whether they (or better) their places, have certain properties and satisfy certain conditions. If we concede everything that *ante rem* structuralism requires, all this would seem to suffice for a response to Benacerraf's dilemma. As regards the applications of a mathematical theory, they would rest on the fact that some systems of non-mathematical objects would instantiate (parts of) certain structures, and this might be secured through appropriate representation theorems.

In closing, let us mention a well-known problem that seems to afflict all versions of non-eliminative structuralism (see Parsons, 2008, p. 107), though it has been raised against *ante rem* structuralism in its most recent formulation: the so-called problem of identity (Keränen, 2001). Many mathematical structures feature an internal symmetry that, in technical terms, is described as a (non-trivial) automorphism: a bijective function from a structure into itself preserving the internal relations (an automorphism is trivial when it is given just by the relation of identity, being thus defined on every structure). An example is given by the group $< \mathbb{Z}, + >$, given by the set of relative numbers $(0, 1, 2, \ldots; -1, -2, \ldots)$ upon which addition is defined in the usual way. An automorphism is defined on this group, corresponding to the change of sign of its elements: if G maps $-h$ to every relative number h (that is: $G(h) = -h$), it is clear that, for any pair of relative numbers h and k, $G(h + k) = G(h) + G(k)$. If we stick to this group only (that is, if we disregard the ordering relation and any other relations or operations that are not defined on $< \mathbb{Z}, + >$), and we describe its elements in purely relational terms, there is no way of distinguishing between 1 and -1. The statement '1 is identical with -1' is thus true. In order to say that it is false, extra-structural truth-conditions must be appealed to. The same holds for any structure Σ upon which an (non-trivial) automorphism G can be defined: in order to say that a statement like 'For every object x of Σ, x is identical with $G(x)$' is false, we need to appeal to extra-structural truth-conditions.

Shapiro has contributed with three articles (2006a, 2006b, 2008) to this debate. The basic point of his reply is that *ante rem* structuralism is not bound to accept any version of the principle of the identity of indiscernibles: it can deny that if x and y are objects (places) of the same structure that share all the structural relations relative to their structure, then x is identical to y (2008, p. 286). Let us come back to the group $< \mathbb{Z}, + >$. Since it is a group, every element of $\mathbb{Z}$ has an inverse relative to +. Take any element a of $\mathbb{Z}$ different from the neutral element. It can easily be proved that the inverse of a is distinct from a. Let us denote it with '$-a$'. It follows that a and $-a$ are distinct, even though $G(a) = -a$ and $G(-a) = a$. This proves that the principle of the identity of indiscernibles does not hold in $< \mathbb{Z}, + >$: a is distinct from $-a$, even though it is indiscernible from it within the structure. Shapiro's point is thus that it is a mathematical fact that certain places in a structure upon which a (non-trivial) automorphism is defined are distinct even though they are indiscernible within the structure. Far from constituting a problem for *ante rem* (and, in general, for non-eliminative) structuralism, this shows that this philosophical view is consonant with mathematical practice, where it is common to concede that some indiscernible objects are distinct: points in Euclidean geometry are a clear example of this.

A different way of expressing the same idea, suggested by Leitgeb and Ladyman (2008, pp. 28–9), consists in acknowledging that "the identity ... of places in a structure is not to be accounted for by anything other than the structure itself", namely that "the identity relation for positions in a structure ... ought to be viewed as an integral component of a structure in the same way as, for example, the successor relation is an integral component of the structure of natural numbers". This means, for instance, that $< \mathbb{Z}, + >$ is equivalent to $< \mathbb{Z}, =, + >$, that is that once $< \mathbb{Z}, + >$ is given, it is possible to define the relation of identity on $\mathbb{Z}$, and thus give conditions of identity for the elements of $\mathbb{Z}$. Every structure is thus well defined only if the identity among the elements of its domain is defined according to the conditions imposed to the structure itself. It follows that if x and y are distinct elements in the domain of a structure, they are discernible within that structure at least because '$x = y$' does not hold.

In a conference held in Paris in June 2009, Resnik noticed that the problem could be solved on the same lines, at least at a very general level, if one argues that certain properties of the entire structure could be taken as structural properties of its places, so that in a structure with an automorphism G all elements a would be distinct from $G(a)$ thanks to a structural property of these elements, this being on its turn a property of the structure itself, as showed above for the case of $< \mathbb{Z}, + >$ and $G(a) = -a$.

5.4 A Second Version of Non-eliminative Structuralism: Parsons and the Role of Intuition

If '$x \neq y$' does hold in a certain structure, we can conclude that x and y are distinct within it. But this tells us nothing of what makes them so. Let us take again $< \mathbb{Z}, =, + >$. Whatever a is (unless it is the neutral element), it is a theorem of the structure that $a \neq -a$. This can be expressed by saying that one among a and $-a$ is positive whereas the other is negative. But not only cannot it be determined within the structure which one is positive and which one is negative; it also cannot be explained what it means to say that one is positive and the other is negative. This might be the reason why Parsons (2008, p. 108) believes that Leitgeb's and Ladyman's answer does not take the identity objection considered at the end of the previous section very seriously, and why he suggests his own (see also Parsons, 2004, ch. 4). His basic idea is that structures involved in mathematics falls under two kinds: some "are assumed in mathematics without the obligation to construct them within other structures" (2008, p. 108); others can be only constructed in this way. The thought is then that the basic structures out of which all other structures required in mathematics are constructed should not have non-trivial automorphisms, whereas those with these sort of automorphisms should be constructed in such a way that the objects that are indiscernible but distinct in them are made discernible by properties of the basic structures upon which they are constructed.

One could doubt that all the structures required by mathematics can be constructed out of more basic structures with no non-trivial automorphism. Parsons himself acknowledges that the Euclidean plane could point to a similar doubt. Still, this response to the identity objection makes clear where Parsons' non-eliminative structuralism and *ante rem* structuralism come apart. In presenting the former, we will follow here the exposition in Parsons (2008), a book containing parts, or revised versions, of papers that Parsons has devoted to the topic since the early 1980s.

According to Parsons, not all mathematical objects are intrinsically places in structures. This is not to say that those that are not do not, or cannot, occupy places in structures, but only that their nature and existence do not depend on their occupying these places. Of course, those that are intrinsically places in structures exist in so far as the structures of which they are places exist. The others exist independently of the structures whose places they possibly occupy, and among these structures, some exist thanks to the existence of these very objects. In other terms, Parsons argues that structuralist mathematics (so to say: this term is not Parsons', we introduce it for the sake of exposition in order to refer to that part of mathematics which deals with objects that are intrinsically places in structures) is possible only in so far as it co-exists with non-structuralist one, whose objects are not intrinsically places in structures.

Concerning the former, Parsons' account relies on a basic idea that is not significantly different from Shapiro's (Parsons, 2008, p. 100): statements of structuralist mathematics can be understood as being about objects, even if this "does not require us to take more as objectively determined about the objects about which it speaks than that language itself specifies". In other words: these statements can speak of objects whose properties and relations are all and only those which they assign to them. Under this condition, these statements must thus be taken at face value, as being about objects. Despite the little weight being put on truths here, the underlying idea is thus close to Frege's thesis. Parsons explicitly hints at this similarity when he argues that, in so far as mathematics is at issue, "the usable general characterization of the notion of object comes from *logic*", which means that "we speak of particular objects by referring to them by singular terms: names, demonstratives, and descriptions" and "reference to objects will typically and standardly occur in the context of sentences, and, therefore, in the use of singular terms, by the application to singular terms of *predicates*" (1982, p. 492; 2008, p. 3). It follows that an account of structuralist mathematics does not entail anything about which objects are outside specific contexts. However, statements of such mathematics speak of objects in structures, and not of structures as such, apart from the obvious cases in which the relevant objects are themselves structures (2008, p. 111).

As we have said, however, structuralist mathematics does not make the whole of mathematics. According to Parsons, some structures – which Shapiro would have rather called 'systems' – might include objects that are determined independently of them. For example, as we shall see later, though the arithmetic stemming from Peano's axioms and from the definition of the usual arithmetical operations is part of structuralist mathematics (and is thus not about natural numbers as such, unless they are taken to be identical with the places in the structure of progression: 2008, p. 101), there might be structures among whose objects are some which play the role of natural numbers in them, and still are not just places in them, and this yield contextually determined identity statements such as '3 is α', where α is one such object.

Parsons describes then a pluralist framework, where different types of structures, conceived in different ways and with different roles peacefully cohabit. This framework needs a more general notion of structure than Shapiro's. Parsons (2004, pp. 62–3; 2008, pp. 111–12) suggests a "metalinguistic" notion, according to which the domain of a structure is specified by a (monadic) predicate, and relations and functions defined on it are specified by other predicate or functional letters (or functors, in Parsons' terminology). He describes two ways for obtaining a structure so conceived (2008, p. 112):

> The most concrete way of giving a structure would be by predicates and functors that are antecedently understood, so that there is

some independent verification of fundamental propositions about the structure (say, axioms for the theory of this kind of structure). Brouwer and Hilbert could be taken to have been attempting something like this in their descriptions of the intuitive basis of arithmetic. A less problematic, but also less philosophically interesting, case is where a type of structure is defined in the abstract and then examples are given from different branches of mathematics, but in this case, the "realizations" of one type of structure are simply found in another structure.

The crucial point in Parsons' proposal is that the first way gives the non-structuralist basis for structuralist mathematics.

This makes clear that the opposition between structuralist and non-structuralist mathematics is reminiscent of Hilbert's distinction between ideal and finitary arithmetic (see § 2.4.3), and Hilbert's description of finitary arithmetic did in fact inspire Parsons' account of the latter. This account relies on a theory of intuition, where the notion of intuition is itself based on the distinction between two types of abstract objects.

Beside the usual abstract objects that are wholly independent of any concrete object, there would be "quasi-concrete" abstract objects, namely objects "of a kind which goes with an intrinsic, concrete 'representation', such that different objects of the kind in question are distinguishable by having different representations" (Parsons, 2008, p. 34). The nature of the relation of representation between these objects and the corresponding concrete objects can differ according to the sorts of quasi-concrete objects considered. Among these Parsons cites (*ibid.*, p. 43) geometrical figures, represented by appropriate diagrams, and sets of concrete objects, represented by these very objects taken together. The cases on which he pauses at more length, however, are expression types and strings of strokes (henceforth, 'strings'). These cases show, among other things, that for a certain sort of abstract objects to be quasi-concrete it is not required that the conditions of identity of such objects are such that if a and b are two of them, represented by a^* and b^*, then $a = b$ if and only if $a^* = b^*$. This seems to hold for sets of concrete objects and, in our opinion, also for geometrical figures, at least to the extent that Euclid's geometry is concerned (Panza, 2012). But it doesn't hold for expression types and strings of strokes. The former are types represented by their tokens, like the logical constant '∃' or the English phoneme /mæθ'mætɪks/. The latter are configurations of signs such as '| | |' or '| | | |', and are represented by tokens of appropriate signs. Clearly, in neither case the conditions of identity of the relevant objects are *ipso facto* the conditions of identity of the concrete objects that represent them. However, it is clear that we access both expression types and strings of strokes through the perception of these concrete objects, and that we distinguish an expression type from another and a string from another because we know how to make appropriate distinctions between those concrete objects. Moreover, at least in the case of strings of strokes, our perception of the concrete objects allows us to obtain knowledge about the corresponding quasi-concrete objects. For instance, by observing the collection of signs '| | |' and '| | | |' we can learn that it is possible to go from the string represented by the first collection to that represented by the second just by the addition of one and only one stroke.

Both these examples are suggested by Hilbert: leaving trivial typographical conventions aside, strings are the objects of Hilbert's finitary mathematics, while his metamathematics is a theory of expression types organized by an appropriate syntax. The crucial role that these quasi-concrete objects have within mathematics is thus made manifest (*mutatis mudandis*) by Hilbert's programme. Parsons confines himself to the examination of strings in their relation with arithmetic.

Before coming to this, we must observe that perception of concrete representations of quasi-concrete objects yields, according to Parsons, an intuition of the latter, that is, an intuition of abstract objects. This is a form of non-propositional intuition: it is intuition of objects (while propositional intuition is intuition that some proposition is true). It is thus cognate with Kant's sensible intuition. But in so far as it is intuition of abstract objects, it extends Kant's sensible intuition beyond its original scope. According to Parsons, some quasi-concrete abstract objects, strings among these, are mathematical objects. The intuition of them extends the sensible intuition to the domain that in Kant's views pertains to pure intuition, which is not intuition of objects at all. Parsons thus offers a conception of pure intuition different from Kant's, though inspired by Kant's sensible intuition (Parsons, 1979–80; 2008, ch. 5, discusses at length the relation between his form of intuition and those of Kant and Husserl). Notice also that according to Parsons, the intuition of quasi-concrete objects comes with a form of propositional intuition: the intuition that the relevant quasi-concrete objects have certain properties and stand in certain relations. Moreover, even though the intuition of quasi-concrete objects is mediated by the corresponding sensible intuition, it provides a direct justification for the corresponding propositions, which are then given to us as true propositions. This propositional intuition thus provides, in its turn, an intuitive knowledge whose content is given by propositions about quasi-concrete objects, these being in many cases mathematical objects. In some cases, this knowledge can also be general (Burgess, 2008, p. 406): a crucial example is the knowledge that every string can be extended by adding another stroke.

Consider now that the relation that quasi-concrete objects bear to their representations make the former being not intrinsically places in structures. This helps in understanding how Parsons conceives the relation between structuralist mathematics and its non-structuralist basis. The latter is given by mathematical theories about quasi-concrete objects organized in structures (or systems, in Shapiro's terminology) whose places are occupied by these objects (which are obviously determined independently of the structure they belong to). These structures are, in their turn, instantiation of other structures that can be obtained by them through abstraction. Parsons only considers the case of the structure formed by strings ordered according to the successor relation, determined by the operation of adding one stroke to a string. Not only does he take it that this is an instantiation of the structure of progression, but also that it is possible to have intuitive knowledge of some relational properties of the objects forming it, knowledge that can be used to obtain the structure of progression itself by abstraction. The kind of abstraction at stake here is what Tait (1986) calls 'Dedekind abstraction' (Parsons, 2008, pp. 47 and 104–5). Suppose that a certain structure has been somehow described. In the present case, it is the structure formed by the string | and by all and only strings | |, | | |, | | | |, etc., obtained from the first by successively adding |. It is then possible to define a new structure "of that type" by requiring that there be an isomorphism between it and the original structure. We obtain thus the structure of progression, namely <N, 0, Suc>, where 'N' designates a monadic predicate fixing the domain of the structure (by designating the property of being a natural number), '0' is an individual constant denoting the number mapped to | by the constitutive isomorphism, and 'Suc' designates the relation of successor that this isomorphism maps to the relation connecting any string to the one obtained by it by adding |. From the fact that the first structure is a structure of quasi-concrete objects it does not follow that the second also is its being obtained by abstraction suggests indeed that it isn't.

It suggests it, but it does not prove it. One could indeed argue that the same structure can also be obtained through intuition of its elements, namely that the natural numbers itself are, as such, quasi-concrete objects. In order to get clear about this, one should specify the conditions of identity that Parsons' metalinguistic notion of structure imposes to structures. He is not clear about this, but devotes the whole of Chapter 6 in his book to argue in favour of the claim that even if it is possible to get to natural numbers in other ways, none of these ways can merely rely on the intuition of quasi-concrete objects. In other words, according to Parsons, natural numbers, as such, are not quasi-concrete.

This does not entail that, *ipso facto*, no intuitive knowledge is involved in arithmetic, to the extent that the latter is, as whole, part of structuralist mathematics. For it could happen that intuitive knowledge of some properties of strings somehow turns into intuitive knowledge of arithmetic. One could argue that this is precluded by the fact that intuitive knowledge of arithmetic would require knowledge that the structure of progression is instantiated by the structure of strings, which in turn could not possibly be intuitive. But Parsons does not accept this argument; instead, he explicitly asks whether a sufficiently comprehensive intuitive arithmetic can be had.

A way of getting it could be that of elaborating enough of a theory of strings so as to obtain a substantial part of arithmetic. Hilbert had little doubt about this: finitary arithmetic is such a theory for him. At a closer look, however, the possibility of a similar theory does not follow from the possibility of having intuitive knowledge of some properties of strings. Hilbert does not go much further in the description of his finitary arithmetic directly in terms of strings, and it is perfectly plausible that such intuitive knowledge suffices only to provide the basis for Dedekind abstraction, but cannot then lead beyond the relation of successor defined on strings, up to the definition on them of the usual arithmetical operations. In order to evaluate whether this is feasible, Parsons considers a weak version of first-order arithmetic, called PRA (Primitive Recursive Arithmetic), suggested by Thoralf Skolem (1923), whose language has no quantification and in which the induction axiom is replaced by a schema of inference rules. PRA is generally considered the best formalization of Hilbert's finitary arithmetic, and if it can be formalized in some way, PRA is the best candidate for this. Parsons then ponders whether from the fact that strings provide a model of PRA it follows that the theorems of PRA express intuitive knowledge. He calls the positive reply to this question 'Hilbert's thesis', and discusses it in Chapter 7 of his book. The thesis is eventually rejected, but it is shown that a large part of PRA – involving all that depends on the operations of addition and multiplication, but not what depends on exponentiation – expresses intuitive knowledge.

If Parsons is right, this provides a response to Benacerraf's dilemma for this part of PRA. In order to know whether this reply holds also for those parts of all versions of arithmetic that correspond to this same part of PRA, in particular for all theorems relying only on addition, multiplication and the order relation, we should enter into complex problems concerning the ontological and epistemological relations between PRA and richer versions of arithmetic. Parsons does dwell on this, simply suggesting that the remaining parts of PRA, and of all other versions of arithmetic as well, should be counted as structuralist mathematics, and that this obviously holds for analysis and set theory. He seems to believe that the sort of connection established by his account between structuralist and intuitive mathematics can provide an appropriate epistemology for the former. Since it is easy (see § 5.3) to define the truth of a statements about places in a structure in such a way that it depends on the properties of the structure, this epistemology would allow, on its own, to reply to the dilemma also for non-intuitive mathematics.

Parsons, however, is not explicit about this. In the last chapter of his book he just sketches how "Reason" is supposed to lead to the edification of arithmetic and set theory. Parsons distinguishes between Reason (with a capital-r) as a faculty or form of intellectual activity from the different reasons (with non-capital-r) to which one appeal in the exercise of this activity. According to Parsons, "Reason is in play when actions, including assertions and related speech acts, are supported or defended by reasons" (2000, p. 299; 2008, p. 317). Parsons points to five features of Reason (2000, pp. 299–304; 2008, pp. 317–25): the reasons involved in it are on their turn supported by reasons; the regress of reasons stop when one meets reasons that are considered "intrinsically plausible"; reasons tend to systematization; their systematization engenders a hierarchy of levels; Reason is the final judge for the formulation of our judgements (even in judgements based on perception, Reason decides whether and to what extent evidence should be credited). The term 'reason' belongs to a Kantian language, but these features point to some kind of phenomenology of Platonic *dianoia* (see § 1.1). Moreover, Reason as just described is supposed to be very close to the intuition that, according to Gödel, is involved in set theory (see § 2.5). If one still wishes to call it 'intuition' and to take some of the reasons involved in it as intuitive, one must distinguish it both from the intuition of quasi-concrete objects and from the propositional intuition based on it. The essential difference is that the intrinsically plausible reasons on which Reason is based do not require any intervention of external events, contrary to what is the case for the perceptions upon which these other forms of intuitions are based (2000, p. 310; 2008, p. 326). Even though there is no explicit admission in Parsons, this seems to suggest that

Gödel's platonism is to be taken as the exaltation of the abilities of competent mathematicians in distinguishing reasons that are intrinsically plausible from others that aren't. If this is so, then appealing to it in responding to Benacerraf's dilemma would have the effect of trivializing the dilemma itself. In order to get a more substantial response both to this dilemma and to Plato's problem, these reasons must be anchored to a different ground, even if this were acknowledged as relevant only thanks to further reasons: this is exactly what, according to Parsons, is made possible by quasi-concrete objects.

It should be clear by now that the strength of Parsons' proposal lies in the connection it establishes between a (non-eliminativist) structuralist conception of mathematics and a theory of the intuition of some abstract objects. This connection not only seems to account for the formal theories of modern mathematics, where a structuralist conception of objects is accompanied by a symbolic practice apparently based on the intuition of expression types; but it also seems to explain, in part at least, the historical evolution that led to a structuralist mathematics through different forms of non-structuralist one. It might also happen to account for a new form of mathematics, less inclined to structuralist constraints, that is growing under the pressure of the computational capacities afforded by the progress in informatics. Should this be eventually the case, Parsons' ideas would just be a starting point for further inquiries, whose philosophical, logical, historical, and even mathematical complexities, are not hard to imagine.

6
The Indispensability Argument: Structure and Basic Notions

Most of the platonist strategies we have discussed so far are based on a clarification of the notion of mathematical objects. The strategy to be discussed in this and the following chapters is different, and is based on the so-called Indispensability Argument (henceforth, IA). IA is neutral on the nature of mathematical objects. It stems form the simple undeniable datum that mathematical theories are applicable, and actually applied to (empirical) science.[1] Its premises are *prima facie* acceptable by nominalists and platonists alike, and it would offer, if sound, plausible grounds for the claims that mathematical statements are true and/or mathematical objects exist, or at least that we are justified in so believing. Not relying on its sleeves on *a priori* reasoning, IA is particularly suitable for supporting platonism even in an empiricist framework. IA is commonly credited to Quine and Putnam, though some precursors have been recognized in Frege, von Neumann, Carnap and Gödel.[2] The basic structure of the argument is suggested in many of Quine's remarks, for it seems to flow naturally from some of Quine's tenets.[3] However, the first explicit formulation is in Putnam (1971). As mentioned earlier, IA can be considered as a non-conservative response to Benacerraf's dilemma, at least insofar as endorsing it is meant to imply that that the truths of mathematics are not *a priori*, what contradicts the second part of condition (*ii*) of conservative responses.

Before reviewing the recent debate in Chapter 7, we offer in the present chapter a recognition of the possible different formulations of IA, together with a brief discussion of the most salient notions and theses that it may involve.

6.1 Four Versions of IA

Suppose that there are true scientific theories, or at least scientific theories that we are justified in holding true. Within those theories, certain properties and relations are attributed to certain objects. In astronomy, for instance, the relations described by the laws of gravitation are attributed to planets. Let us grant that a scientific theory is true only if the objects it speaks of exist. Many accepted scientific theories make abundant use of portions of mathematics, thus including, beyond statements about physical objects, statements about mathematical objects like numbers and sets. Suppose this happens for some among true (or justifiably held true) scientific theories. Since it is natural to conclude that the theories in question could not be true if these mathematical objects did not exist, it follows that these objects exist, or that we are justified in holding that they exist. This argument might be rejected by arguing that, however helpful, the use of mathematics in the relevant scientific theories is dispensable: everything these theories say about the physical world and its objects could also be said without appealing to mathematical objects. To avoid this rejoinder, one needs to make sure that this use is indispensable.

Roughly, this is the basic thought underlying IA.[4] To see how this argument could be stated more precisely, let us suppose a mathematical theory to be putatively about some objects, and talk of these as the objects of this theory. Let S and M be a scientific and a mathematical theory, respectively. For short, let us use '$O_{[M]}$' to denote the objects of M.

The argument could then be firstly stated in more precise terms as follows, either in general, or with respect to an indeterminate instantiation (namely, an instantiation relative to two indeterminate theories S and M):

[1a] *Non-epistemic IA for Platonism*

(*i*) There are true scientific theories / S is true.
(*ii*) Among them, some are such that some mathematical theories are indispensable to them / M is indispensable to S.
(*iii*) The former are true only if the latter are themselves true / S is true only if M is true.
(*iv*) A mathematical theory is true only if its objects exist / M is true only if the $O_{[M]}$ exist.

(*v*) The objects of the mathematical theories mentioned in (*ii*) exist / The $O_{[M]}$ exist.

In this version, the argument is indubitably valid, but it might not be sound, since some of its premises might be false. For example, one could doubt that scientific theories, even the best ones we have, are true. A way of weakening the argument accordingly, without undermining its validity, is to state it in epistemic terms, as being merely concerned with our justification in believing in the truth of the relevant theories or in the existence of their objects:

[1*b*] *Epistemic* IA *for Platonism*

(*i*) We are justified in believing that some scientific theories are true / We are justified in believing that S is true.
(*ii*) Among these theories, some are such that some mathematical theories are indispensable to them / M is indispensable to S.
(*iii*) We are justified in believing that the former are true only if we are justified in believing that the latter are themselves true / We are justified in believing that S is true only if we are justified in believing that M is true.
(*iv*) We are justified in believing that a mathematical theory is true only if we are justified in believing that its objects exist / We are justified in believing that M is true only if we are justified in believing that the $O_{[M]}$ exist.

(*v*) We are justified in believing that the objects of the mathematical theories mentioned in (*ii*) exist / We are justified in believing that the $O_{[M]}$ exist.

Stating IA this way, or anyway presenting it in an epistemic form, has not only the advantage of avoiding the highly controversial premise 1*a*(*i*) (or other similar premises in other non-epistemic versions of the argument). It is also fully in agreement with the view that ontology cannot but be normative in character – that is, merely aiming at telling us what we are justified in believing to exist, rather than what exist (Colyvan, 2001, p. 11), which is a view endorsed by many. These are certainly the main reasons why IA is often stated in an epistemic form.

Both [1*a*] and [1*b*] concern particular mathematical objects only in their instantiated versions, and once it is supposed that the instantiation is obtained by considering a pair of particular theories. In their general or indeterminately instantiated versions, neither [1*a*] nor [1*b*] specify which mathematical objects they are concerned with, and thus leave open the possibility that they are endorsed without pointing to any mathematical objects in particular. Tough generic, their conclusion would still suffice to undermine nominalism, and to support platonism.

A similar sort of indeterminacy might also affect particular instances of the argument. We have seen (in § 3.1) that multiple reductions of arithmetic to set-theory are available. Suppose that M is arithmetic, and that its objects are taken to be sets. Through [1a] and [1b] we can conclude that some sets exist, or that we are justified in believing that they exist, without being able to specify which they are. Furthermore, as observed by Alan Baker (2003a, pp. 55–8), it is enough to notice that arithmetic can be reduced both to set theory or to category theory (see MacLane, 1997), to understand that [1a] and [1b] can be endorsed without being able to discriminate whether the relevant objects are sets or categories.

Another important feature of [1a] and [1b] is the following. Both conclusions 1a(v) and 1b(v) concern the existence of mathematical objects, whereas the premises 1a(i)–1a(iii) and 1b(i)–1b(iii) only concern the truth of mathematical and scientific theories. The step from the latter to the former is licensed by premises 1a(iv) and 1b(iv), both of which could be plausibly questioned. Rejecting them is nonetheless not enough to refute IA, since the remaining premises in both [1a] and [1b] are still enough to validly deliver significant conclusions. But, insofar as these conclusions do not concern existence, IA, when thus amended, will not be an argument for platonism any more, but merely one for semantic realism (see *Introduction*, § 1.7). Here is how it could be stated, then:

[2a] *Non-epistemic IA for Realism*

(i) There are true scientific theories / S is true.
(ii) Among them, some are such that some mathematical theories are indispensable to them / M is indispensable to S.
(iii) The former are true only if the latter are themselves true / S is true only if M is true.

(iv) The mathematical theories mentioned in (ii) are true / M is true.[5]

[2b] *Epistemic IA for Realism*

(i) We are justified in believing that some scientific theories are true / We are justified in believing that S is true.
(ii) Among these theories, some are such that some mathematical theories are indispensable to them / M is indispensable to S.

(*iii*) We are justified in believing that the former are true only if we are justified in believing that the latter are themselves true / We are justified in believing that s is true only if we are justified in believing that M is true.

(*iv*) We are justified in believing that the mathematical theories mentioned in (*ii*) are true / We are justified in believing that M is true.

One might deny that it is plausible to claim that a mathematical theory is true without thereby making its truth depend on the existence of some objects (for instance, by arguing that any non-literal interpretation of mathematical statements is untenable: see Burgess and Rosen, 1997). On this view, IA for realism ([2*a*] or [2*b*]) could not be endorsed without also endorsing IA for platonism ([1*a*] or [1*b*]). Still, insofar as [2*a*] and [2*b*] are neutral as to whether the truth of mathematical theories depends on the existence of suitable mathematical objects, they represent desirable versions of IA for those claiming for the objectivity and truth of mathematics, but not qualifying themselves as platonists (for instance, Hellman and Putnam: see § 4.3, and § 7.2.2, respectively). It is helpful to keep [2*a*] and [2*b*] distinct from [1*a*] and [1*b*]), then.[6]

We thus have four versions of IA: a non-epistemic and an epistemic version for platonism, and a non-epistemic and an epistemic version for realism. We take them to be minimal versions of the argument. By calling them 'minimal' we mean that these versions feature the least or least controversial theoretical ingredients required to state IA, either epistemically or not, and either for platonism or for realism, respectively. These versions seem to us to provide, then, the basic structure of the argument, or better, four basic schemas, on the basis of which different arguments can be obtained, so as to get most of the versions of IA that are available in recent discussions.

Arguments [1*a*], [1*b*], [2*a*] and [2*b*] are better understood as schemas of arguments, rather than arguments proper. This is so in more than one sense. First, in all four of them, 's' and 'M', can be substituted, respectively, with names of particular scientific and mathematical theories. Second, the meaning of 'indispensable' can be specified in them in different ways, so as to get strictly different arguments according to which specification is chosen. Third, the same holds for the notions of justification and truth (provided they are preserved under indispensability, however the latter is spelled out). Lastly, [1*a*] and [1*b*] involve the notion of aboutness, and this can also be specified in different ways, so as to provide strictly different arguments.

Before reviewing the debate on IA (in Chapter 7) it seems then helpful to us, to have a look to the main ingredients of the argument.

6.2 The Quine-Putnam Argument and Colyvan's Argument

It will help to introduce this discussion by considering three other versions of IA, often discussed and considered as paradigmatic.

6.2.1 The Quine-Putnam Argument

Putnam (1971, p. 347) contains what is unanimously seen as the first clear formulation of IA, building upon Quine's suggestions, and the starting point of any subsequent discussion:[7]

[3] *Quine-Putnam Indispensability Argument*

> ... quantification over mathematical entities is indispensable for science, both formal and physical, therefore we should accept such quantification; but this commits us to accepting the existence of the mathematical entities in question.

This appears as an argument for platonism on the lines of [1*a*] or [1*b*],[8] with premise 1*a*(*ii*) or 1*b*(*ii*) specified in terms of quantification (see § 6.3.2, above). As a matter of fact, Putnam presents his argument as "an argument for realism", however. This deserves clarification.

The term 'realism' admits many attested readings. In the sense in which we used it above, it refers to the thesis that statements of a certain sort are true, irrespective of what makes them true (and possibly independently of our cognitive activity), which could be better called 'semantic realism'.[9] Sometimes this term is also used to denote the thesis that certain (kinds of) entities exist, which could be better called 'ontological realism'.[10] In this latter sense, realism in mathematics is the same as platonism (Field, 1989a, p. 228).

Now, Putnam appears to waver, at different places, between considering [3] an argument for platonism or for semantic realism (see § 7.2.1, note 13). Be that as it may, the quotation above has been traditionally seen as expressing an argument for ontological realism, or platonism – given also that Putnam continues by observing that "this type of argument stems, of course, from Quine", who surely deemed IA an argument for platonism (see also note 8 above).

6.2.2 Colyvan's Argument (and its Enhanced Version)

Another often-discussed formulation of IA is the one advanced by Colyvan (1998, p. 40; 2001, p. 11):

[4] *Colyvan's Indispensability Argument*

(*i*) We ought to have ontological commitment to all and only those entities that are indispensable to our best scientific theories.
(*ii*) Mathematical entities are indispensable to our best scientific theories;

(*iii*) We ought to have ontological commitment to mathematical entities.

Arguably, premise 4(*i*) should be read as a quantified biconditional: 'For any entity x, we ought to have ontological commitment to x if and only if x is indispensable to our best scientific theories'. The 'only if' implication is expressed by 'only' (and provides a necessary condition: we ought not to have ontological commitment to non-indispensable entities); the 'if' implication is expressed by 'all' (and provides a sufficient condition: we ought to have ontological commitment to indispensable entities). This allows Colyvan (2001, pp. 12–13) to claim that his formulation makes explicit the role played by two important theoretical ingredients of IA coming from Quine's views,[11] – which do not occur, at least not explicitly, in our four versions – since he believes that premise 4(*i*) "follows from the doctrines of *naturalism* and [confirmational] *holism*". We will question the role of these ingredients in § 6.7. Here we just want to notice that the 'only if' direction is useless for the validity of the argument (whose logical form is of a *modus ponens*). This is acknowledged by Colyvan himself (2001, p. 12), which observes that premise 4(*i*) "is a little stronger than required", since "all that is really required . . . is the 'all', not the 'all and only'".

Be that as may be, it is clear that [4] is an epistemic argument for platonism, and is then on the same lines of [1*b*], so that premise 4(*ii*) is to be intended as equivalent to premises 1*b*(*ii*) (cf § 6.3.2 below).

Arguments [3] and [4] are representatives of two different stances on IA. The first is concerned with logico-syntactical features of scientific theories and their expressive power: what is at stake is whether some particular vocabulary is necessary in order to state some given scientific laws (see Putnam 1971, § v–vii). The second widens the scope by taking into account general concerns in the philosophy of science, especially as regards the relation between philosophy and science and the way in which empirical evidence supports scientific theories.

There is yet a third stance on IA, suggested by the acknowledgement that the so-called theoretical entities of physical theories, such as strings or subatomic particles, raise problems analogous to those raised by mathematical objects: how should we justify their existence, since they are (possibly even in principle) unobservable?

Inference to the best explanation (henceforth, IBE) is often used in this connection (see Lipton, 2002): we get a justification in believing in the existence of some such entities if their postulation provides the best explanation available for a given class of (observable) phenomena. IBE is a powerful tool for scientific realists: those who claim that our best scientific theories describe the way the physical world is, even when it comes to unobservable entities – as opposed to those, like Bas van Fraassen (1980), who recommend agnosticism in the latter case (see §§ 7.3.2, 7.4.2). Many have suggested, then, as partly also Quine (1981c, pp. 149–50) did,[12] that this use of IBE in support of scientific realism parallels IA for platonism. On this basis, Field first suggested that IA could be stated in terms of indispensability "for explanations" (1989a, pp. 14–15): if we take a certain theory, that includes some statements entailing that certain (sorts of) objects exist, to provide the best explanation of a given set of (arguably empirical) phenomena, we "have a strong reason to believe" in the truth of the statements in question, and then (possibly) in the existence of the relevant (sorts of) objects.

As a result of this and of the recent vast debate on scientific explanation (Kitcher and Salmon, 1989; Salmon, 1989; Woodward, 2009), and in particular on the role of mathematics in it,[13] Baker (2009, p. 613) has offered the following 'enhanced' version of [4]:

[5] *Baker's Enhanced Indispensability Argument*

(*i*) We ought rationally to believe in the existence of any entity which plays an indispensable explanatory role in our best scientific theories;
(*ii*) Mathematical objects play an indispensable explanatory role in science.

(*iii*) Hence, we ought rationally to believe in the existence of to mathematical objects.

6.3 (In)dispensability

We are now ready to begin our discussion of the main ingredients of IA, beginning, as it is natural, with the notion of indispensability.

Concerning this, two questions must be answered: (*a*) what is to be said to be indispensable, and to what?; (*b*) what does it mean that the former is indispensable to the latter?

6.3.1 What Is Indispensable to What?

As concerns the second half of question (*a*), the natural response is that IA is concerned with indispensability to scientific theories (but see § 7.4.4). For IA to have a point at all, moreover, it must be granted that the relevant scientific theories contains statements involving some mathematical vocabulary.[14]

The first half of question (*a*) is trickier. We can distinguish at least five candidates that could be said to be indispensable to a scientific theory S: (*i*) other theories; (*ii*) quantification over a certain domain of entities, or better putative objects;[15] (*iii*) certain putative objects, as such; (*iv*) reference to such objects; (*v*) a given vocabulary.

Option (*iii*) is suggested by argument [4]. Let us first consider this one. What does it mean that certain entities are indispensable to S? If these are objects that S is putatively dealing with, it is natural to require that they be suitably characterized within S itself, or within an auxiliary related theory (especially if they are abstract objects, for which ostension is unavailable). Let T be this auxiliary theory or the part of S characterizing these objects. One can then say that T is putatively about these objects and then, according to option (*iii*), that the $O_{[T]}$ are indispensable to S. This cannot but mean that S indispensably includes some statements that involve either terms that purportedly denote the $O_{[T]}$, or variables that purportedly range over them, which means in turn that S makes indispensable use of a vocabulary fixed by T. The problem raised by question (*b*) would then be to understand what 'to make indispensable use of' is to mean in this case.

The remaining options do not differ essentially from (*iii*).

As Colyvan himself points out (2001, p. 7), reference to putative objects, or apparent reference cannot but amount to the use of singular terms purportedly denoting some objects. Option (*iv*) reduces then to option (*iii*), since those objects are just the $O_{[T]}$. The same holds for option (*ii*), provided that the place of those singular terms is taken by appropriate predicate constants.[16]

Options (*i*) and (*v*) differ from options (*ii*), (*iii*) and (*iv*) since they do not mention any putative objects. It seems natural, however, to understand also these two options as involving the consideration either of an auxiliary theory related to S, or of an appropriate part of S itself. Let again T be either the latter or the former. Option (*v*) can then be specified in a way similar to options (*ii*), (*iii*) and (*iv*), provided that the relevant vocabulary is the one fixed by T: that this vocabulary is indispensable to S means then that S makes indispensable use of it. The same holds for option (*i*), if this is specified as we shall suggest in the next section. Indeed, according to this specification, that a theory T is indispensable to S means again that S makes indispensable use of a vocabulary fixed by T.

This being said concerning question (*a*), it is important to notice that the difference between options (*ii*), (*iii*), (*iv*) and options (*i*), (*v*), due to the fact that the former do mention objects while the latter do not, is not merely a matter of emphasis. For nothing forces a partisan of options (*i*), (*v*) to admit that the vocabulary fixed by T has an ontological import. These two options are then more congenial to IA for semantic realism, at least if endorsing the latter is not meant to go together with the endorsement of IA for platonism.

6.3.2 What Is (In)dispensability?

We can now rephrase question (*b*) as follows: what does it mean that a scientific theory S makes indispensable use of a vocabulary fixed by an auxiliary theory related to it, or by an appropriate part of itself? Insofar as a part of a theory is itself a theory, answering this question merely requires us to specify what it means that S makes indispensable use of a vocabulary fixed by another theory T either included in it or not. In short, let us agree that saying that this is so is the same as saying that T is indispensable to S (we do not see, indeed, what else saying that T is indispensable to S could mean).

It is clear then that if S did not include statements employing the vocabulary of T, T would be certainly dispensable from S. Let us then suppose that S includes such statements, and let us call these 'T-loaded' statements. The basic idea is that T is dispensable from S if and only if S can be rephrased into a theory S* that does not include any T-loaded statement, that is, if and only if S is equivalent, in some sense to be qualified, to such a theory S*. T will then be indispensable to S if and only if it is not dispensable from it (in § 4.1 we have already seen how Field argues for the dispensability of analysis from Newton's Gravitational Theory along these lines). In order to specify what it is for T to be indispensable to S, we have then to select a suitable equivalence relation among scientific theories.[17]

6.3.2.1 *The Way of Paraphrase*

An option, suggested by Quine, is to conceive of some constants – those which purportedly refer to abstract objects, or designate properties of them or functions with arguments or values on them – as introduced through contextual definitions, and to use these definitions the other way around, in order to eliminate these constants (see Quine, 1939, and many other subsequent works).

Contextual definitions fundamentally differ from explicit definitions. In the latter, a new symbol or expression is introduced as an abbreviation of an already available expression of the same syntactical type. Think of the neo-logicist definition of '0' in Frege arithmetic (see § 5.1). Contextual definitions, instead, license the replacement of certain statements (or systems of statements) with others (usually simpler, in some sense) in which a new expression or symbol occurs that has no correlate of the same syntactical type in the former statements. Think of the definition of numerals through numerical quantifiers (see § 4.1). They license the replacement of statements in which such quantifiers occur but no terms for natural numbers occur, with other statements in which such terms occur: accordingly, '$\exists_2 x(x$ is a moon of Mars)' is replaced by 'Mars has 2 moons'. We can take the new statements, replacing the original ones in agreement with a contextual definition, as paraphrases of these latter, and *vice versa*.

This points to a very simple equivalence relation between theories. Call it '*p*-equivalence': two theories are *p*-equivalent if for any statement of each of them, the other includes either this same statement or a paraphrase of it, and *vice versa*.

In most contexts, constants introduced by contextual definitions behave syntactically as other constants of the same type. But, at least in some cases, it is natural to understand their use as a *façon de parler* (Quine, 1939, pp. 708–9). Hence, in case of singular terms, it is natural to admit that the existence of the objects they purportedly denote is not a necessary condition for the truth of the statements in which they occur. For example, for 'Mars has 2 moons' to be true it is enough that '$\exists_2 x(x$ is a moon of Mars)' be true, and this does certainly not require that there exists an object to which '2' refers. These singular terms are then intended to denote nothing at all. The same holds, of course, *mutatis mutandis*, for the quantification over constants to which predicates introduced in this way apply.

This makes *p*-equivalence suitable for specifying indispensability. Suppose one wants to eliminate the constants belonging to T's vocabulary from the language of S. One can select appropriate contextual definitions for these constants, and then rely on them to replace any T-loaded statement of S with its non-T-loaded paraphrase, thereby obtaining a new theory S* *p*-equivalent to S. If this is possible, T is dispensable from s, according to the previous general definition of dispensability with p-equivalence as the relevant equivalence relation.

But how could this be impossible? After all, one may think, if the vocabulary of s is rich enough, it should be relatively easy to devise contextual definitions for any constant of T's vocabulary included in it. Even if other difficulties are set aside (connected with the proper understanding of the very notion of paraphrase), the main problem is that, for this strategy to be effective, these definitions have to be somehow plausible: it should be plausible to assign the same sense to the original statements and their paraphrases as obtained through those definitions. Quine himself, who initially adhered to a radical nominalist programme based on this strategy (Goodman and Quine, 1947), soon realized the impossibility of finding plausible contextual definitions for many mathematical constants included in the vocabulary of our scientific theories and abandoned this programme, by moving towards platonist views.[18]

6.3.2.2 Observational Equivalence

A different option goes as follows (see Colyvan, 2001, pp. 76–7, though this option is very common). Suppose s's vocabulary can be separated into two complementary sub-vocabularies: Ω, called 'observational', containing all and only those constants that purportedly refer to observable objects or design properties of them, or functions with arguments and values on them; Θ, called 'theoretical', containing all other constants of s's vocabulary. Let us call 'observational' the statements of s in which only constants of Ω occur, and say that s and s* are o-equivalent if they include the same observational statements.[19]

Suppose now that all observational statements of s are non-T-loaded. We can say that T is dispensable from s if and only if s is o-equivalent to some appropriate theory s* in which no T-loaded statements occurs.

Up to now, we have taken for granted that a theory is a set of statements. If all observational statements of s are non-T-loaded, it is then enough to take all and only the observational statements of s in order to get a theory s_Ω o-equivalent to s in which no T-loaded statements occurs. Still, s_Ω might not be appropriate, since some T-loaded statements of s might be required in order to deduce observational statements of s, so that these last statements would turn out to be underivable, and then unjustified, in s_Ω. Consider also that if this were the case, and s were taken to be the deductive closure of some of its statements, including some T-loaded ones, it would be impossible, by eliminating these latter, to specify the set of s's observational statements, and then s_Ω itself.

Considering Craig's theorem (Craig, 1953, 1956; see also Putnam, 1965) makes the situations clearer. Skipping over details, the theorem states that if whatever (formal) theory V (satisfying some very minimal and natural conditions) is recursively enumerable,[20] and its vocabulary is anyhow split into two complementary vocabularies Φ and Ψ, then there is a sub-theory V_Φ of V, including all and only the statements of V employing Φ, which is recursively axiomatizable.[21] Hence, should the requirement that S_Ω be recursively axiomatizable be enough for S_Ω to be appropriate in the sense required above, any theory S (satisfying the very minimal and natural conditions required by Craig's theorem) would admit an appropriate theory S* o-equivalent to it all of whose statements are non-T-loaded, and T would then be always dispensable from S. Thus, the fact that S_Ω is recursively axiomatisable cannot by itself make it appropriate in the required sense.[22]

In order to use o-equivalence as the relevant equivalent relation for specifying indispensability, some different appropriateness condition on S* must then be fixed. Field (1980, pp. 8 and 41) suggests that S* be "reasonably attractive". Colyvan (2001, p. 77) that it be "preferable" to S. But when is a theory attractive or preferable to another? In answering this question, one should appeal to theoretical virtues. One could require, for example, that S* admit for an adequate axiomatization, or that it be formally elegant or simple, or that it provide a suitable explanation for the phenomena S is dealing with, or that it allow prediction of these phenomena, and so on.

Once an appropriate criterion of virtuosity is fixed, one could then use it, together with o-equivalence, to specify indispensability: T would then be said to be dispensable from S if and only if S were o-equivalent to some theory S* complying with this criterion, in which no T-loaded statement occurs.

This would not solve all the problems connected with the use of o-equivalence to specify indispensability, however. One major unsolved problem would be the following. Despite being o-equivalent to S (and complying with the selected criterion of virtuosity), S* may not include all empirically relevant statements of S. In other terms, the empirical content of S may not be exhausted by its observational statements, namely by its statements employing only (non-logical) constants from its observational sub-vocabulary. Some of its statements might unavoidably employ a non-observational vocabulary, namely a mathematical one, and still have empirical content, insofar as they say something about the empirical world: remember the statement 'Gravitational acceleration at poles is equal to 9,823 m/s$^{2'}$, mentioned in § 4.1. In this case, despite being o-equivalent to S, S* might have an empirical content

The Indispensability Argument: Structure and Basic Notions   209

poorer than s's,[23] and this would threaten the tenability of the specification of indispensability using o-equivalence.

Three routes seem to be available here. One could reject such a specification of indispensability (*pace* its being common). One could cash out the criterion of virtuosity so that s* could comply with it only if it preserves s's empirical content. One could limit the applicability of the notion of indispensability, so specified, to the case of scientific theories whose empirical content is exhausted by their observational statements (as we will see in § 7.1, this is Field's strategy).

6.3.2.3 (In)dispensability and Explanation

Neither p-equivalence, nor o-equivalence seem suitable for providing a specification of indispensability adequate for argument [5].

Consider the case raised by Baker himself (2009, p. 614; see also Baker 2005) to exemplify this last argument. It concerns a particular subspecies of cicadas, whose lifecycle is of either thirteen or seventeen years. It has been suggested that an explanation for this is that lifecycles of a prime number of years minimize the chance of overlapping with lifecycles of predators. To conclude from here, according to argument [5], that natural numbers exist, it is not enough to hold that this is a good explanation. One has also to take arithmetic to be explanatorily indispensable for the relevant biological theory.

But what does this mean? A natural response is this: T is explanatorily dispensable from s if and only if s is as equally explanatory as some appropriate theory s* in which no T-loaded statement occurs. In this way, claiming that T is explanatorily indispensable to s means that no such theory s* is available, which means in its turn (under some further appropriate clauses) that T provides the best available explanation for the relevant phenomena. In order to make this fully clear, what is required is both a specification of what is meant by 'equally explanatory', and the identification of a suitable criterion of appropriateness.

6.3.2.4 A General Clarification of (In)dispensability

The foregoing examples suggest a way of getting a better clarification of the notion of (in)dispensability involved in IA. Let M be a mathematical theory. Then:

[*(In)dispensability*]
M is dispensable from S if and only if:

(a) S is ε-equivalent to a theory S* all of whose statements are non-M-loaded, provided ε-equivalence is an appropriate equivalence relation;

(b) s* comply with an appropriate criterion of virtuosity.

M is indispensable to s if and only if it is not dispensable from s, that is, if and only if no theory s* meeting conditions (*a*) and (*b*) is available.

A noteworthy consequence of this definition is that the ordinary way of talking of indispensability in general is partly inaccurate. No theory (theorem, term, entity, or whatever you like) is indispensable *tout court* to a theory s, but only relative to a certain equivalence relation. We should better speak of ε-indispensability, rather than indispensability *simpliciter*. The general, unspecified, notion of indispensability is then an essentially relative notion,[24] with the result that IA can have different philosophical significance, according to the choice of the equivalence relation entering a specification of this notion.

6.4 Quine's Criterion of Ontological Commitment

Any version of IA for platonism has to include a premise stating that certain scientific theories are true, or we are justified in believing that they are true, only if some mathematical objects exist, or we are justified in believing that they exist (in versions [1*a*] and [1*b*], these premises are split in two: premises 1*a*(*iii*) and 1*b*(*iii*) allow us to pass, respectively, from the truth and the justification in believing in the truth of some scientific theories, to the truth and the justification in believing in the truth of some mathematical theories; premises 1*a*(*iv*) and 1*b*(*iv*) then allow us to pass, respectively, from the latter to the existence and the justification in believing in the existence of some mathematical objects). This premise is usually argued for on the basis of a criterion of ontological commitment of scientific theories: a criterion for establishing what objects such a theory says exist. The criterion that is usually appealed to is Quine's (1939, 1948), which is often evoked through the well-known slogan: "to be is to be the value of a variable" (1939, p. 708; 1948, p. 253).[25] The seeming simplicity of this slogan hides at least three conditions.

First, the relevant theory has to be stated in "canonical notation" (1960, ch. V), which is given for Quine in a first-order predicate language with identity; if it is not already presented in this form, it has to be so rephrased. According to Quine, only a first-order predicate language is suitable for displaying the ontological implications of a theory, since it is ontologically neutral, as opposed to other languages, like second-order predicate language or modal language (see Chapter 2, note 15, and § 4.3).

The Indispensability Argument: Structure and Basic Notions 211

Second, the first-order predicate language to be used has to be reduced by eliminating all singular terms[26] and all the remaining non-logical constants that can be eliminated through the procedure described in § 6.3.2.1, so as to put it in its ontologically most elementary form. Notice that our decisions on how to reach this elementary form will be guided by previous philosophical views on which objects we are keen to accept in our ontology.[27]

Third, if the relevant theory is informal, rephrasing it in such an elementary language will be effected by taking its existential statements (like 'there are *F*s that are so-and-so' or 'there is an *F* that is so-and-so') as faithfully rendered by statements of the form '$\exists x[\mathcal{A}(x)]$', where '$\mathcal{A}(x)$' is an appropriate formula in which x occurs free.

Putting these conditions together, we can say that, according to Quine's criterion, in order to determine the ontological commitment of a theory T, one needs first to rephrase it in the form of its ontologically most elementary canonical reformulation $T_0^{[P1]}$ (which, hopefully, could be univocally determined). The ontological commitment of T will then coincide with that of $T_0^{[P1]}$.

Quine's slogan applies to $T_0^{[P1]}$. It states that a theory T says that an object a exists if and only if a satisfies the open formula '$\mathcal{A}(x)$' occurring in some consequence of the form '$\exists x[\mathcal{A}(x)]$' of its ontologically most elementary canonical reformulation $T_0^{[P1]}$ of T.[28] Quine's criterion of ontological commitment can then be stated as follows:

[*Quine's Criterion of Ontological Commitment*]

The ontological commitment of T is the domain formed by all the objects that purportedly satisfy the open formulas '$\mathcal{A}(x)$' occurring in all the consequences of the form '$\exists x[\mathcal{A}(x)]$' of its ontologically most elementary canonical reformulation $T_0^{[P1]}$.

The role of a criterion of ontological commitment within IA for platonism depends on the natural assumption that a theory T is true, or we are justified in believing it true, only if there actually exists, or we are justified in believing that there actually exists, everything that T says exists, from which it follows that T is true, or we are justified in believing that T is true, only if there actually exists, or we are justified in believing that there actually exists, any object included in T's ontological commitment. Now, the role of the premise concerning indispensability in IA for platonism is to warrant that the ontological commitment of the relevant scientific theories includes mathematical objects. The idea is that, if S and M are, respectively, a scientific and a mathematical theory such that M is indispensable to S, then the $O_{[M]}$ are certainly part of the ontological commitment of S. But is this actually so?

The answer depends on the equivalence relation and the criterion of virtuosity which are appealed to in order to specify the notion of indispensability, and on the relation that S is taken to bear to its ontologically most elementary canonical reformulation $S_0^{[P1]}$. In particular, the answer is in the affirmative if and only if the unavailability of a theory S* meeting conditions (a) and (b) of [(In)dispensability] is enough for the $O_{[M]}$ to be among the objects that purportedly satisfy the open formulas '$\mathcal{A}(x)$' occurring in the consequences of the form '$\exists x[\mathcal{A}(x)]$' of $S_0^{[P1]}$.

This makes clear that no accurate diagnosis on the soundness of a version of IA for platonism can avoid considering the connection between the criterion of ontological commitment that this version of IA appeals to (be it Quine's or any other) and the equivalence relation and the criterion of virtuosity that are appealed to in specifying the notion of indispensability.

In IA for realism, no ingredient analogous to the criterion of ontological commitment occurs. Still related problems also arise, since the role of the premise concerning indispensability is there to warrant that the truth or the justification in believing to the truth of the relevant scientific theories entails the truth or the justification in believing to the truth of some mathematical theories, the basic idea being that this certainly happens if the latter theories are indispensable to the former. But, once again, whether this is so or not depends on the way the notion of indispensability is specified (and on the conception of truth that it is adopted).

6.5 Naturalism

Naturalists, broadly speaking, require philosophical inquiries to be informed by the results and methodology of natural sciences. This broad attitude can be variously elaborated (Papineau, 1993, 2007; De Caro and Macarthur, 2004, 2010; Maddy, 2007; Ritchie, 2009). A radical form of naturalism would require philosophical disputes to be entirely reduced to, and replaced by, scientific arguments. A milder form would acknowledge genuinely philosophical disputes as legitimate, but would ask that the evidence for their solution is sought in empirical findings and scientific methods. Various intermediate forms can be thought of.

Quine was a fierce naturalist, for instance in his conception of epistemology (Quine 1969b). More generally, naturalism consisted for him in the "abandonment of the goal of a first philosophy" (1981b, p. 72), in "the recognition that it is within science itself, and not in some prior philosophy, that reality is to be identified and described" (1981a, p. 21). Following the empiricist tradition, he believed that empirical evidence is the only kind of evidence on which knowledge is to be based, and that it is science – and not *a priori* philosophical arguments – that delivers such evidence and explains how it can be acquired. He believed this to hold also for existential knowledge, to the extent that "ontological questions . . . are on a par with questions of natural science" (Quine, 1951a, p. 45), since any object – be it concrete or abstract – is to be conceived of as a posit of our theories (Quine, 1960, ch. 7; 1981a). Science is then, for Quine, the sole ground for our judgements in matter of ontology.

From this and Quine's criterion of ontological commitment, it follows that legitimately claiming that there exists an object a should be the same as claiming that a satisfies the open formula '$\mathcal{A}(x)$' occurring in a consequence of the form '$\exists x[\mathcal{A}(x)]$' of the ontologically most elementary canonical reformulation $s_0^{[p1]}$ of an accepted scientific theory s. This is why Quine took his criterion to point to "essentially scientific reasons" for answering ontological questions (Quine, 1969c, p. 97).[29]

This is not the same as arguing that only scientific theories can be true or that we are justified in believing as true only scientific theories.[30] Quine's point is rather that the verdict on the truth of non-scientific theories has to depend on a verdict on the truth of scientific theories to which the former are indispensable. It follows that the reasons for declaring a non-scientific theory to be true must ultimately depend, for Quine, on IA itself or on structurally similar arguments.

This should be enough to make clear how much Quine's naturalism is tied to his views in matters of ontology and to IA, in particular. Still, naturalism in general can be held independently of these conceptions.

To remain at a sufficient level of generality, we can identify two aspects of naturalism, respectively concerned with truth and existence, that can be singled out by the two following theses:

[*Semantic Naturalism*]
(Since scientific theories are the only source of genuine knowledge), the only theories we are justified in holding true are either scientific theories, or theories that are indispensable to scientific theories.[31]

[*Ontological Naturalism*]
(Since scientific theories are the only source of genuine knowledge), we are justified in believing in the existence of only those objects that are part of the ontological commitment of theories that we are justified in holding true according to [*Semantic Naturalism*].[32]

6.6 Confirmational Holism

Confirmational holism is a thesis concerning how empirical evidence confirms scientific hypotheses. In Quine's famous slogan, it is the thesis that "our statements about the external world face the tribunal of sense experience not individually but only as a corporate body" (1951a, p. 41).[33]

The basic idea (see for instance Quine, 1990, pp. 9–16) is that no hypothesis can be experimentally tested in isolation, but it is always tested in a quite larger context, possibly a whole theory or even the whole web of our theories. According to the hypothetico-deductive conception of confirmation (see § 7.4), an hypothesis h is confirmed if $h \Rightarrow \gamma$ and γ is experimentally verified. Though open to many difficulties (see § 7.4.1), this conception is usually assumed when confirmational holism is discussed in connection with IA. This is then presented as the thesis that no experiment can verify a consequence of a single hypothesis, for what happens, at best, is that $(\varphi \wedge h) \Rightarrow \gamma$, where φ is a large conjunction of auxiliary hypotheses. Hence, verifying γ would not confirm only h but rather the conjunction $\varphi \wedge h$. Analogously, verifying $\neg \gamma$ would not disconfirm only h but rather the conjunction $\varphi \wedge h$ (thus leaving open which conjunct or conjuncts in φ are to be rejected).

Different forms of confirmational holism differ for the extension to be ascribed to φ. Quine's view is that φ coincides in fact with the whole web of our theories, including mathematical and logical ones, which are then submitted to experimental (or more generally empirical) confirmation and disconfirmation.[34] Still, just as for any hypothesis, these theories cannot be confirmed or disconfirmed as such, but only together with the whole web they belong to.

To present it in a brief formulation, confirmational holism can be identified, in general, with the following thesis:

[*Confirmational Holism*]
(Since empirical evidence does not confirm scientific hypotheses in isolation, but rather whole theories) confirming an hypothesis goes together with confirming a whole theory that this hypothesis is taken to belong to.

If we assume, as it is natural to do, that confirmation of a theory contributes to our justification in holding it true, we can easily understand how this thesis can be taken to be relevant to argument [4], even though confirmation is not explicitly mentioned in it. The idea is that confirmation of a scientific theory s entails confirmation of any theory M indispensable to s, and this provides justification in believing M true, or at least contributes to providing such justification. Still, in order to make this idea suitably clear, one should at least shed light on the relation between confirmation and justification in general, or at least offer a version of IA which involved confirmation directly rather than justification in general.

6.7 The Dispensability of Naturalism and Confirmational Holism

Both naturalism and confirmational holism are epistemic theses. They can then be plausibly considered as relevant only for epistemic versions of IA. However, neither argument [1b] nor argument [2b] make an explicit mention of either naturalism or confirmational holism. Moreover, neither theses seem required as background assumptions for arguing in favour of the relevant premises. This suggests that at least some (valid) versions of IA are not forced to appeal to either naturalism or confirmational holism. This is also stressed by Putnam himself (2012, pp. 183 and 188), when he claims to have "never subscribed . . . in . . . [his] life" to the 'only' implication in premise (*i*) of argument [3] – the implication that Colyvan takes to follow from naturalism – and "never claimed that mathematics is 'confirmed' by its applications in physics".

That confirmational holism might be unnecessary for IA has already been suggested on various grounds by many authors (see, for example: Resnik, 1995; Chihara, 2004, § 5.6; Dieveney, 2007; Azzouni, 2009; and Colyvan himself, 2001, p. 37). The basic idea is that there is no need to agree with such a very strong thesis about confirmation in order to endorse that truth or justification in believing in the truth of scientific theories transmits to the mathematical theories which are indispensable to them, since, if appropriately understood, indispensability should be enough for warranting this transmission, also from a non-holistic perspective. One might even argue that appealing to indispensability implicitly clashes with confirmational holism. Indeed, according to the latter, confirming, or in general justifying, s entails confirming or justifying any other theory somehow related to s though not indispensable to it, with the result that the supporter of confirmational holism could endorse an argument similar to IA but involving no appeal to indispensability.

Naturalism, on the contrary, is usually perceived as a necessary assumption of any (epistemic) version of IA, both by supporters and critics.³⁵ A possible explanation for this is that epistemic IA is usually understood as delivering not merely a sufficient reason for asserting its conclusion, but rather the only legitimate reason on whose ground this conclusion could be argued for. For this to be so, the indispensability of a mathematical theory to a scientific theory that we are justified in believing true is, indeed, not only to be considered as a sufficient condition for our being justified in believing that the former theory is true and its objects exist, but also as a necessary condition for this, which is just what naturalism prescribes. Understanding epistemic IA in this way has two shortcomings. First, it fits poorly with the fact that IA's conclusions are often conceived as claiming that there are true mathematical theories and that there exist mathematical objects, or, more specifically, that some mathematical theories (those to which the argument has been applied) are true and their objects exists, and not that these particular theories are the only true ones or that these particular mathematical objects are the only which exist. Second, if these are the conclusions that supporters of IA have in mind, the preceding understanding of IA simply asks for much more than is actually required in order to get them.

Beyond that, this way of thinking is open to obvious objections. Indeed, binding the justification in the truth of a mathematical theory and in the existence of its objects to the indispensability of this theory to a scientific theory that we are justified in believing true entails both that such a justification relies on grounds as contingent as those on which our justification for scientific theories relies,³⁶ and that unapplied mathematical theories are in principle excluded from the possibility of benefiting from it.³⁷

7
The Indispensability Argument: The Debate

As Chapter 6 should have made clear, IA can be questioned on various grounds, depending on the target version and on the notions this is meant to involve. In this chapter we discuss some of the major objections that have been levelled against IA, and some possible replies.

7.1 Against Indispensability

Whatever version of IA we are considering, let us call 'indispensability premise' the premise that claims that some true, or justifiably held true scientific theories are such that some mathematical theories are indispensable to them. The major attack to this premise has been mounted by Field, some of whose views have been expounded in § 4.1. Here, we shall focus on his attack to the indispensability premise.

Field considers IA a valid argument; he even sees it as the only non-circular argument available for platonism (Field, 1980, p. 4). But he takes it to be unsound, for he does not take the indispensability premise to be true. In § 4.1, we have seen that "good" mathematics, namely the acceptable one, has to be, for Field, conservative over any consistent body of nominalistic statements. Let, then, S be a consistent, axiomatized, and deductively closed first-order theory (this restriction is motivated by Shapiro's objection to Field expounded in § 4.1). Let us divide the set A of S's axioms into two complementary sets, A_N and A_M, these being, respectively, the set of all and only nominalistic axioms of S, and the set of M-loaded axioms of S, where M is an acceptable mathematical theory whose vocabulary is included in S's (and is the only non-nominalistic vocabulary included in S's). Since M is required to be conservative over any consistent body of nominalistic statements, a nominalistic statement φ is a theorem of S if and only if it can be proved within the sub-theory s_N, whose axioms are those belonging to A_N. It follows that s_N includes all and only the nominalistic consequences of s. If an *o*-equivalence relation is defined to the effect that two scientific theories are *o*-equivalent if (and only if) they have the same nominalistic consequences, s and s_N will then be *o*-equivalent. Insofar as all the consequences of s are provable within s_N and the axioms of s_N are just the nominalistic axioms of s, it seems natural to concede that s_N complies with some acceptable criterion of virtuosity. The sub-theory s_M, whose axioms are those belonging to A_M, and *a fortiori* the theory M itself, are then dispensable from s, according to a specification of [*(In)dispensability*] in which ε-equivalence is replaced by *o*-equivalence and that criterion of virtuosity is adopted.

One might think, then, that it follows from the conservativeness of acceptable mathematics over consistent bodies of nominalistic statements that mathematics is *ipso facto* dispensable from science. Quite to the contrary, Field notices that acceptable mathematics could turn out to be indispensable to science even if it is conservative. Here is what he writes (1982, p. 50, emphasis in the original; 1989a, p. 44):

> There are two *prima facie* ways that mathematics might be useful, despite its conservativeness: it might be useful in *facilitating deductions* (between nominalistic premises and nominalistic conclusions), and it might be useful in being *theoretically indispensable* (i.e., needed in the premises of some important theory).
>
> The conservativeness of mathematics does not in itself show that there can be no reason to believe mathematics: as I have repeatedly stressed from the time of my book *Science without Numbers*, to combat the argument for platonism one must also show mathematics dispensable.

A reason for taking the conservativeness of acceptable mathematics not to be enough to ensure its dispensability from science could be that the foregoing argument does not apply to some relevant cases, since: (*i*) it concerns only first-order scientific theories; (*ii*) it assumes these theories to be axiomatized and deductively closed; (*iii*) it requires that the set of s's axioms can be divided into to complementary sets of statements such as A_N and A_M. Still, this could hardly be Field's reason, for: (*i**) in § 4.1, we observed that Field replies to Shapiro's objection motivating restriction to first-order theories; (*ii**) every theory can be conceived as axiomatized,[1] actual scientific theories are surely deductively closed, and this latter condition could in any case be omitted by appropriately modifying the foregoing argument; (*iii**) Field shows no qualms on the separability of nominalistic and mathematical sub-vocabularies in scientific theories. Field's reason is rather voiced in the parenthesis in the following quote (1989b, p. 243; see also 1982, pp. 50 and 55–6):

> Mathematics . . . seems to be theoretically indispensable, that is, needed among the premises of important theories, including empirical theories. (This is not ruled out by the conservativeness of mathematics: the conservativeness of mathematics tells you something about how you can use mathematics when you have a nominalistic theory, but the issue here is whether sufficiently powerful nominalistic theories are available).

How should we understand the possibility that there are no "sufficiently powerful nominalistic" scientific theories?

To begin with, notice that Field's argument for the conservativeness of mathematics can be generalized by replacing bodies of nominalistic statements with bodies of generically non-mathematical statements. Hence, if s is a consistent, axiomatized, deductively closed, first-order scientific theory, the set of whose axioms can be divided into two complementary sets A_H and A_M of non-mathematical and M-loaded axioms respectively, we can reason as above and conclude that the sub-theory s_M of s, whose axioms are those belonging to A_M, and *a fortiori* the theory M itself, are dispensable for s according to any specification of [*(In)dispensability*] where ε-equivalence is replaced by an equivalence relation such that two scientific theories are equivalent if and only if they have the same non-mathematical statements.

If Field envisages the possibility that sufficiently powerful nominalistic scientific theories may not be available, this is thus not because he believes that in our scientific theories occur, beside mathematical statements, other non-nominalistic non-mathematical statements (for instance, statements about theoretical entities or metaphysical statements).

Field's point rather hinges on something we already pointed out in § 4.1 and discussed in § 6.3.2. Among the empirically relevant theorems of a scientific theory, some might occur that are not nominalistic: the empirical content of a scientific theory will not, in this case, be exhausted by its observational statements, and the relation of *o*-equivalence will not be appropriate for a specification of [*(In)dispensability*].

However, if our (best) scientific theories turned out to be rephrasable in such a way that their empirical content is entirely given by nominalistic statements, then acceptable mathematics, insofar as it is conservative over bodies of nominalistic statements, would also be dispensable from those theories, by a specification of [*(In)dispensability*] in terms of *o*-equivalence (defined as above, to the effect that two scientific theories with the same nominalistic statements are *o*-equivalent).[2] Assuming our (best) scientific theories are those that we are justified in holding true, this would be enough to deny the indispensability premise in every epistemic version of IA, and to make at least unwarranted the same premise in every non-epistemic version of IA.

After having argued that acceptable mathematics is conservative over any consistent body of nominalistic statements, in order to reject IA's indispensability premise, Field needs then only to show that our (best) scientific theories can be suitably rephrased so that their empirical content is exhausted by nominalistic statements.

In § 4.1 we gave a broad-brush description of Field's attempt at showing that things are so through the example, which he takes as paradigmatic, of Newton's Gravitational Theory (in a modern version of it), which is shown to be rephrasable in a purely nominalistic language. Here it is however important to notice that the rephrased version he offers is, as it were, a global reformulation: it does not contemplate that each theorem of the original Newtonian Gravitational Theory where mathematical constants occur is replaced by a theorem of the new version using only nominalistic vocabulary and counting as a paraphrase of the former (Field, 1980, p. 1; 1989a, pp. 6–7). Field thus does not share Quine's view that the reformulation of a scientific theory should be obtained by paraphrasing each of its theorems. In order to ensure that the new theory is equivalent to the original one in some appropriate sense, and that it counts as a reformulation of it, Field rather appeals to representation theorems, which establish that some suitable structural properties of the system allegedly formed by both concrete and mathematical objects that is described by the original theory, also count as structural properties of the new system described by the new theory, which is only composed of concrete objects. By means of these theorems, Field wants to show that the ontological commitment of the new theory, provided by concrete objects only, forms a system that is structurally equivalent (isomorphic) to the system formed by the ontological commitment of the original theory, provided by both concrete and allegedly mathematical objects.

Hence, rephrasing the original theory as Field suggests does show that the empirical content of the theory can be captured by nominalistic statements only, but not that mathematics is dispensable from Newton's

Gravitational Theory, since nothing in this rephrased version requires the new theory to be wholly free of mathematical theorems. What matters is that these latter do not concur to provide the empirical content of the theory. Field himself shows (1980, p. 60) that not all properties that are enjoyed in the original theory by the temperature function (this being a mathematical function defined on $\mathbb{R}^4$ and with values in $\mathbb{R}^4$, see § 4.1) are expressed through nominalistic statements in the new version of the theory: only those "of any physical importance" are.

Field seems thus to be implicitly appealing to two different equivalence relations in his argument against IA: *o*-equivalence, on whose basis the (in)dispensability of mathematics for appropriately rephrased scientific theories can be assessed; and another relation, linking Newtonian Gravitational Theory to his reformulations of it, which is ensured once appropriate representation theorems can be proved.[3] This latter equivalence relation is somehow related to explanation: the relevant reformulation must be able to explain every empirical phenomenon already explained by the original Theory of Gravitation. According to Field, the original explanation is "extrinsic", since it employs a mathematical language, whereas the explanation given by its reformulation is "intrinsic", since it only employs a nominalistic language (and thus, says Field, makes no appeal to entities that do not contribute to the constitution of those facts that Newton's Theory is meant to explain). Field's aim is just to show that for any extrinsic explanation, an intrinsic one can be found (1980, pp. 41–4; 1989a, pp. 14–20). This is all the second equivalence relation is intended to ensure.[4]

It should then be clear that Field offers in fact two independent sub-arguments. The first – which is intended to show that acceptable mathematics is conservative over any consistent body of nominalistic statements – is general in character: if it is correct, it applies to all possible cases and establishes its conclusion once and for all. The second – which is intended to show that our (best) scientific theories can be appropriately rephrased so that their empirical content is exhausted by nominalistic statements – includes, on the contrary, both a particular argument, which is only concerned with Newton's Gravitational Theory, and an hypothetical generalization to every scientific theory that might be relevant to IA, assuming that Newton's Gravitational Theory can be considered paradigmatic.[5] Hence, even conceding to Field both that acceptable mathematics is conservative over any consistent body of nominalistic statements, and that Newton's Gravitational Theory can be rephrased as he suggests, it is then enough to come up with an example of a scientific theory which involves a mathematical part that cannot be rephrased so as to make its empirical content wholly exhausted by nominalistic statements, to have a counter-objection to Field's criticism to IA.

This can be obtained in two ways: either by considering the case of an actual scientific theory, or by appealing to an *ad hoc* theory, which is then hypothetically considered as accepted. The first strategy has been pursued by David Malament (1982), who considers the hard-to-nominalize example of Quantum Mechanics. We cannot enter here into any details of this objection (Balaguer, 1996a, 1996b, tries to reply to Malament's objections; see also Meyer, 2009, and Leng, 2010, chs 3 and 9). The second strategy is implicitly at work in Melia (2000), which discusses the case of a theory using a first-order nominalistic language of mereology but also includes ZF set-theory, and is meant to have at least a consequence that does not follow from its nominalist part and that, despite it not being itself nominalist, is certainly empirically relevant, since it concerns the "concrete world" (*ibid.*, p. 458).

Let us look better at Melia's argument (*ibid.*, pp. 459–61).[6] The theory under consideration, call it 'Mer[ZF]', includes two sets of axioms. The first, call it '$A_N^{[Mer]}$', includes infinitely many non-logical axioms. Two of them serve to introduce a binary predicate constant '*Par*', designating the relation of being part of (notice that every object is part of itself), and establish that it determines a partial order over the objects of the theory. Another axiom introduces the individual constant '*u*', denoting the mereological sum of all the objects, establishing that for every object x, $Par(x,u)$. Two other axioms introduce the unary predicate constant '*At*', designating the property of being an atom, and establish that an atom has no parts, and that every object has (at least) an atom as a part. In addition, there are infinitely many axioms of the following kind: one establishes that there are distinct objects x and y such that $At(x)$ and $At(y)$; another establishes that there are distinct objects x, y, z such that $At(x)$, $At(y)$ and $At(z)$, and so on. This establishes the existence of infinitely many atoms with no recourse to a mathematical language. Finally, $A_N^{[Mer]}$ includes a comprehension axiom schema warranting for the existence, and fixing the properties, of mereological sums different from u. Intuitively speaking, this axiom schema establishes that if F is a monadic property, and there are Fs, then there is the region[7] s_F containing all the Fs, and such that anything that has a part in common with s_F has also a part in common with an F.

The second set of axioms of our theory, call it '$A_M^{[Mer]}$', includes all the axioms of ZF plus infinitely many bridge-laws including: (*i*) all instances of the comprehension axiom schema included in $A_N^{[\text{Mer}]}$ where F is a property designed by predicate constants in ZF's vocabulary; (*ii*) an axiom establishing that there exists the set of all the non-mathematical objects of Mer[ZF] (namely, all the objects denoted by individual constants included only in the nominalist vocabulary of Mer[ZF], and thus not included in the vocabulary of ZF); (*iii*) infinitely many axioms establishing that there are infinitely many atoms that are not sets (urelements).

Melia's point is the following. It can be proven (2000, pp. 475–6) that the axioms of $A_N^{[\text{Mer}]}$ have a model containing "a denumerable number of atoms" and "an infinite number of regions", but such that "every region either contains only a finite number of atoms, or is such that its complement contains a finite number of atoms" (*ibid.*, p. 460) This is not, however, a model of Mer[ZF], since it follows from its axioms altogether that there are regions including infinitely many atoms and such that their complement includes infinitely many atoms, regions that Melia calls "infinite and co-infinite". The proof of this last claim is quite easy (*ibid.*, pp. 460–1). Since it is possible to define in ZF the ordinal numbers, and the non-mathematical objects of Mer[ZF] form a set, there exists a bijective mapping from this set to the ordinals. Thus, given the comprehension axiom schema extended to the whole language of Mer[ZF], there exists a region s_{EN} containing just all the non-mathematical objects of Mer[ZF] that are mapped to an even finite ordinal number, and this region is clearly infinite and co-infinite.

However it is defined, the predicate 'EN' is certainly not included in the nominalistic vocabulary of Mer[ZF]. Even if it follows from $A_N^{[\text{Mer}]}+A_M^{[\text{Mer}]}$, the statement asserting the existence of s_{EN} in the language of Mer[ZF] is thus not a nominalistic consequence of this theory. Now, from this statement another one follows by existential generalization, call it 'ψ_{ICI}', asserting in the language of Mer[ZF] that there exists a infinite and co-infinite region not otherwise specified. It is hard to deny that this statement provides part of the empirical content of this theory, since it asserts the existence of at least a region of concrete objects that is specified simply by saying that it is infinite and co-infinite. It is clear, however, that however it is formulated, ψ_{ICI} includes constants beyond those of the nominalist vocabulary of Mer[ZF], since this latter vocabulary has no sufficient resources for defining a predicate expressing the condition that a certain region is infinite and co-infinite. We must then conclude that ψ_{ICI} follows from $A_N^{[\text{Mer}]}+A_M^{[\text{Mer}]}$ but not from $A_N^{[\text{Mer}]}$ alone: it is not a nominalistic consequence of Mer[ZF], but provides part of the empirical content of Mer[ZF].

Melia's argument does not then contradict the conservativeness of the (impure) mathematical theory whose axioms are those belonging to $A_M^{[\text{Mer}]}$ over the body of the nominalistic consequences of the axioms belonging to $A_N^{[\text{Mer}]}$. It offers then no counter-example to Field's thesis on the conservativeness of mathematics. Nonetheless, it presents us with a theory that includes an (impure) mathematical theory and uses this theory in order to draw consequences that express part of its own empirical content and cannot be obtained otherwise.

Tough doubts might be raised on the real significance of this example, Melia takes himself to have established that IA cannot be attacked by attacking the indispensability premise. He indeed proves (*ibid.*, pp. 476–8) that the nominalistic consequences of Mer[ZF] are just the consequences of $A_N^{[\text{Mer}]}$, and concludes from this that a nominalist cannot deny IA's indispensability premise by resting content of what he calls the "trivial strategy", namely by arguing that a scientific theory employing a mathematical one can always be replaced by the theory formed by the body of its nominalistic consequences: this could, as a matter of fact, lead to a theory poorer in empirical content than the original one. In other words, Melia takes his example as denying that the dispensability of a mathematical theory for a scientific theory can be established by a specification of [*(In)dispensability*] in terms of *o*-equivalence (so defined that two scientific theories with the same nominalistic theorems are *o*-equivalent). We leave the reader to evaluate the strength of Melia's argument; let us notice, however, that the possibility of reaching its conclusion seems implicitly adumbrated when Field feels the need of accompanying his argument for the conservativeness of mathematics with the second argument concerning Newton's Theory of Gravitation.

7.2 Against Ontological Commitment

A second possible objection to IA for platonism concerns the premise asserting that if a given mathematical theory M is indispensable to a given true or justifiedly held true scientific theory, then its mathematical objects exist, or we are justified in believing they exist. Most criticism has not been addressed directly to this claim, however, but rather to the criterion of ontological commitment used to argue for this premise, namely Quine's.

These objections have followed two major strategies. Some have been launched directly against Quine's criterion; others have focused on some of the assumptions presupposed by Quine's criterion, namely: (*a*) that there is only one sense in which it can be said that an object exists; (*b*) that all statements of an informal theory having the form 'there is (at least) one *F*', 'there are *Fs*', etc., can and must always be rendered in a first-order predicate language by statements of the form '$\exists x[\mathcal{A}(x)]$'.[8]

In this section we will only discuss those objections that we take to be relevant to the debate on IA. Objections explicitly addressed to Quine's criterion should not be confused with a number of other arguments which have already been mentioned. More specifically, the following do not count as objections to Quine's criterion: (*i*) all proposals according to which mathematics should be so reinterpreted that the meaning of mathematical statements is not given by their surface grammar, to the extent that, despite appearances, no mathematical statement would be existentially quantifying over a domain of mathematical objects (see, for instance, Hellman's proposal, § 4.3, Chihara, 1990); (*ii*) all those views denying that mathematical objects are abstract and their existence is independent from the cognitive subject, being rather conceived either as mental constructions (as in Brouwer's intuitionism, see § 2.4.5) or as concrete objects (as already suggested by Mill, 1843, and again by Maddy, see § 4.3). The former are compatible with Quine's criterion, and simply lead to the conclusion that the ontological commitment of scientific theories is free of mathematical objects. The latter are views about the nature of mathematical objects, which are allowed to occur in the ontological commitment of mathematical theories. Supporters of these theories might well reject Quine's criterion, but nothing in their views mandates that they do so.

Let us then consider some of the major criticisms that have been addressed to Quine's criterion, and to its assumptions (*a*) and (*b*) in particular.

7.2.1 Existence and Ontology: Carnap's Objection

A well-known objection against assumption (*a*), advanced independently of discussion of IA, is Carnap's (1950a). It is based on the thesis (already discussed in § 2.4.1) that answering to questions about the existence of a particular kind of entities makes sense only within a particular linguistic framework.

Take the framework of arithmetic as an example. A statement like 'There exist prime numbers greater than 2' is, within it, an appropriate answer to a question concerning the relevant theory, and only depends on a proof in this theory. The existence of prime numbers greater than 2 is then to be considered, in this sense, as nothing but a consequence of the axioms of this very theory. Thus understood, the assertion makes no genuine ontological claim. And in Carnap's view, it would be illegitimate to take it as making one such claim, for that would be tantamount to confuse an assertion within a given framework with an assertion external to any particular framework.

Externally to any framework, we can only say that there are reasons for adopting some frameworks, given some circumstances and aims. But this has nothing to do with "the reality of the entities" (*ibid*, p. 250) allegedly involved in these frameworks: it is only a matter of contingent and pragmatic considerations.

Carnap advanced these views a few years after the publication of Quine's "On What There Is" (Quine, 1948), where Quine most explicitly proposed his criterion of ontological commitment. Sometimes Carnap seems to be mediating between his own views and those of his former student. Not only he claims (1950a, p. 249, fn. 4) that "W. V. Quine was the first to recognize the importance of the introduction of variables as indicating the acceptance of entities", but, although he notices that Quine does not accept his distinction between internal and external questions, he also adds (*ibid.*, p. 250, note 6):

> With respect to the basic attitude to take in choosing a language form (an "ontology" in Quine's terminology, which seems to me misleading), there appears now to be agreement between us: "the obvious counsel is tolerance and an experimental spirit". [Carnap is quoting from Quine, 1948, p. 38].

The real contrast is however stronger than Carnap might want us to think. In order to agree with Quine it is not enough to acknowledge the role of bound variables as "indicating the acceptance of entities". This can at most lead us to conclude that, given any theory, its ontological commitment is to be determined as Quine's criterion prescribes. The crucial point of Quine's proposal is that his criterion does not simply lead to "accepting some entities" within a given framework, but rather to establishing that they exist in the only sense in which it can be said that an object exists. There is no room, in Quine, for any weak sense of understanding the expression 'accepting an entity'. The tolerance Quine suggests in the passage quoted by Carnap is mainly concerned with the attitude one should take in exploring alternative theories with different ontological commitments. But once it is established (provided it can be done[9]) which ones we are justified in holding true, there is no way of denying "the reality of the entities" over which a theory quantifies, as Carnap would like.

7.2.2 Objectivity Without Objects: Putnam on Equivalent Descriptions

An objection to assumption (*b*) can be found in Putnam (1967). Putnam begins by recalling the well-known wave/particle duality in the explanation of the behaviour of elementary particles: their behaviour can be described both in terms of waves and in corpuscular terms, even though these two descriptions of physical reality are apparently associated with two incompatible images of reality. According to Putnam, we can assign the same explanatory power to the two descriptions, although we cannot assign to each statement pertaining to one of them the same meaning as that of a statement pertaining to the other. Moreover, we can neither say that one of them is more fundamental than the other, nor that one can be reduced to the other. In short, we are faced with two "equivalent descriptions" (*ibid.*, p. 8; the expression is taken from Hans Reichenbach, 1951, pp. 131–2), with different meanings.

According to Putnam, a similar situation also occurs in mathematics. In giving an account of mathematics, we can evoke two different images: that of "mathematics as set theory", and that of "mathematics as modal logic" (*ibid.* p. 9; one could also mention category-theory as a further option, but this was not among the options Putnam was considering). According to the former, mathematical statements describe a domain of objects and their properties; according to the latter, they simply tell us "what follows from what" (*ibid.*, p. 11).

Putnam considers the example of Fermat's last theorem, according to which, if *n* is a natural number greater than 2, there is no triple of positive natural numbers giving solutions for the equation $x^n + y^n = z^n$. At the time of the publication of Putnam's essay, this was just a conjecture; a proof has been given later (Wiles, 1995). The form of Putnam's exposition depends on the possibility that it might have turned to be no theorem at all, but Putnam's point is wholly independent of this. We will be faithful to his exposition, then.

Let ¬*Fermat* be the following statement of first-order Peano arithmetic:

$$\exists x, y, z, n \, [n > 2 \wedge x \neq 0 \wedge y \neq 0 \wedge z \neq 0 \wedge x^n + y^n = z^n]$$

where variables are meant to range over natural numbers. This statement says that there is (at least) one triple of positive natural numbers giving a solution to Fermat's equation for *n* > 2, and thus that there is a counter-example to his last theorem. If ¬*Fermat* happened to be a theorem of first-order Peano arithmetic, it could be proved from a sub-set of its theorems that could easily be singled out.[10] Let 'AX_{AP^1}' stand for the conjunction of first-order Peano axioms; Fermat's last theorem would then be false, namely it would have a counter-example, if and only if the following were to hold:

$$\Box[AX_{AP^1} \Rightarrow \neg Fermat]$$

If this statement were true, Putnam says, then also the following would:

$$\Box[AX^* \Rightarrow \neg Fermat^*]$$

where 'AX^*' and '$\neg Fermat^*$' would stand for the formulae obtained by AX_{AP^1} and $\neg Fermat$ when the constants belonging to the vocabulary of first-order Peano arithmetic are replaced with appropriate schematic letters. The first of these statements would indeed be true if and only if the implication '$AX_{AP^1} \Rightarrow \neg Fermat$' were necessarily true, and its necessity were preserved under the indicated substitution. But insofar as the second statement has only schematic letters, it is in fact a schema of statements, and is then true if and only if it is a truth of first-order modal logic (thus neither of arithmetic, nor of any other particular theory).

According to Putnam (*ibid.*, p. 10), the "mathematical content" of '$\Box[AX^* \Rightarrow \neg Fermat^*]$' is the same as that of $\neg Fermat$. However, "the pictures in the mind called up by these two ways of formulating what one might as well consider to be the same mathematical assertion can be quite different". For, while $\neg Fermat$ suggests that mathematics describes some particular "external" objects, '$\Box[AX^* \Rightarrow \neg Fermat^*]$' suggests that it deals with entailments among statements, and thus concerns no other objects than statements. Still, "each of these two ways of looking at mathematics can be used to clarify the other". If one has qualms concerning the use of modal operators, one could be comforted by the idea that the same mathematical statement can be interpreted in set-theoretical terms. If one agrees with the objections to platonism advanced in Benacerraf (1965), one could be comforted by the idea that natural numbers can be seen as "things one can quantify over", but that any "fact" concerning them could be seen as a fact concerning any progression. From this it would follow that natural numbers exist, but "all this comes to, for mathematics anyway", is that it is possible that there are progressions and that there are necessary truth of the form 'if α is a progression, then . . .'.

All this could be made to square with Quine's criterion of ontological commitment. Putnam concedes the possibility that in '$\Box[AX^* \Rightarrow \neg Fermat^*]$'

the symbol '□' be interpreted as a predicate that applies to statements, so that this statement could in fact be taken as saying that "a certain object, namely the statement '$AX^* \Rightarrow \neg Fermat^*$' has a certain property". Still, on this understanding, Putnam observes, "the only 'object' this commits us to is the statement '$AX^* \Rightarrow \neg Fermat^*$' (*ibid.*, p. 11; notation slightly adapted). Hence, if one accepts Putnam's argument, one cannot endorse assumption (*b*),[11] and this for at least two reasons.

On the one side, the informal statement asserting that there are four natural numbers providing a counter-example to Fermat's last theorem, can indeed be rendered either by $\neg Fermat$, which is the negation of an existential statement of a first-order language, or by '$\Box[AX^* \Rightarrow \neg Fermat^*]$', which is rather a modal statement. Beyond this, Putnam takes his example to show that the same "mathematical fact" (*ibid.*, p. 9) admits different equivalent descriptions leading to different ontological commitments, with the result that a mathematician asserting that there are four natural numbers providing a counter-example to Fermat's last theorem would not be forced to commit her/himself to the existence of some objects rather than others.

Putnam's proposal raises several issues, the most serious of which being that it is not clear in which sense Putnam is speaking of mathematical facts, mathematical content, or equivalent descriptions of a mathematical fact.[12] Another issue is that Putnam does not elaborate his conception of mathematics as modal logic beyond the example we have reported concerning first-order Peano arithmetic (even though he suggests how this conception could be extended to set theory: *ibid.* pp. 19–22). However, his suggestion has inspired Hellman to his fully worked-out proposal (see § 4.3).

How can this conception of Putnam's be reconciled with his adoption of IA (see § 6.2.1)? A way of making the two things compatible is to claim that mathematics is indispensable to our scientific theories whatever description of the relevant mathematical facts is chosen among all the equivalent ones available. By itself, this would make Putnam's views compatible only with IA for semantic realism, however. If they have to be compatible with IA for platonism too, this should be supplemented with the claim that all these descriptions are somehow associated with the same ontological commitment where this includes abstract objects, or at least with different ontological commitments such that each of them includes mathematical objects, and this clashes with what Putnam says about the ontological commitment of '$\Box[AX^* \Rightarrow \neg Fermat^*]$'. However, it is natural to read argument [3] in § 6.2.1 as an argument for platonism. When he gives this version of IA,

Putnam neither discusses nor rejects his theory of equivalent descriptions. Burgess and Rosen (1997, p. 201) submit that if Putnam had also considered this theory on that occasion, he would have accepted the indispensability of mathematics for science, but not at the same time that the truth of mathematical theories entails the existence of mathematical objects, thus rejecting IA for platonism. A similar view, suggested also in Putnam (1975a), is openly advanced in Putnam (2004, p. 102), and yet more explicitly in Putnam (2012).[13]

7.2.3 The Easy Road for Rejecting IA.

In recent years, several authors have raised objections to IA for platonism that are to some extent based on criticism of Quine's criterion. For reasons of space, we cannot go into details, but we want to mention them at least cursorily for the sake of completeness. All these proposals have a common feature (which they share with Carnap's view): their objections to the thesis that the indispensability of a mathematical theory to a true or a justifiedly held true scientific theory make the objects of the former theory exist, or we are justified in believing that their existence is not based, as it is for Field, on attempted reformulations of the latter theories in purely nominalistic terms. These authors believe that even if statements of the form '$\exists x[\mathcal{A}(x)]$', where the variable range on a domain of putative mathematical objects, occur in the ontologically most elementary canonical reformulations of these mathematical theories, this does not entail that these objects exist, or that we are justified in believing that they exist. They do not thus reject the indispensability premise in IA, but rather believe that indispensability does not entail that there exists mathematical objects, nor that we are justified in believing that they exist.

Since their objections are not based on any attempt at nominalistically reformulating scientific theories, bur rather on more general philosophical arguments, some refer to these views (as well as to Yablo's, which we will consider in § 7.2.4) as those opting for an "easy road" (Balaguer, 2008) in defence of nominalism. Similar proposals are for instance advanced by Melia (1995, 2000) and Azzouni (1997a, 1997b, 1998, 2004). Balaguer (1996b, 1998, 2009) has maintained a version of fictionalism that has much in common with these views, even though it is not directly based on criticism to Quine's criterion. Colyvan has questioned that any easy road for supporting nominalism along these lines is tenable (see Colyvan, 2010 and *Forthcoming*; the latter replies to papers by Azzouni, Bueno, Leng, Liggins and Yablo, which were not available at the time of writing but are focused on this debate).

7.2.4 Fictionalism as an Objection to IA

According to fictionalism, once the truth of a mathematical statement (be it pure or impure) is appropriately conceived as truth-in-the-fiction of mathematics, it has no ontological consequence outside the fiction, and this alone suffices to block IA.

Let us begin by considering Field's fictionalism. We already noticed (§§ 4.1 and 7.1) that Field's objection to IA, over and beyond the intrinsic interest of his fictionalism, is not as much based on the latter, but rather on his rejection of the indispensability premise. It is then important to note that Field's position is wholly compatible with Quine's criterion and with its assumptions (*a*) and (*b*).

Like Quine, in accordance with assumption (*a*), Field does not distinguish among different ways in which it can be said that an object exists. For even acknowledging that there are two ways of speaking of the truth of a mathematical statement – one according to which it is not true, since it is not true *simpliciter*, and another according to which it is true, since it is true-in-the-story – it does not follow that we are appealing to two different notions of existence. The predicates 'true *simpliciter*' and 'true-in-the-story', despite their sharing important features, should be taken as two different predicates. Therefore, when a fictionalist claims that the statement 'There are [or, equivalently, there exist] infinitely many natural numbers' is not true *simpliciter*, but is true-in-the-story of arithmetic, (s)he need not be taken as ascribing to natural numbers some sort of existence-in-the-story as opposed to existence in reality.[14] According to Field, natural numbers don't exist; the fact that 'There are infinitely many natural numbers' is true-in-the-story of arithmetic is irrelevant to their existence: it is just a fact about the statement, due to its being entailed by the axioms of arithmetic.

Field also accepts assumption (*b*), and for this reason maintains that mathematical statements of the form 'There is (at least) an *F*', 'There are *F*s', etc. are not true *simpliciter*. As we saw in § 4.2, if he believes that these statements are true-in-the-story, this is not because he thinks they should be understood under some alternative interpretation.

Can we say that Field's fictionalism contrasts with Quine's criterion? Since, as we already said, the only plausible argument for the truth of mathematics would for Field be IA, and since Field rejects IA by rejection of the indispensability premise, he can conclude that there is no reason for holding the statements of mathematics true. In this argument, the thesis that mathematical statements can be held true-in-the-story has no real role to play. Fictionalism can indeed be adopted once one has independently shown that there is no reason for holding those statements true simpliciter. Therefore, not only Field's fictionalism is not really essential to his objection to IA; it also is no objection to Quine's criterion, nor need it be, since this criterion does not require that M-loaded statement be part of the ontologically most elementary canonical reformulation of a scientific theory for which a mathematical theory M is dispensable. Field's fictionalism is just a way of claiming that there is some sense in which statements of mathematics are (non-vacuously) true, and it comes conceptually after an independent argument, showing that they are not true *simpliciter*, has been put forward.

Field's fictionalism appears thus more as an auxiliary component of his overall view, which is helpful in accounting for some unpalatable features of it, but not to reject IA.

On the contrary, Yablo does not really question IA's indispensability premise, and presents his fictionalism as a criticism to Quine's criterion. Let us see how he proceeds.

Consider again the statement 'The number of Martian moons is 2' – call it 'φ'. In § 4.2 we said that, according to Yablo, the literal content of this statement is that the natural number 2 is the number of Martian moons. Let us grant that this statement entails the other 'There is an object that is a natural number and is the number of Martian moons', which, according to our assumption (*b*), we should write as '$\exists x[NN(x) \land x = \#(MM)]$', where '*NN*' is the predicate '(being a) natural number', *MM* is the predicate '(being) a Martian moon', and '#' is, as in Frege arithmetic (see § 5.1), a functional operator associating to any monadic predicate the number of things that fall under the concept expressed by this predicate. Let us also suppose that both the predicate '*NN*' and the operator '#', or at least the terms that one gets by applying it to a monadic predicate, are not eliminable through appropriate contextual definitions. We can thus derive '$\exists x[NN(x)]$', and if φ is a theorem of an informal theory, we should say that the ontological commitment of that theory includes natural numbers.

A way of blocking this argument without rejecting assumption (*b*) is implicit in Yablo's own proposal: it is enough that we admit that it is the real content, and not the literal content, of φ that we need to consider when we are to determine the ontological commitment of a theory in which φ occurs.

A similar argument could apply to statements of a form other than that of φ, for instance statements straightforwardly existential like 'There exists an object that is a natural number and is the number of Martian moons', or to pure mathematical statements like 'There is one and only one even prime number'. In these and other cases, Yablo's point seems to be that if Quine's criterion is applied to the real content of mathematical statements, it does not entail that there are mathematical objects.

The problem with this view is how the real content of all mathematical statements can be singled out. This suggests that the major part of the work to be done on a theory in order to determine its ontological commitment is not made by the criterion itself, but rather by a prior evaluation, largely due to the adoption of some philosophical views rather than others, which decides on the nature of the ontologically minimal canonical reformulation of the theory. What is put into question, then, is the hope of finding such a reformulation in an univocal way, and thus of giving a tool for conducting ontological disputes independently of any philosophical assumption. Yablo (2001, pp. 232–3) seems to point to this problem when he stresses the difficulties we encounter in finding a sharp boundary between figurative and non-figurative speech in ordinary language and informal theorizing (see Leng, 2010, § 6.2, for discussion).

It remains that (as observed in § 4.2) Yablo does not believe that for any mathematical statement it is possible to determine its real content. Should this be the case, his objection to IA could simply consist in denying the indispensability premise and offering a nominalistic reformulation of mathematical theories: his fictionalism would then be an auxiliary aspect of a general nominalistic proposal, as in Field's case.[15] But why is it not always possible, for Yablo, to determine the real content of a mathematical statement? The reason is that in many cases the use of a metaphorical or figurative language (like that of mathematical fiction) seems not only helpful, but also necessarily required for the communication of contents: these are about the real world, but could not be communicated otherwise.

Ineliminable metaphors are pervasive in natural language and especially poetry, Yablo argues. Consider the statement 'Juliet is the Sun': there is no single statement that we can offer as a paraphrase of it that wholly captures its meaning. But there are also less recondite examples. Suppose that a scientific theory (say, demography) contains the statement 'The average family has 2.3 children' (see Yablo, 1998, pp. 250–1; 2001, pp. 93–7). When literally taken, this statement, being of the form '$F(a)$', would entail that there exists an object, a family, which has the odd properties of being average and of having 2.3 children. This is obviously not what we intend to communicate, since no one believes this. What is meant is rather that the number of families, divided for the number of children, is 2.3. According to Yablo, to express the real content of this statement without recourse to any mathematical vocabulary, we should then use an infinite disjunction of statements with numerical quantifiers: 'There are twenty-three families and ten children, or forty-six families and twenty children, or . . .'[16] Clearly, it is not possible to communicate (nor to think) with infinitely long statements. The introduction of a figurative language in which it is possible to speak of average families, allows us to express a content that could not be expressed otherwise.

However, the examples offered by Yablo, especially as regards scientific theories, do not seem to delineate a wholly cogent position. First of all, if we accept that the real content is what we really assert, in those cases where metaphorical or figurative language is ineliminable there would be no way of specifying what we really assert. Even though this seems to be acceptable in some contexts, like poetry, it is hard to believe that it is so, in the context of scientific theories. Second, despite its being founded on a stimulating insight, Yablo's view should be extended and shown to be applicable to any mathematical theory whatsoever.

Nonetheless, Yablo's point, if it is accepted, could indeed serve as a criticism to Quine's criterion. As said above, Yablo (1998) sees ineliminable metaphors as a symptom of the impossibility of finding general criteria for individuating and eliminating figurative aspects of language, both ordinary and scientific. If this is so, the very possibility of determining the ontologically most elementary canonical reformulation of a scientific theory is put in jeopardy.

According to Yablo, assertions made through statements that indispensably employ the figurative language of a mathematical fiction would thus be performed, both in ordinary and scientific discourse, within a game of make-believe with an attitude of simulation (see § 4.2). We already noticed that the distinction between the propositional attitude of simulation and that of belief echoes Carnap's distinction between internal and external existence. This seems to suggest an objection to assumption (*a*): we would have two different ways of understanding a single existential statement, one in which its literal content is genuinely believed to be true, and the other in which it is just simulated. We pointed earlier to the problems raised by this view. Notice also that this point only seems to concern Quine's criterion when it is understood – as it sometimes is, but differently from the way we have understood it here – as a criterion for establishing not what a statement is committed to, but what a speaker takes herself to be committed to in asserting it, which is an interpretation of Quine's criterion that, despite its relevance for ontological debates, seems to be of minor relevance for IA.[17]

Be that as it may, Yablo's argument suggests a possible epistemic position according to which we are justified in holding true a scientific theory S for which a mathematical theory M is indispensable, even if we are not justified in believing that M is true and that the $O_{[M]}$ exist. It would be enough to argue that the M-loaded statements of S, whatever their real content may be, are only quasi-asserted, in order to conclude that the $O_{[M]}$ are not included in the ontological commitment of S.

7.3 Against Naturalism and Scientific Realism

In § 6.7 we argued that naturalism is, at best, tied with epistemic versions of IA, and suggested that even these versions can be freed by this assumption, unless they are intended to deliver the only legitimate reason on whose ground its conclusion could be argued for. Despite this, naturalism is usually considered as an essential ingredient of IA, and is the one that has undergone less criticism. Possibly, a reason for the former fact is that IA is perceived as a typically Quinean argument – and since Quine surely championed a naturalist philosophy, it is taken for granted that naturalism is relevant to IA – while one for the latter fact is that naturalism is today widely accepted and discussed also independently of IA.

7.3.1 Maddy's Objection and the Recent Debate on Naturalism

An objection to those versions of IA which (are considered to) include naturalism has been raised by Maddy (1990a, 1992, 2005). She first notices that the idea, advanced by Quine, that those mathematical theories that do not find application in science should be considered as "mathematical recreation", and their objects should be deprived of any ontological right (see note 37, Chapter 6), clashes with mathematical practice.

Maddy agrees with Quine in claiming that ontological questions should be answered by appeal to scientific inquiry, and that in case of conflict between scientific findings and philosophical claims it is philosophy that must go. She also agrees with Quine that one major source for justification of mathematical truth and existence is application of mathematical theories in empirical sciences. What Maddy disagrees with is that this is the unique source.

According to Maddy, the naturalistic philosopher, even if willing to support IA, cannot neglect that mathematics is to be (as generally it is) regarded as a model of rigorous inquiry, and that, despite this, the methods adopted by mathematicians in proposing their theories and the reasons by which they accept or reject them have often little, if anything, to do with applications. But those who adopt versions of IA that (necessarily or not) explicitly rely on naturalism, as it is in the case of

[4], would be bound to reject this practice as illegitimate on a par with other metaphysical reasonings. And this should be unacceptable, to a naturalist in particular.

One might still support versions of IA with naturalism once these implausible aspects are avoided (Maddy, 1992, pp. 279–80):

> We could argue, first, on the purely ontological front, that the successful application of mathematics gives us good reason to believe that there are mathematical things. Then, given that mathematical things exist, we ask: By what methods can we best determine precisely what mathematical things there are and what properties these things enjoy? To this, our experience to date resoundingly answers: by mathematical methods, the very methods mathematicians use; these methods have effectively produced all of mathematics, including the part so far applied in physical science.

This quotation suggests a version of epistemic IA for platonism, as it were, in two steps. Initially some version of IA, giving only sufficient conditions for its conclusion, could be used to derive the very general claim that we are justified in believing that there are mathematical objects, because of the "successful application of mathematics", or, better, of the indispensability of mathematics for science. Once this is ensured, the naturalist should maintain that, in order to decide which mathematical objects exist, one should better look at mathematical methods, or even only at them. The naturalist tenet to be codified by IA would then roughly be that the indispensability of mathematics for science provides us with the general justification in believing in the existence of mathematical objects (and possibly in believing in the existence of the objects of those mathematical theories whose indispensability to science has been proved); but then it is by accepted mathematical standards, even independently of considerations of indispensability, that we can establish which particular mathematical objects exist. As we shall see in § 7.4.3, Maddy does not actually endorse this argument, since she considers its first part unsound (for reasons that have no direct connection to naturalism). Others may well accept it, of course. But let us leave this aside, and focus on the part of the argument that depends on naturalism, conceived of as Maddy suggests.

By this tenet, the naturalist should acknowledge that the practice of pure mathematics can play its own role in the justification of our ontological beliefs concerning mathematics (and thus on our beliefs about the truth of mathematical theories), but credit this power to mathematical practice only insofar as the general claim of the existence of mathematical objects had been previously warranted by empirical considerations pertaining to the application of mathematics in natural sciences. This position is wholly consistent with Maddy's views presented in § 4.4, and has been defended by Maddy on several occasions (1992, 1994, 1997, 2005) as an integral part to her naturalist conception of mathematics.[18]

Some, however, for instance Burgess and Rosen (1997, p. 212), would still see Maddy's naturalism as too mild, arguing that "a thoroughgoing naturalist" should take the fact that mathematics commonly and "conveniently" makes reference to abstract objects as "sufficient to warrant acquiescing in their existence". Following this line of thought, Burgess and Rosen have suggested a naturalist argument for platonism (Burgess and Rosen, 2005, pp. 516–17). It runs as follows.

First come some preliminary assumptions: standard (pure and applied) mathematics abounds in existence theorems saying that certain mathematical objects exist; competent scientists and mathematicians accept these theorems "without conscious silent reservations" and rely on them in their theorizing and in applications; these theorems are to be taken literally; accepting them "without conscious silent reservations" and relying on them in theorizing and in applications is just to believe them literally true. It follows that competent scientists and mathematicians believe in the existence of the objects these existential theorems are about, with the result that they would be wrong, if nominalism were correct. Add then the following further assumptions: these theorems are acceptable by mathematical standards; they are also acceptable by scientific standards; as far as these theorems are concerned, there is no good philosophical argument for rejecting or questioning the mathematical and scientific standards of acceptability. It follows that belief in the existence of the objects these theorems are about is justified.

All these assumptions are controversial, and Burgess and Rosen discuss various nominalistic strategies that derive from rejecting one or more of them.[19] Some of these premises can easily be seen to express [*Semantic Naturalism*] and [*Ontological Naturalism*] as we stated them in § 6.5. If all premises are accepted, Burgess' and Rosen's argument establishes something we may call 'platonism by default'. They equate nominalists to sceptics who deny the existence of entities to which their own language, when literally understood, is instead committed to, and reject the tenability of this sceptical challenge, arguing that nominalists are asking more than our accepted standards of justification allow us to infer. In their view, mathematical existential theorems are "innocent until proven guilty", as they are justified by default by our only available standards of justification.

More space would be needed to discuss this position. What is relevant to us is that this sort of naturalist, contrary to those who follow Quine, would not take indispensability as delivering the unique grounds for belief in the truth of mathematical theories and in the existence of their objects; moreover, they should not give IA much consideration, unless they understand it intra-mathematically, as it were, as arguing that there are all those abstract objects that are indispensable to the theories that are accepted by the mathematical community.

Nonetheless, Burgess' and Rosen's argument offers a good example of how platonism can be argued for by relying on (some version of) naturalism, but not on any version of (non-intra-mathematical) IA appealing to it, nor on Maddy's idea that mathematical standards can justify belief in the existence of some mathematical objects only after an argument based on the indispensability of mathematics to science has been offered for the general claim that mathematical objects exist.[20]

For the sake of completeness, let us add that Maddy's most recent conception of naturalism goes somewhat in a different direction from what we have presented above.

In more recent works (2007, 2011), she describes at length the naturalist position she attributes to what she calls the "second philosopher" (roughly, a naturalist philosopher who rejects, following Quine, the idea of a first philosophy superior to science). Maddy delineates three different alternative positions on the ontology of mathematics, and ponders which suits the second philosopher's views. "Robust Realism" is the view that mathematical objects exist as mind-independent abstract objects; the robust realist is not guided, in his/her verdicts on ontology, as the second philosopher would like, by methodological considerations internal to mathematics or science (for example by considering whether the adoption of some new axiom for set-theory can improve the theory's expressive power, independently of its ontological import), but rather by metaphysical considerations which the second philosopher sees as unsupported by evidence (Maddy, 2011, pp. 57–9, 63). "Thin Realism" is the view that mathematical objects exist, but they are only taken to have the properties ascribed to them by the relevant theories, and to lack the properties that those theories and natural science in general "ignore as irrelevant" (Maddy, 2007, p. 369). "Arealism" is the view that mathematical objects do not exist.

While Maddy takes Robust Realism to be mistaken, she believes that, despite their different ontological stances, both Thin Realism and

Arealism can account equally well for those aspects of mathematics that are relevant, from both an internally methodological and an historical point of view, to mathematical and scientific practice.[21] These are a conglomerate of facts at which Maddy gestures by reference to various examples from the history of mathematics, and which she globally refers to as "the facts of mathematical depth" (2011, p. 137). This view still awaits a thorough elaboration (see Shapiro, 2009). What is important to stress here is that it sets aside Maddy's earlier proposal of a cognitive basis for platonism (see § 4.4, note 2), and gives hardly any role to her improved two-step version of IA described above: on this basis, no form of platonism seems required in order for the second philosopher to explain the truly significant facts of mathematical depth, whatever this come to in details.

7.3.2 Against Scientific Realism

Even though IA can (and maybe should) be formulated without naturalism, it must necessarily include a premise asserting that some scientific theories are true or we are justified in holding them true. One need not be a naturalist in order to endorse such a premise, but this surely requires some form of scientific realism.

Strictly speaking, [*Semantic Naturalism*] does not mandate that we are justified in holding our scientific theories true, or even that there are (or can be) scientific theories that we are (will be) justified in holding true: it states necessary conditions for this justification, but does not by itself claim that these conditions hold. Nevertheless, denying that these conditions hold would entail the sceptical conclusion that, as a matter of fact, scientific inquiry, even if it represents our only possible source of justification, cannot justify any of our judgements and beliefs on the truth of our theories. And this is clearly opposite to the spirit of naturalism.

Scientific realism is all but a clear-cut position. To a first approximation, we can say that a scientific realist maintains that our scientific theories aim at giving a (literally) true description of reality, and that accepting a theory is tantamount to believing it true, namely to believing true both its theoretical and observational statements (and not only the latter). More precisely, Psillos (1999, p. xvii) defines realism as the conjunction of three claims, a metaphysical one, an epistemic one, and a semantic one. According to the first, "the world has a definite and mind-independent natural-kind structure". According to the second, scientific theories, taken at face-value are "truth-conditioned descriptions of their intended domain, both observable and unobservable", and are "capable of being true or false"; "the theoretical terms featuring in theories have putative factual reference" so that if they are true, "the unobservable entities they posit populate the world". According to the third, "mature and predictively successful scientific theories" are "well-confirmed and approximately true of the world" so that "the entities posited by them . . . do inhabit the world".

All these theses are controversial. Some of the views discussed in Chapter 4 (for instance, Field's and Yablo's fictionalism) can be seen as rejecting the second thesis. Other objections stem from anti-realist positions such as that advanced by van Fraassen (1980; see also § 7.4.2 below). Other reasons for objecting to scientific realism could be the following.

One could observe that the notion of mature scientific theory is rather vague, in so far as it hinges on the possibility of drawing a suitable distinction between good and non-good scientific theories. This is crucial to IA: those willing to assert that IA is sound must acknowledge that it is possible, at least in principle, to draw this distinction; alternatively, they must assume that there certainly are true (or at least justified) scientific theories even if we are not in a position to identify them.

One could also claim that our experimental observations, however rigorous, are necessarily influenced by subjective and sociological facts, or by metaphysical presuppositions or conceptual schemes imposed by us, so that these observations cannot serve to justify theories that purport to be true or at least "approximately true" of a mind-independent world. This view can take various forms, some of which can seem more or less plausible to those who believe that science has many virtues nonetheless. Among the latter, we could mention Karl Popper, for his notion of "influential metaphysics" (1934, on which see Watkins, 1958), and Lakatos, for his notion of "positive heuristics" (1970).

One could also point to the fact that scientific theories appear to be under-determined by empirical evidence: there could be (and in some cases there are, as a matter of fact) several different and incompatible scientific theories, all consistent with the empirical data. Elaborations on this point might make one doubt that our mature theories could even be said to be approximately true (see Duhem, 1908; Quine, 1975; Putnam, 1983; van Fraassen, 1980).

A similar conclusion could also be reached by arguing that IBE – which is typically appealed to by scientific realists in order to justify their beliefs in the existence of unobservable entities – is not a reliable form of inference (see Lipton, 2002).

Finally, many examples could be made of theories that were once thought of as empirically adequate, and have later been rejected as blatantly false. The so-called 'pessimistic meta-induction argument' (see Laudan, 1981) also suggests that the theories that we now take as our best ones, never matter how mature (in some sense) or justified, will eventually share the same unfortunate destiny of their predecessors. Therefore, we have no grounds for thinking that we have now, nor that we can have in principle, even approximately true descriptions of the world.

These brief remarks cannot even sketchily account for the impressively vast debate on scientific realism in the philosophy of science (surveys can be found in Boyd, 1991, 2002, and Psillos, 1999). What is important here is that there are several ways of questioning scientific realism, and many of them can be levelled as objections to the necessary premises of IA asserting that there are true, approximately true, or justifiedly held true or approximately true, scientific theories. One might reply that the claims according to which we are justified in holding some scientific theories true or approximately true are sufficiently weak to allow sidestepping this objection, at least if the notions of justification or/and approximate truth are spelled out appropriately; making them too weak, however, would have the unwelcome result of putting some other necessary premises of IA into question (for instance, premises (*iii*) in arguments [2*a*] and [2*b*]).

7.4 Against Confirmational Holism

In §§ 6.6 and 6.7 we saw to what extent and under which conditions confirmational holism might be relevant to IA, more precisely to epistemic versions of it. Here we will consider criticisms of IA coming from criticisms of the holistic model of confirmation (framed within an hypothetico-deductive model of confirmation). We can classify them in three categories, according to whether they focus, respectively, on: (*a*) the hypothetico-deductive model of confirmation, endorsed by Quine himself; (*b*) the plausibility of an holist conception of confirmation; (*c*) the possibility of formulating IA with no appeal to confirmational holism.

7.4.1 Objections to the Hypothetico-Deductive Model of Confirmation

We will not devote much space to point (*a*), since the discussion on confirmation of scientific theories has been extremely vast, and independent from IA.

It is important to stress, however, that already in Quine's times the hypothetico-deducitve model was subject to strong objections, mainly raised by Carl Gustav Hempel (1945). It is indeed well-known that this model, at least the naïve version we have mentioned in § 6.6 – according to which an evidence e confirms an hypothesis h if e follows logically from h – gives rise to various paradoxes. In trying to solving these, more sophisticated models have been proposed (surveys are in Glymour, 1980, and Earman and Salmon, 1999).

Quine was well-aware of these problems. Even though he never intended to advance any original theory of confirmation (as he acknowledges in 1981d, p. 453), he attempted at least to appropriately amend the hypothetico-deductive model in order to prevent the emergence of paradoxes (see Quine, Ullian, 1978; objections to these attempts are in Chihara, 1981).

It is a fact, in any case, that the holistic nature of confirmation does not hinge on its satisfying the hypothetico-deductive model. Clearly, this model does not imply that confirmation is holistic; *vice versa*, *confirmational holism* can also be defended if that model is not endorsed (for instance, as Hellman, 1999, has argued on the basis of a probabilistic model of confirmation).

7.4.2 Sober's Contrastive Empiricism

The most important objection against IA related to point (*b*) – namely, to the denial of the holistic nature of confirmation – has been raised by Elliot Sober (1993). According to him, even granted that there are mathematical theories indispensable to some scientific theories, empirical evidence does not allow confirmation of the latter to transmit to the former, nor to deliver any justification in the existence of the objects of the former.

Sober's position is explicitly inspired by the "constructive empiricist" version of scientific anti-realism advanced by van Fraassen (1980). It will help to sketch at least the basics of this view.

According to the latter, the main aim of our scientific theories is not to attain a generally true (even only approximately true) description of the world, but just to account for observable phenomena. In order to do this, we can posit unobservable physical entities, or appeal to alleged mathematical objects, but what we say about these entities cannot in any way entail that we are justified in believing them to exist. Accepting a scientific theory, in van Fraassen's sense, is not the same as believing in the truth of every statement of the theory literally understood, but merely reduces to believing true what the theory says about observable entities and phenomena (that is, to believing, as it is common to say, that the theory is empirically adequate, what does not imply that it is globally true).

It is not easy to spell this view out in detail. If we come back to a previous example (see § 4.1), the statement 'The acceleration of gravity at poles is equal to 9,283 m/s^2' seems to say something both about an observable phenomenon (tested through appropriate experiments), and about a rational number (and also about something whose nature can be variously described, such as the acceleration of gravity). Suppose we accept the theory to which it belongs, that is, we take it to be empirically adequate. Should we say that this statement is true (insofar as it tells us something about observable entities, like terrestrial poles), or rather that it is not so (insofar as it tells something us about a rational number)? How is it possible to separate between a part of the statement's content that is about observable entities and that we could plausibly hold true, from one that is not about them? Van Fraassen, it goes without saying, has answers to offer to these questions, though we cannot enter into this here.[22] What is relevant is that he cannot accept IBE as a valid form of inference, at least insofar as it leads to justifying the existence of unobservable physical entities or mathematical objects.[23] He accepts that theories in which unobservable entities and/or mathematical objects are posited can successfully explain some phenomena, but he believes that this is not enough for being justified in taking these entities or objects to exist (he rather counsels agnosticism about them).

According to this view, unobservable entities and mathematical objects are epistemically on a par. There is then at least one thing in common between constructive empiricism and confirmational holism. Justification for the existence of unobservable physical entities and for mathematical objects work similarly for both: as the holist takes the existence of both to be justified when the theory is confirmed, so van Fraassen believes that the confirmation of the theory gives no justification for believing in the existence of either.

Sober's contrastive empiricism differs from van Fraassen's views on this very last point, and suggests that it is wrong to consider the justification for the existence of unobservable physical entities and of mathematical objects to be on a par with respect to the theory's confirmation. According to Sober, while IBE is valid when the existence of unobservable physical entities is at stake, it is not so when it is used to justify the existence of mathematical objects.

Like others (for instance, Carnap, 1950b, 1952, or Jaakko Hintikka, 1968), Sober has a probabilistic conception of confirmation (for detailed presentations, see Lakatos, 1968; Howson and Urbach, 1989; Kuipers, 1997). According to him, confirmation of a theory is regulated by the

"Likelihood Principle": observation o favours hypothesis h_1 over hypothesis h_2 if and only if $P(o/h_1) > P(o/h_2)$, namely if and only if the probability that h_1 confers on o (the probability of o given h_1) is greater than the probability h_2 confers on o (the probability of o given h_2), which could also be expressed by saying that, given o, h_1 is likelier than h_2.

Two given hypotheses h_1 and h_2 can be such that the Likelihood Principle can discriminate between the two on the basis of a certain observation o, but this is not the same as saying that this principle allows discriminating in general between an hypothesis h and any other rival hypothesis, for it might happen that confrontation between two alternative hypotheses is impossible on the basis of it, whatever the available observation may be. A simple example (echoing the well-known paradox of confirmation signalled by Goodman, 1954) is the following. Let h_1 be 'Andrea, and not Marco, killed the man', h_2 be 'Marco, and not Andrea, killed the man', and h_3 be 'Andrea did not kill the man, even though all available evidence will make it appear that he did'. A given set of observations will allow discriminating between h_1 and h_2, but it will be impossible to discriminate on its basis between h_1 and h_3.

If we grant, with Sober, that the Likelihood Principle is the fundamental principle governing confirmation, it follows that we never confirm a single hypothesis (or a theory) as such, but only relative to one or more rival hypotheses (or theories).

According to Sober, within this model of confirmation it is possible to confirm (or disconfirm) hypotheses that assert the existence of unobservable physical entities, but it is impossible either to confirm or to disconfirm mathematical theories. Quite the opposite: in this model "indispensability is not synonym for empirical confirmation, but its very antithesis" (1993, p. 44).

In order for it to be the case that two mathematical theories M_1 and M_2 are respectively confirmed and disconfirmed according to the Likelihood Principle it should happen that $P_A(o/M_1) > P_A(o/M_2)$, where A is a set of background assumptions, and $P_A(x/y)$ is the probability of x given y, under the assumptions of the assumptions in A. But, according to Sober, this is wholly implausible, and this suffices to show that mathematical theories are not the kind of theories that can be confirmed on this model.

Nevertheless, Sober does not go into many details in order to support his claim. He confines himself to arithmetic and argues that: (*i*) there is no alternative to arithmetic that could be contrasted with it; (*ii*) even if there were one, it would not confer to observation o a probability different from that which is conferred to it by arithmetic itself (whatever the assumptions in A are). If (i) and (ii) are accepted, Sober concludes, the Likelihood Principle becomes "unrealistic" (*ibid.*, p. 45) when applied to two alternative mathematical theories. As a consequence, it makes no sense to speak of empirical confirmation of a mathematical theory, contrary to what the confirmational holist suggests.

It is however far from clear how (i) and (ii) should be understood exactly. As regards (i), Sober gives no hint, for example, as to what a theory alternative to arithmetic – and, *a fortiori*, to any other mathematical theory – might be. This should be made clear, however, since, in a very weak sense, it is clearly possible to find alternatives to any mathematical theory simply by discarding or modifying one or more of its axioms or definitions (provided consistency is preserved). A clear example is given by non-Euclidean geometries. But the same seems also to hold for arithmetic. Let, for example, PA^1 be first-order Peano arithmetic (understood as usual). It will be enough to replace the axioms (or definitions) introducing usual associative addition in it with others introducing a non-associative addition, in order to obtain a different, in some sense alternative, theory. Call it PA^{1*}. Now let s_1 and s_2 be two scientific theories obtained by conjoining the same body of nominalistic (or generally non-mathematical) statements with PA^1 or PA^{1*} respectively. One might suppose that an observation o, in conjunction with the same background hypotheses, would confer different probability to s_1 and to s_2, and would thus favour s_1 over s_2 and, indirectly, PA^1 over PA^{1*} (or vice versa). In order to deny that this is possible, Sober might deny that PA^{1*} is a genuine alternative to arithmetic, or else he should give an argument for showing that no observation can, under the same background assumptions, confer different probability to PA^1 and PA^{1*}.[24] However, Sober does not elaborate this point further, and also leaves unsettled how his position could generalize beyond arithmetic.[25]

7.4.3 Maddy and the Scientific Practice Objection

As we saw in § 7.3.1, endorsing an epistemic version of IA for platonism involving naturalism would be, according to Maddy, in contrast with scientific practice, if naturalism were understood *à la* Quine, as in [*Ontological Naturalism*]. Accordingly, she envisages an amended version of IA, based on a different understanding of naturalism and fitting, according to her, with mathematical practice. Maddy (1992, 1997), however, takes also this modified argument to be unsound.

Maddy's further objections, all falling within category (*b*), are motivated by the fact that even in its modified version, IA for platonism does appeal to confirmational holism, and this seems to give no faithful description of how and when scientists actually take their scientific theories and their component parts as confirmed.

Quine (1955, p. 247) suggests that a scientific theory should be evaluated according to five criteria: simplicity, familiarity of principles (the theory uses laws that are already known from other theories, in order to explain phenomena that would otherwise need the introduction of new laws), scope (the ability to explain with a single theory phenomena that would otherwise require several theories; this is usually called 'unificatory power'), fecundity (the property of facilitating further extensions of the theory), and finally the fact that "such testable consequences of the theory as have been tested have turned out well". According to Maddy (1997, p. 133), these criteria are those we appeal to in order to justify our scientific theories. For short, let us call 'Q-justified' a theory which is justified according to them. Maddy also seems to accept (plausibly in connection with the fifth criterion) that if a theory is Q-justified it should also count as confirmed.

This being granted, she tries to show, through examples from the history of science, that confirmational holism contrasts with scientific practice. These examples concern significant cases of theories that were considered confirmed and even Q-justified, but still included parts, even indispensable ones, not equally considered as confirmed.

One of these examples (see Maddy, 1997, pp. 135–46) concerns the attitude of the scientific community with respect to the development of the atomic theory of matter, especially in connection with the existence of atoms. Maddy notices that during the nineteenth century, through the development of chemistry (thanks to works by Dalton, Berzelius, Dumas, Cannizzaro and others), many chemical hypotheses were corroborated by empirical evidence, also by appeal to the atomic theory, and the latter was also applied in other fields, for instance in physics for the kinetic theory of health. She then adds that by the end of the nineteenth century atomic theory was generally thought to be indispensable for several Q-justified theories. This, however, did not prevent many from remaining sceptical towards the existence of atoms (for instance, Duhem, Poincaré and Mach), or even to deny their existence. Such a sceptical attitude was motivated by the request of a "direct verification" of atomic theory, making the existence of atoms somehow observable. This attitude was rather common until atomic theory admitted only of an indirect confirmation as an indispensable part of other Q-justified (and then confirmed) theories, and was only dispelled between 1908 and 1913, when Perrin's experiments allowed for a direct confirmation of this theory.

Leaving many important details aside, what is essential in Maddy's reconstruction is that she takes it to show that a theory considered indispensable for Q-justified and confirmed scientific theories can be, and indeed was, considered as non-confirmed, or at least as not justified in a way appropriate for inferring that some entities posited though it could be justifiedly held to exist.

It might be retorted that all that Maddy's reconstruction in fact shows is that the five criteria singled out by Quine do not suffice to establish when a theory is to be taken as appropriately justified; her objection should thus be addressed to these conditions, and not to confirmational holism, nor to (epistemic versions of) IA assuming it. But Maddy could counter-argue that, whatever one might think about these criteria, the example of atomic theory shows that the requirement, that the kind of empirical evidence confirming a scientific theory should also allow justifying all the theories that are indispensable to it, does not fit with actual scientific practice.

Maddy (1995; 1997, pp. 143–57) offers other examples and arguments (see also Cheyne, 2001, pp. 161–5). One among the latter is particularly relevant. It relies on the acknowledgment that many idealizations and approximations commonly used in science (frictionless planes, liquids with infinite depth, etc.) should be considered as indispensable to the scientific theories in which they occur, since no suitable paraphrases for replacing statements in which these occur could be devised. Still, no one believes that these statements are literally true.[26]

On this basis, Maddy (1992, p. 281) concludes then that "we must allow a distinction to be drawn between parts of a theory that are true and parts that are merely useful", even if these latter are indispensable. Hence, for her, "the indispensability of mathematics in well-confirmed scientific theories no longer serves to establish its truth", nor the existence of its objects.

7.4.4 Resnik and the Pragmatic Indispensability Argument

Let us now come to considerations relative to point (c). Many seem to accept that epistemic versions of IA can avoid appealing to confirmational holism. As mentioned above, Colyvan himself admits this (2001, p. 37), and others, like Dieveney (2007) or Azzouni (2009) have argued for it.[27] In the same vein, Michael Resnik (1995; 1997, ch. 3) has suggested an epistemic version of IA intended to do entirely without confirmational holism.[28]

According to Resnik (1995, pp. 168–9), it is a fact that scientists assume that the mathematical theories they use are true and their objects exist, and that they appeal to the properties of these objects in describing empirical phenomena. This is true, Resnik claims, for any scientific theory, also when the theory is not true, nor even confirmed. This is something which pertains to scientific practice, not to the truth or confirmation of scientific theories. Resnik wants to show that this practice is in good standing. In order to do this, he offers a version of IA that is independent of the fact that a given scientific theory is true, held true, or even confirmed; it relies on purely pragmatic considerations, and it does not assume confirmational holism nor scientific realism (Resnik, 1997, pp. 46–7). The argument is divided into different parts, but for the sake of simplicity, we shall present them together, with minor modifications (7(*i*), 7(*ii*), 7(*iv*) and 7(*v*) are premises, 7(*iii*), 7(*vi*) and 7(*vii*) are conclusions):[29]

[7] *Pragmatic IA*

(*i*) In stating its laws and conducting its derivations science assumes the existence of many mathematical objects and the truth of much mathematics.
(*ii*) These assumptions are indispensable to the pursuit of science; moreover, many of the important conclusions drawn from and within science could not be drawn without taking mathematical claims to be true.

(*iii*) So we are justified in drawing conclusions from and within science only if we are justified in taking the mathematics used in science to be true.
(*iv*) We are justified in doing science.
(*v*) The only way we know of doing science involves drawing conclusions from and within it.

(*vi*) So, we are justified in taking that mathematics to be true.

(*vii*) So, mathematics is true.

This argument differs from all those considered so far in many respects.[30] One is that it does not concern scientific theories as sets of statements, but rather scientific practice and our justification in doing science in a certain way. As Resnik claims (*ibid.*, p. 171), it does not establish that "the evidence for science (one body of statements) is also evidence for its mathematical components (another body of statements)", but rather that "the justification for doing science (one act) also justifies our accepting as true such mathematics as science uses (another act)".

Resnik takes the first part of the argument to be by itself sufficient to oppose the views of those who deny the truth of mathematical statements. Conclusion 7(*iii*) tells us that denying truth to our mathematical theories would make our scientific practice unjustified. It is unclear, however, what inferential relations hold between 7(*i*), 7(*ii*), and 7(*iii*). At most one could argue that 7(*iii*) follows by contraposition from the second part of 7(*ii*), the one following 'moreover'. Resnik seems to be implicitly assuming that things are as this second part of 7(*ii*), and thus also 7(*iii*), say they are but then it is unclear which role premise 7(*i*) is supposed to play, even though Resnik (1992, 1995, p. 169; 1997, pp. 43–4) pauses at length on it, offering examples from the history of biology and astronomy in its support.

Apart from these difficulties concerning the first part of the argument, notice the conditional character of its conclusion 7(*iii*): by itself it does not say that we are justified in holding true the mathematical theories that are indispensable to our scientific practice. One might indeed claim that we have no justification for "drawing conclusions from and within science". Premises 7(*iv*) and 7(*v*) ensure that we have this justification. From this, together with 7(*iii*), it follows that we are justified in taking the mathematical theories employed in science to be true, as asserted by conclusion 7(*vi*). The passage from 7(*vi*) to 7(*vii*), namely the third part of the argument[31] is justified, according to Resnik, by naturalism, which he briefly defines (1995, p. 166; 1997, p. 45) as the thesis that "natural science is our ultimate arbiter of truth and existence". Naturalism, according to Resnik, rules out scepticism about the justification and truth of our scientific theories. Still, it is not clear how, just by itself, it can warrant the passage from the justification of a theory to its truth.

Even though the formulation of the argument raises these and other problems,[32] Resnik has the merit of having first suggested, in the recent debate, that IA could be offered in versions free of the assumption of confirmational holism.

Concluding Remarks

For its countenancing an everlasting domain of abstract objects which mathematics is supposed to describe, platonism has often been charged with generating two undesirable gaps. The first gap separates both the activity of mathematicians – their elaborating theories and conducting proofs – and the result of their activity – the system of those theories, namely mathematics as a discipline which evolves in human history thanks to human beings – from mathematics' subject matter – what it is about and what it describes. This is the gap signalled by Benacerraf's dilemma and, more generally, by the problem of access, which the dilemma raises in all its meaningfulness. The responses discussed in Chapter 5 try to fill this gap in various ways, while those discussed in Chapter 4 deny it, either by rejecting its very premise – platonism – or by giving a more palatable reading of it. The second gap separates mathematics itself – which, despite its being the outcome of a human activity, is often described by the platonist as embodying the sort of necessity stemming from the independence of the objects this activity describes – from the roundabout, contingent and always perfectible attempts science makes in order to account for the material and corruptible world which surrounds us and to which we belong. This is the gap IA intends to fill by appeal to the notion of indispensability. Beyond the difficulties this argument might encounter, discussed in Chapters 6 and 7, it remains that even though indispensability was conceived as the symptom of an intrinsic cohesion between mathematics and science, so that its mere acknowledgement would provide, together with some more or less additional premises, an argument for platonism (and/or semantic realism), this would still leave the first gap untouched. Scientific practice itself would disclose the existence of the objects mathematics is about, and would grant for the truth of its theorems; nonetheless, it would not say a word about the nature of these objects, nor would it explain that truth, leaving the philosopher facing the most difficult task. Once the nominalist option is ruled out by considerations of indispensability, the platonist would be confronted with an inescapable duty: to explain what it means exactly that mathematical objects exist, that mathematical theorems describe them, and that these are true.

If our book has hit its target, it should by now be clear that this is, according to us, the most challenging task for platonists. Articulating platonism requires, in our opinion, much more than the production of an argument in favour of the existence of mathematical objects, or for the inadequacy of nominalism. It requires that the nature of these objects is explained in a way that makes the claim of their existence plausible, showing how one can appeal to it in order to account both for the peculiarity of mathematical reasoning and its results, and for the reliability and successful application of these latter. It is not enough, to answer the challenge, to recall, maybe on behalf of Kant, that existence is not a property, and that it is not predicated in different ways of different entities. Insistence on the fact that mathematical objects exist in the same sense in which we say of a star or a town that they exist, is certainly no solution to the problem. Rather, it is a way of making it even more pressing and, maybe, to foreclose the possibility of a number of answers. The case of IA is paradigmatic. Its basic idea is that mathematics and observation inextricably co-exist in science, so that speaking of existence and truth in mathematics is no different from speaking of existence and truth in empirical domains. But does not this, by itself, already raise an ineluctable question? Does this not require that this univocality of sense is explained, justified, made sufficiently plausible?

Supporters of IA may react by noticing that the intimate connection between mathematics and empirical inquiry that they call 'indispensability' is not the outcome of sophisticated and questionable arguments, but rather a fact that intellectual honesty would make anyone able to certify. But this just is where the problem lies. If this is a fact, it requires an explanation. It is not enough, in order to account for it, to say that it follows from it that there cannot be different senses for speaking of existence and truth in mathematics and science respectively. Quite to the contrary, it is from the explanation that this result may rather follow.

Still, raising the problem is not the same as surrendering to it. If we accept that existence can be understood in different ways, platonism might have resources for reaching a plausible explanation.

There is a quite simple, way of accounting for some applications of arithmetic to science, which has been mentioned in various places. Many scientific theories, maybe all of them, use statements in which numerals occur. We say, for instance, that Mars has 2 moons or that some subspecies of cicadas have a lifecycle of 17 years. We have observed that it is possible to rephrase these statements by means of numerical quantifiers, thus eliding any reference to abstract objects such as natural numbers, and showing that the occurrence of numerals in them is nothing but a useful and innocuous way of saying. Innocuous as it may be, this way of saying signals the presence of an abstraction. Acknowledging that with numerals we can say the same things that can be said without them entails acknowledging that the use of naturals is governed by appropriate stipulations, which fix the sense to be assigned to them and allow us to claim that the statements in which they occur are true. Thus fixing the sense of certain singular terms brings about a new domain of objects of which we can speak in order to say true things of other objects. Once this is conceded, it is of little help to say that these new objects exist in our imagination, or just out of our will and capacity of abstraction, and thus do not exist at all. What really matters is that such imagination or abstraction has not just made a useful job, but also suggests the possibility of extending our stipulations, and adding new ones to them, so as to allow speaking of these new objects as such. This is actually what leads to arithmetic: a theory of abstract objects delivered by the acceptance of suitable stipulations which idealize a concrete procedure such as counting, define the structure of progression, and allow proving theorems concerning this structure as such, theorems which have no counterpart in statements which merely express results of counting.

It still remains, however, that nominalists are on the right track when they claim that the mere possibility of appealing to paraphrases with numerical quantifiers is enough to show that arithmetic, as a theory of abstract objects, is not indispensable in many scientific context. The position just outlined, which closely resembles Shapiro's structuralist platonism, thus comes out as compatible with a nominalist account of certain applications of arithmetic to science. But this does not mean that all applications of mathematics to science are of the same kind.

Imagine someone says that a given plot of land is square, and its diagonal measures $\sqrt{2}$ kilometres. There is no way of understanding this statement as expressing the result of counting. Numerical quantifiers will not allow eliminating the real number $\sqrt{2}$. And possibly no other eliminativist strategy could be applied in this case. What is being said here cannot be said without real numbers. Nonetheless, it is not clear that in saying what we say, we are directly speaking of an empirical object like our plot of land. What is said here is not the outcome of an empirical procedure that has been applied to this land, but rather the outcome of a mathematical argument that gets applied to a square, of side with assigned length of 1 kilometre, which we assume as an ideal model of the plot of land. In other words, our statement only indirectly concerns the land, insofar as it is accepted that the square and its diagonal give an appropriate ideal representation of it. Mathematics cannot be eliminated here: it really is indispensable. But the statement that the plot of land is squared and its diagonal is $\sqrt{2}$ kilometres is not true because the land is in fact like this, but rather because it has been settled, on the basis of appropriate empirical procedures, that the land is suitably represented by a square whose side is assigned a length of 1 kilometre.

It should be clear that IA does not apply to any of those cases. Not to the first case, since the relevant mathematical theories are not indispensable for the relevant scientific theories; not to the second one, since the mathematical theories are indispensable but are not used to get an empirical description, but rather to provide an abstract model of the configuration of the empirical objects under scrutiny.

We do not want to claim that all cases of application of mathematics to science fall under the two categories just exemplified. The first example shows that conceiving of arithmetic as a theory of abstract objects generated through appropriate stipulations idealizing an empirical procedure, is compatible with a nominalist account of some applications of arithmetic to science. The second example seems rather to show that other applications of mathematics to science ask for a different treatment. This could, in turn, be compatible with the idea that the relevant mathematical theories stem from other suitable stipulations, of a different origin. It would be enough to admit that these stipulations can deliver adequate representations of what science intends to speak.

This will certainly not suffice to answer the request for an explanation that we hope our book has engendered. But it should at least suggest that there exists a viable route for those platonists who accept that mathematics derives, in ultimate analysis, from appropriate stipulations, attributing existence (in some suitable sense) to abstract objects, and permitting us to say true things of them. The two main problems of this approach would be, on the one side, to explain what legitimates the adopted stipulation and makes it possible that the existence they attribute to the relevant objects, though different in nature from that of stars and towns, is not a mere fiction either, like the one a novel attributes to its characters; and on the other side, to explain how these objects can provide suitable representations of what science is about. These, it will be claimed, are, with due changes, the same problems to which a more traditional platonism promises to answer by appeal to the independent existence of abstract objects. To a large extent, this is indeed so. But, at the end of our exposition, we hope at least to have shown that invoking the existence of such objects will not suffice. Doing so simply amounts to denying Plato's problem. The main purpose of this book has been to show, on the contrary, that this problem is still pressing for whoever attempts to account in an acceptable way for what we are doing when we do mathematics.

Notes

Introduction

1. For a survey, see Rosen (2001).
2. Throughout the book we will use the term 'arithmetic', without further qualification, to refer to the theory of natural numbers.
3. See, for instance, Dummett (1978b).
4. In what follows we will use, unless otherwise specified, the adjective 'true' *tout court*, and its derivatives, implicitly assuming that statements considered as true are not vacuously such, or not only vacuously such (when appropriately interpreted).

1 The Origins

1. This change is all but explicit. However, we do not see how Plato's metaphor could be interpreted differently, especially in light of what has been said above concerning *Theaetetus*, *Meno* and *Phaedrus*.
2. Plato is here using the term *'logismos'*. A more literal translation would render it as 'calculation'. We prefer, however, to translate it as 'logistic' since it seems to us that Plato is here not merely referring to a mathematical procedure or practice, bur rather to a branch of mathematics, namely to that branch that he elsewhere calls *'hē logistikē'* (or *'hē logistikē technē'*), a term that we shall render as 'logistic' too. Logistic was the art of calculating with numbers, and was distinct from the science about them. We will come back to this point in what follows.
3. The terms *'dianoia'* and *'nous'* occur elsewhere in Plato's dialogues, and with different meanings. For example, they occur in the passage quoted above from the *Phaedrus* (247c–d), where they are rendered, respectively, as 'mind' and 'intelligence'.
4. Here too, Plato insists on the indispensability of earthly knowledge (342d–e): "For unless a man somehow or other grasps the four of these, he will never perfectly acquire knowledge of the fifth."
5. Plato uses *'nous'* here, but with a different meaning in respect to the meaning that the term has in the aforementioned metaphor of the divided line. See also *Timaeus*, 51d–52a.
6. This conception will be codified in definitions VII.1 and VII.2 of Euclid's *Elements*.
7. This second interpretation is close to the classic interpretation given by Jakob Klein (1934–6, pp. 17–25) who, however, translates the relevant passage differently from us, having "with reference to *how much* either happens to be" where we have "with the odd and the even, however many they are".
8. Technically, a progression is the structure common to every system of elements that are linearly ordered and include a first element relative to this

order, starting from which they can all be reached by progressing according to the order.
9. Aristotle is using here the plural neuter *ta mathēmatika*. Leaving aside the unorthodox literal translation of this expression as 'mathematical things', we remain with two other possible translations – 'mathematical objects' and 'mathematical entities' – among which we select the former, which squares nicely with modern terminology.
10. His confutation of option (*i*) hinges on various arguments presented at the beginning of the second chapter of Book XIII, which in turn take up those presented at the end of the second chapter of Book III.
11. In *Physics*, Aristotle is crystal clear that time – defined as "number of movement in respect of the before and after" (219b 2) – exists only in so far as movement (218b, 21) and the soul that enumerates it (223a, 21–9) exist.
12. This reading of Euclid's *Elements* is suggested in Panza (2012).
13. Proclus alludes to this argument in passing, however, already in the *Commentary*'s first Part (Chapter VI, 13, 9–11).
14. We use here on purpose the somewhat vague term 'thought'. A proper account of how Aristotle conceives of the higher faculties of the soul would take us too far afield.
15. With 'sensibility' we mean, here and in what follows, the faculty of having perceptual sensations.
16. Points would need a separate treatment, but this would not raise any further difficulty.
17. For a more detailed exposition of what is in this section, see Panza (1997). Among the many expositions of Kant's philosophy, see Cassirer (1918–21) and Strawson (1966). On Kant's philosophy of mathematics, see Friedman (1992).
18. As a consequence, knowledge is not, for Kant, true and justified opinion; it is rather a state of mind involving concepts and intuitions.
19. This does not seem to require any special explanation as long as geometrical schemas are concerned. As regards arithmetical ones, consider that, according to Kant, there is a concept for each number, and not only the concept of number, unit, even and odd. If there has to be multiplication of entities at all, this occurs at the level in which concepts, and not schemas, are generated.

2 From Frege to Gödel (Through Hilbert)

1. Where a and b are two real numbers, a complex number is a binomial $a + ib$, where i is the square root of -1 (namely, it is such that $i^2 = -1$). A suitable definition of reals yields thus a suitable definition of complex numbers.
2. For a comprehensive exposition of Frege's philosophy of mathematics, see Dummett (1991). General introductions to Frege's philosophy can be found in Dummett (1973), Kenny (1995).
3. The acronym 'FA' (for 'Frege arithmetic') is used today for a different theory, which will be discussed in § 5.1.
4. All we shall say on Frege in the present section is referred to the first part of *Grundgesetze*, §§ I.1–I.52.

5. A formal system is generally also called 'formal theory'. By the conventions presented in the *Preface*, however, a theory is a system of statements; it does therefore include neither a language nor deductive rules, even though its statements are obviously expressed in a certain language and can be derived by certain rules. A sentence of a given language (be this formal or not) provides a way for expressing a statement, but it is not necessary that it does. It can simply be taken as a concatenation of symbols or words in a given language. This is common in logic, where sentences of a formal system are studied as such and for their deductive relations, independently of any other function they might happen to have. Strictly speaking, by our terminological conventions, a formal system should not be taken as a theory, even though it includes (or might include) a (formal) theory.
6. Today, we have non-axiomatic formal systems in which the role of axioms is played by suitable inference rules, namely introduction and elimination rules. In our presentation we will only consider axiomatic systems, however.
7. Notice that interpreting a sentence is not *ipso facto* equivalent to turning it into a statement: it does not entail that the force of asserting something is attached to it, but only that it has been fixed what the sentence would assert were that force to be attached to it. On the other hand, a statement can be distinguished by the sentence that expresses it even if no interpretation has been fixed. Thus one might assert, for instance, that $2 + 3 = 5$, by uttering the sentence '$2 + 3 = 5$' with assertoric force, without having established yet what the terms '2', '3', and '5' mean. If assertoric force is attached to a sentence, this turns it into a statement. If the sentence is interpreted, then also the statement counts as interpreted; it thus asserts something specific, and can be either true or false.
8. Things are by far less clear for the extension of this system presented in the second volume for the purpose of defining real numbers. We shall come back to this point in § 2.1.4.
9. As is well known, and as we shall see later, Frege's whole system is inconsistent and this is clearly enough to deny that is a logical system. The relevant question lies elsewhere, however, since different ways of emending Frege's system so as to free it from (apparent) inconsistency are known (and will be mentioned later). The relevant question is whether the system so emended can or cannot be taken as logic. And this is where the real controversy lies. We shall come back to this question in § 5.1, where we shall consider a way of recasting Frege's definition of natural numbers whose proponents (known as 'neo-logicists') take to remain within bounds of logic (or analyticity, at least).
10. We will use modern notation, in order to simplify the exposition. This allows us to simplify things significantly by avoiding considering the function $—\xi$, which plays a crucial role in Frege's system, since it is supposed to occur as a component in five of the seven fundamental functions we shall consider here (namely negation, implication, the two universal quantifies, and the value-range). Frege calls this function 'horizontal', and defines it by stating that its value is T for T as argument and F otherwise. Avoiding considering this function prevents us entering into some important details concerning the nature of Frege's formal language and the way he understands it.

258 *Notes*

On this and other matters concerned with Frege's formal language, we refer the reader to Landini (2012).

11. More precisely, the second-order quantifier should be denoted by '$\forall P[\mu_\xi(P(\xi))]$', where the subscript in 'μ_ξ' indicates that in $P(\xi)$ the symbol 'ξ' is not a placeholder for empty places for objects: the argument of the functions over which 'P' ranges could for instance be held fixed, like in '$\forall P[f(P(\Gamma))]$', or bounded by another quantifier, like in '$\forall P \forall x[f(P(x))]$'.

12. Frege's use of the symbol '|—' should not be confused with the modern use of the '⊢', which indicates that the formula following the symbol has to be proved, or is a theorem.

13. Frege considers the symbol '|—' as composed by '|' and '—' and takes the former (which he calls 'judgement-stroke' to express the act of assertion and the latter to designate the function horizontal (see note 10 above). It follows that, for Frege, this function occurs in any statement.

14. So far, we have only talked about names of objects. Frege also believes, however, that functions have names, like '$f(\xi)$'. The difference between the two cases is crucial, but we cannot pause on similar aspects of Frege's philosophy of language, despite their relation to his philosophy of mathematics. The reader might find various texts treating these issues, among which we recommend Dummett (1973).

15. Following Quine, some believe that second-order logic is not properly logic, but rather set-theory in disguise, and are thus not disposed to concede that the fifth of Frege's fundamental functions presented above corresponds to a logical constant. The reason is that for Quine, second-order variables should be taken as ranging over sets (whose elements are individuals over which first-order variables range, or couples, etc. of individuals). Beside Quine's, there are several possible interpretations: see Quine (1970, pp. 66–8), Boolos (1975, 1984, 1985), Linnebo (2003, 2008), Shapiro (2005b).

16. For reasons that will be clear later, it is important to notice that Frege stresses the analogy between two of these axioms by denoting them with the same roman numeral. These are Laws II*a* and II*b*. The three others are Laws I, III, and IV (the latter of which is actually redundant).

17. By 'implicit definition' one generally means a definition that is obtained not by laying down what is the *definiendum*, or by introducing a convention that stipulates how some words or symbols should be understood, but rather by laying down certain conditions that *definiendum* must satisfy. When it is taken as a definition of the *Num.*(–) operator, HP is an implicit definition, since it just stipulates an identity condition for the values of this operator. An operator is just a suitable function: if it is applied to certain entities taken as its arguments, it delivers other entities as values. In § 5.1 we will explain why the operator implicitly defined by HP does not only deliver natural numbers, but cardinal numbers more generally. In a formal language, an operator is thus designated by a functional constant.

18. Kant has a similar definition of the extension of a concept also in the *Logik Jäsche* (§ 1.7). In both cases, he contrasts the extension of a concept to its "content", namely its characteristic marks. In *Logik Blomberg* (§ 204) he defines in a similar way the *sphaera* of a concept, and then appeals to the notion of extension in order to make is definition clearer: "The multitude of things . . . that I can think under the *conceptus communis* constitute the *sphaera conceptus*".

19. Notice that Frege agrees with Kant that geometrical truths are synthetic *a priori* (§ 89).
20. For more on what follows in this section, see Dummett (1991, pp. 201–8).
21. See note 16.
22. According to modern notation and conception of predicate logic, BLV should rather be written as follows: '$(\dot{\varepsilon}[F(\varepsilon)] = \dot{\varepsilon}[G(\varepsilon)]) \Leftrightarrow \forall x[F(x) \Leftrightarrow G(x)]$'. Here '$F$' and '$G$' are (schematic) predicate letters, and so '$\dot{\varepsilon}[F(\varepsilon)] = \dot{\varepsilon}[G(\varepsilon)]$', '$F(x)$' and '$G(x)$' are sentences, rather than names of objects, which motivate the use of '$\Leftrightarrow$' instead of '='. This is due to the abandonment of Frege's idea that truth-values are objects, that is, items of the same type of those denoted by individual constants.
23. We are used today to use the term 'Peano axioms' to refer to different axiom systems, among which some are first-order and some are second-order (respectively called 'first-order Peano axioms' and 'second-order Peano axioms'). In the former case, these systems include an axiom schema, replacing the second-order induction axiom, and contain thus, properly speaking, an infinite number of axioms. Frege clearly proves them in a second-order version.
24. On the notion of logic in Kant and Frege, see MacFarlane (2002).
25. Just to give an example of how Frege is reasoning here, let us see how this proof works for the function $\xi = \zeta$. If we take both the arguments of these functions to be value-ranges of certain functions, then BLV allows us to determine whether these functions have the same value-range or not. If yes, the function takes T as value, if not it takes F as value. If we take an argument to be the value-range of a certain function, and the other a truth-value, then the functions take either T or F as value according to whether the latter has been supposed to be the same value-range as the former or not (that is, according to the foregoing supposition, whether the former is the same value-range as that of $\xi = (\xi = \xi)$, if the latter is T, or whether the former is the same value-range as that of $\xi = \neg\forall x[x = x]$, if the latter is F, or not). If we take an argument to be the value-range of a certain function and the other an object other than a truth-value, then the function takes F as value.
26. Today, we would not say that a name of function has a reference, unless we were willing to claim, *contra* Frege, that a function is an object. On the problems related to the extension of the notion of reference to constants different from singular terms, see for instance Dummett (1973, ch. 7) and Wright (1998c).
27. A recursive criterion is a criterion that applies to some cases if and only if it is established whether other cases pertaining (in some sense to be specified) to a lower level satisfy it. It thus applies only to cases satisfying a hierarchy of levels, and requires a starting point, or basis. The same can be said for any recursive rule or stipulation. A typical case of recursive definition is the following definition of natural numbers: 0 is a natural number; if n is a natural number, then $s(n)$ is a natural number; nothing else is a natural number. If we have introduced the function $s(-)$, called 'successor function', then it follows from the first and second clause that $s(0)$ is a natural number, and then that also $s(s(0))$ is a natural number, and so on, by the third clause, for all and only natural numbers.
28. On the source of the contradiction, see also Boolos (1993) and Dummett (1994).

29. Russell generally talks of classes, but he distinguishes between "class as many" and "class as one", a distinction that closely resembles the modern one between class and set. More on this issue can be found in any good introduction to set theory, such as Quine (1963), Suppes (1972), Potter (2004, especially Appendix C). Here, we will use 'set' in its current meaning.
30. But see note 15 above.
31. This idea echoes Frege's context principle, but it is radically different from that expressed by it: Frege maintained that the meaning of words should be found in the context of the statements in which they occur, but he did not claim that they have a meaning only in such contexts, or even worse that some words could contribute to the formation of meaningful statements despite their being meaningless.
32. What follows is merely a rough discussion of the rise of set theory. Many detailed texts are available though, like Hallet (1984) or Ferreirós (1999). Cavaillès (1938) is a classical exposition.
33. If we take a finite set, it is easy to see that if it has n members, the set of its subsets has 2^n. *Mutatis mutandis*, this holds for infinite sets too. If α is the cardinal of an infinite set, the cardinal of the set of its subsets is 2^α.
34. On the notion of progression, see Chapter 1, note 8. Here it is enough to notice that a set satisfies second-order Peano axioms if and only if it forms a progression under a successor relation which is appropriately defined on it (at least insofar as these axioms are not taken to include axioms for order, addition and multiplication, as it is sometimes the case).
35. Actually, the issue is far more complex: the precise relations between this model and Peano axioms depend on the way in which these latter are formulated. If they are formulated in a second-order language (thus being finite in number), they have a single model, up to isomorphism (namely, any two distinct models of these axioms are isomorphic to each other: there is a bijection among their elements preserving all the relations that make them count as models of the relevant axioms). When this happens, a theory is said to be categorical. If they are formulated in a first-order language (thus being infinite in number since they necessarily include an induction axiom schema: cf. note 23 above, not all of their models are isomorphic: some (called 'non-standard') include some additional elements. This is a consequence of the upward Löwenheim-Skolem theorem. Here are some logic manuals in which these issues can be explored further: Mendelson (1964), Kleene (1967), Shoenfield (1967), Boolos and Jeffrey (1989), Ebbinghaus, Flum and Thomas (2004). On model theory, see. Chang and Keisler (1973), Hodges (1993), Rothmaler (2000).
36. See § 2.5. For more on this, see Cohen (1966), Woodin (2001a, 2001b), Kanamori (2004).
37. We cannot but give a very general formulation of this tenet here, but we hope it will suffice for a sufficiently clear idea of the point at issue.
38. The presence of paradoxes will of course be unacceptable even from Carnap's point of view. The logical point then lies in the relation between impredicative definitions and the paradoxes. Poincaré (1906) saw in the former the main cause of the latter; we now know that many impredicative definitions are perfectly consistent (an instance is HP, used for a definition of natural numbers: see § 5.1). For more on this, see Chihara (1973), Heinzmann (1985), Feferman (2005).
39. But see § 4.2 and § 7.2.4.

40. Some of Hilbert's and Bernay's articles can be found in Hilbert (1932–5, vol. III). Translations in collections can be found in van Heijenoort (1967), Largeault (1992), Ewald (1996), Mancosu (1998). The most mature version of Hilbert's and Bernays' foundational proposal is in Hilbert and Bernays (1934–9). Many of the things we will say in what follows do not square with the ideas presented in this latter book, where finitary mathematics is seen as a formal theory. Our exposition will be closer to earlier versions of that proposal. The difference between the two approaches largely depends on a change in the conception of intuition supposedly involved in finitary aritmetic. On this, see Mancosu (1998), pp. 149–88). On Hilbert's programme, see Giaquinto (1983) and Detlefsen (1986).
41. Hilbert makes use of inverted commas neither around his number-signs nor around the signs he uses for the purpose of communication. For the former, this could be justified by the fact that these signs are not supposed to denote anything else but themselves, with the result that the object $1 + 1$ is here the same as the object '$1 + 1$'. But for the latter, things are different. In this case, we prefer then to introduce inverted commas according to the conventions widely accepted today.
42. According to Parsons' terminology (see Parsons 1998, p. 252; 2008, pp. 241–2), this means that 'concrete' should here be taken as synonymous with 'quasi-concrete' (see § 5.4).
43. Today we know, thanks to Gödel's theorems themselves, that this prevents T from being a second-order theory satisfying other requisites that are generally required by a good mathematical theory.
44. Gödel actually proves a slightly weaker theorem than this. The stronger version we have stated has been proved by John Barkley Rosser (1936).
45. Brouwer (CW) contains Brouwer's original works (with English translations of Dutch texts). Translations can also be found in many collections (see also note 40). Among the many works on Brouwer, see Stigt (1990), van Dalen (1999–2005) and van Atten (2004).
46. Let us take variables to range over the relevant objects. If one wants to prove '$\exists x(F(x))$', for some property F, one proves that the negation of this statement, '$\neg\exists x(F(x))$'(which is equivalent to '$\forall x \neg F(x)$'), leads to contradiction, from which '$\neg\neg\exists x(F(x))$' (or '$\neg\forall x \neg F(x)$') follows by *modus tollens*. By double negation elimination, one then gets the statement to be proved.
47. On this, see Tieszen (2005) and van Atten (2006). On Husserl's philosophy of mathematics, see Tieszen (1995) and Haddock and Hill (2000).
48. Gödel provides a model of ZF (provided by the proper class L of constructible sets), and shows that it satisfies the Axiom of Choice and the Continuum Hypothesis. Clearly, all we will say here about CH is relative to classical mathematics.

3 Benacerraf's Arguments

1. Other accounts of the debate which will occupy us in the following chapters can be found in Engel (1995), Shapiro (2000a, pp. 281–9), Bostock (2009). For more detailed discussions, see the essays in Shapiro (2005a).
2. Benacerraf distinguishes between what he calls 'transitive' counting, by which objects are counted, and what he calls 'intransitive' counting, consisting in

uttering the names of natural numbers in their right order. According to Benacerraf, one cannot count transitively if one hasn't mastered intransitive counting, since to count transitively consists in associating objects to the names of natural numbers uttered in their correct order.
3. Benacerraf (1965, p. 51, note 3) acknowledges that not everyone concedes that the possibility of defining the usual arithmetical operations on the elements of a progression, and that of using these elements in counting objects, are necessary conditions for taking these elements to be the natural numbers. Many believe (Benacerraf mentions Quine, 1960, § 54) that natural numbers can be identified with the elements of any progression whatsoever. This point is relevant under many respects, but it does not affect the argument we are discussing, since for the latter it is enough that Zermelo's and von Neumann's progressions are considered as adequate. Benacerraf (1996a, Appendix; 1966b) will later admit that the decidability of the ordering relation on the elements of a progression is not essential, arguing that it is always possible to define a decidable ordering relation on a progression. This issue is not settled, however: see, for example, Halbach and Horsten (2005), who maintain that the decidability condition is essential.
4. See note 3.
5. On the various theories of truth, see for instance Künne (2003); Burgess and Burgess (2011).
6. For general discussions on the epistemological challenges to platonism raised by Benacerraf's dilemma, see for instance Wright (1983, §§ 11–12); Hale (1987, ch. 4); Lewis (1986, pp. 108–15); Field (1989a, pp. 25–30), (1989b, § 2); Burgess and Rosen (1997, pp. 35–60); Divers and Miller (1999), Cheyne (2001), Liggins (2006, 2010b), Linnebo (2006), Callard (2007), Azzouni (2008), Kasa (2010).

4 Non-conservative Responses to Benacerraf's Dilemma

1. As a matter of fact, Field (1984) suggests that it is still possible to speak of mathematical truth even without denying that knowledge requires truth. However, mathematical knowledge is not given, according to him, by the theorems of mathematics (as complying with clause (*i*)); it is, rather, either knowledge that certain theorems have some (modal) logical properties, or knowledge of the fact that some mathematicians accept certain theories or assumptions.
2. Granted that a nominalistic statement is a statement formulated in a nominalistic language, nothing prevents it being necessarily true. If '*t*' is a term of such a language, the statement 'For any *x*, if *x* is identical to *t*, then it is identical to *t*' is a case in point, and also is a consequence (given any minimal logic) of any theory some of whose axioms are statements in which '*t*' occurs. If this happens with a bridge-law of a nominalistically impure mathematical theory, the statement is a consequence of this theory.
3. One could doubt that it is possible, given this condition, to have nominalistically impure mathematical theories that have nominalistic consequences non-necessarily true. Still, Field (1982, pp. 48–9) gives an example of such a theory. He shows how to obtain a (consistent) version of set-theory, allowing for "sets with non-sets as members" and for the existence of "a set

of all non-sets", where the axiom schema of separation and replacement admit of instances involving "the empirical vocabulary that is ordinarily used to describe non-sets", and in which the axiom of infinity is dropped and replaced by its negation. This theory would have "consequences about the physical world that conflict with most current physical theories", and are then, obviously, non-necessarily true. But notice that the "most current physical theories" to which Field refers here are theories implying – or so understood as to imply – the "existence of an infinite discrete linear ordering of non-mathematical entities". His point is, indeed, that in such a theory, together with the described version of set-theory, one could "define a function symbol mapping some x in the ordering into 0, its immediate successor into 1, the successor of that into 2 and so on", which, together with the existence of a set of all non-sets and the replacement axioms-schema extended in the mentioned way, entails "the existence of a set containing 0, 1, 2 and so on, in violation of the negation of the axiom of infinity". One could have, however, good reasons for arguing that such a "physical theory" – or at least such a theory so understood – is not genuinely physical, after all.
4. This would be the case, for example, with the version of set-theory mentioned in note 3, if it were taken to be a pure mathematical theory, namely if the non-sets that are allowed to be members of sets were taken to be some sort of mathematical entities (under the supposition, of course, that some mathematical entities could be non-sets).
5. Intuitively speaking, an uninteresting (mathematical) theory is certainly not good, whatever condition it might meet. Still, Field's limitation to interesting mathematical theories has also a technical aim: we shall come back to this in § 6.3.
6. In order to understand why this condition is equivalent to the preceding one, one can reason as follows. Suppose that $\neg\varphi$ is a nominalistic, non-contradictory, statement, and that M+φ is inconsistent. Then, $\neg\varphi$ would be a consequence of M that is not necessarily true (since φ is not contradictory), and thus M would have nominalistic consequences that are not necessarily true. It follows that $\neg\varphi$ would be a consistent nominalistic statement, whereas M+$\neg\varphi$ would be inconsistent.
7. In Field (1980) the conservativeness of mathematics is not defined as we have defined it earlier following Field (1982), but rather directly through the condition just stated. Field (1980, pp. 12–13) thus has a slightly modified version of the first part of the argument given earlier to show that "our mathematical theories" are conservative, and presents (*ibid.*, p. 16–19) two other more technical arguments, specifically addressing the case of set theory (which is considered as a mathematical theory *par excellence*).
8. Notice that the issue here is whether these terms and variables can be eliminated from a not purely mathematical language. The easy solution adopted here cannot thus serve to define natural numbers as such and to give an extra-mathematical foundation for arithmetic.
9. Field (1980, p. 21) defines these numerical existential quantifier on the basis of other quantifiers: $\exists_{\geq i} x[F(x)]$, to be read as 'There are at least i xs that are Fs'. It is easy to pass from one definition to the other; we give only one of these for the sake of simplicity.

10. Field (1985) contains replies to Shapiro's objections (see also Field, 1984, for more on the second-order issue), but these do not have the last word on the matter.
11. In the sense that is relevant here, the language of mereology is based on a fundamental relation, that of (being) part of, that gets applied to concrete objects. If these objects do not have parts, they are called 'atoms'. A concrete object that has parts is said to be the 'mereological sum' of its parts. These latter can be either atoms or other mereological sums. A mereological sum is thus a concrete object constituted by its parts (atoms or sums), these being concrete in their turn. These parts need not necessarily be contiguous in space (we can thus have the mereological sum of Mount Everest and Big Ben). This language clearly involves quantification over concrete objects only (atoms or sums). For an introduction, see Varzi (2009).
12. Burgess and Rosen (1997, § II.A) show how this argument could be carried out in detail.
13. The distinction was originally suggested by Burgess (1983), refined in Burgess and Rosen (1997, pp. 6–7). For a more detailed exposition, see Burgess, Rosen (2005, pp. 515–18). For discussion, see Chihara (2005, pp. 483–9) and Hellman (1998).
14. Burgess and Rosen interchangeably use expressions like 'content', 'what a statement says' and 'meaning'. In order to avoid undue qualifications, we will only use the term 'content'.
15. We say 'propositional attitude' in order to make clear that the term 'attitude' is used so as to denote a propositional mental state. But we do not, nor do the authors considered here, want to imply that attributing a particular propositional attitude entails the existence of propositions as abstract objects.
16. Hellman (1998, pp. 342–4) has argued against Burgess' and Rosen's distinction, claiming that it is not generally among the aims of nominalists, not even those qualifying as revolutionary, to suggest that nominalistic versions of mathematical and scientific theories should replace traditional ones in practice. Nominalists would rather be putting forward theories on the epistemology of mathematics fully compatible with the actual use of traditional theories.
17. Yablo seems to have modified his views lately, elaborating suggestions presented in Yablo (2006, 2009). We present them in the form in which they have been discussed until recently.
18. See Rayo, Yablo (2001).
19. See, on this, McGee (1997, pp. 56–62).
20. Though Yablo (2002, pp. 179–85) suggests how to extend his view concerning the real content of mathematical statements to set-theory, it is not clear whether he can claim that the real content of set-theoretical statements is given by logical truths, nor if this can be claimed for any mathematical theory.
21. Yablo has suggested in unpublished work that a statement like 'There are infinite prime numbers' should be interpreted as 'There are natural numbers, and they are such that there is an infinite number of them which are prime'. 'There are infinite prime numbers' should then be considered analogously to a physical law like 'Objects with no net force on them move with constant

velocity' (to be rendered as 'There are bodies with no net force on them and they move with constant velocity'), and its real content would express a condition that natural numbers would satisfy, if they existed. But what is this condition, if it is to be external to arithmetic?
22. For a defence of Field's programme on similar lines, see also Liggins (2006).
23. Leng (2010) defends a fictionalist view that closely resemble Yablo's. Leng also has detailed discussions of many issues related to recent debates on mathematical fictionalism.
24. The first component is called 'hypothetical' since the translation scheme transforms any statement of a given mathematical theory in a modal conditional statement. By 'hypothetical component' one can mean, depending on the context, both the component of the overall interpretation procedure, or, for any specific modalized version of a theory, the collection of statements translated in accordance with that procedure.
25. To these assumptions the axioms of second-order logic must be added (where the comprehension schema is appropriately modified in modal terms, see Hellman, 1989, p. 24; 2005, pp. 553–4), together with those of the S5 system of quantified modal logic, without the Barcan formula (for details, see any good introduction to modal logic, such as Hughes and Cresswell, 1996).
26. To be precise: neither of actual objects, nor of possible but non-actual objects, the so-called *possibilia*. These latter are avoided thanks to a particular restriction of the second-order comprehension schema (Hellman, 2005, pp. 553–4).
27. Parsons is commenting on a suggestion by Putnam (1967) on the possibility of interpreting set theory nominalistically over concrete domains, a suggestion that Hellman (1989, p. 71) recalls, before offering a modal structuralist interpretation that is not nominalist (1989, ch. 2). Parsons' objection can however be addressed to Hellman's modal structuralism.
28. This seems to go against Zalta's idea (see § 5.2) that an abstract object is an object that is not concrete in any possible world, an object that could not exist for a nominalist.
29. For discussions on the various kinds of modality, see Plantinga (1974), Kripke (1980), Lewis (1986), Hale (1999), Fine (2005), Mackie (2006). For various uses of modality in mathematics, see Putnam (1967), Field (1984, 1989b), Parsons (1983b), Hale (1996b), Chihara (2008), Leng (2010, § 3.2), Friedman (2005).
30. For more on this point and its consequences for the interaction between cognitive sciences and the philosophy of mathematics, see Doridot and Panza (2004). For a similar objection, see Parsons (2008, p. 211).
31. Maddy has subsequently rejected the indispensability argument (see §§ 7.3.1 and 7.4.3). In her 1997 work (p. 152, note 30), Maddy still takes it that set theory has cognitive grounds, but claims for set-theoretical platonism no more on the basis of an argument as that just presented, but on the basis of her conception of naturalism (see § 7.3.1). Recently, Maddy has also argued for a form of mathematical realism that leaves unsettled whether platonism about sets is true (see § 7.3.1.1); according to her latest views, her argument for set-theoretical platonism on cognitive grounds could turn out to be "unnecessary" (Maddy 2011, p. 72, cf. also p. 118).
32. Two simple examples might help appreciating the difference: both the notion of empty set and the distinction between a set E and the sets

{E} and $E \cup \{E\}$, which are crucial in set-theory, have no clear analogue in any system of concepts suitable for an appropriate account of perceptual experiences.

5 Conservative Responses to Benacerraf's Dilemma

1. Several essays by Hale and Wright are collected in their (2001) work. See also Hale (1987); Hale and Wright (2002); Cook (2007). For a survey, see MacBride (2003).
2. This name is common today, and comes from a passage in Wright (1983, p. 51). For a more detailed formulation, see MacBride (2003, p. 108).
3. A relation R is an equivalence relation if it is reflective, symmetric, and transitive, namely if for every x, y, and z, respectively: $R(x,x)$; if $R(x,y)$, then $R(y,x)$; if $R(x,y)$ and $R(y,z)$ then $R(x,z)$.
4. Zalta (PM) presents a version of the theory updated to 1999.
5. Using a typed language not only ensures greater rigour, but also allows that what we will later say about mathematical objects can be said about higher-order entities, such as monadic properties, relations, etc.
6. It is important to understand that the sheaf of properties of which Nessy is a reification is not given by the two properties of being a monster and living in Loch Ness, but only by the property of being a monster living in Loch Ness and by all properties, if any, that an object has if it has that property. Taken separately, the two properties of being a monster and of living in Loch Ness do not satisfy the formula '$\forall y[P(y) \Leftrightarrow [Mst(y) \wedge LcNs(y)]]$', indeed.
7. In the philosophy of language, by 'proposition' one usually means the content of a statement or what a statement expresses. In mathematics the same term is used instead to denote theorems or, more generally, statements, and this usage is for the sake of simplicity often adopted also in the philosophy of mathematics as well. Here the term is used with the former meaning.
8. The term 'second-order Peano arithmetic' is used to denote different variants of second-order arithmetic, all involving a finite number of axioms suggested by those advanced by Peano himself in his seminal work (1889). In the variant considered by Zalta, these axioms are five and include a monadic predicate constant designating the property of being a natural number (this is necessary, since the theory has to be immersed within Object Theory, and the objects the former characterizes must then be distinguished from other possible objects of the latter; in other presentations, this constant can be avoided, since no individuals distinct from natural numbers are involved) and no dyadic functional or predicate constant used to designate addition, multiplication and order (which makes this variant merely characterize the structure of progression; see Chapter 1, note 8).

6 The Indispensability Argument: Structure and Basic Notions

1. On the applicability of mathematics, see Steiner (1989, 1998, 2005), Pincock (2012).

2. We mentioned Gödel (1947) in § 2.5. For the others, see Frege (1893–1903, § 91), on which Garavaso (2005); Carnap (1939, p. 64), on which Psillos (1999, pp. 10–11); von Neumann (1947, p. 6).
3. See Quine (1948, pp. 16–19; 1951a, pp. 44–6; 1954, pp. 121–2; 1966b, p. 244; 1969c, p. 97–8; 1981c, pp. 149–50).
4. Though typically endorsed under the assumption that mathematical objects are abstract, IA is independent of the nature of mathematical objects, and then in principle acceptable by nominalists who believes that there are mathematical objects, but they are concrete. For the sake of generality, we will merely speak of mathematical objects.
5. Those who believe the first premise to be too harsh can still accept a weaker formulation in which that premise is discharged and the conclusion is conditional in form: if there are true theories / if S is true, then . . .
6. On this issue, see also Pincock (2004), Paseu (2007).
7. Mark Steiner (1978, pp. 19–20) argues, however, for a different interpretation of Quine's views, suggesting a transcendental version of IA.
8. While Putnam's formulation suggests an epistemic reading, his overall discussion suggests a non-epistemic one.
9. A close, but distinct, reading of 'realism' is the one famously introduced by Dummett, according to whom the term refers to the thesis "that statements of [a] disputed class possess an objective truth-value, independently of our means of knowing it: they are true or false in virtue of a reality existing independently of us" (1978b, p. 146).
10. In §§ 6.2.2, 7.3.2 and 7.4.2 we shall pause on another related view, usually called 'scientific realism'.
11. Notice that not all of Quine's views that might be relevant to mathematics are always considered as relevant, or even compatible for IA. Bueno (2003) has argued, for instance, that Quine's views on ontological relativity (Quine, 1968) may clash with his alleged endorsement of IA. We will here consider only those aspects of Quine's views that bear relevance to IA.
12. But see Busch and Sereni (2012).
13. Baker's argument has been offered in the context of a discussion concerning the role of mathematical explanation in science and its relations to IA. Here we can only point to the relevant works, among which are Baker (2001, 2005), Colyvan (2002), Melia (2002), Batterman (2010), Pincock (2011), Saatsi (2011), Bueno and French (2012). On the partly related issue of explanations in mathematics, see Mancosu (2008b, 2011).
14. Here and in what follows we call 'vocabulary' the set of non-logical constants (namely, singular terms and predicate or functional constants) of a language.
15. We use the adjective 'putative' in order to stress that the indispensability to S of either quantification over a certain domain of entities, or these very entities, or the reference to them, should not be taken to entail by itself that there is a certain relation between S and some existent objects, for this would of course make IA question-begging.
16. Option (*ii*) is suggested by argument [3] and is particularly faithful to Quine's view. Notice that, if only quantification has to be considered, and not also singular terms, then the vocabulary fixed by T is to be restricted to predicate constants alone. As suggested by Quine (1948) himself, this could be done by introducing for any singular term a corresponding predicate which is

intended to apply only to the referent of the original term, and substitute all occurrences of the former accordingly: for instance, 'Pegasus flies' would become 'There is one and only one x such that x pegazises and x flies'.

17. See Chapter 5, note 3.
18. See Quine (1960, ch. 7, note 5). Quine's doubts on the nominalist programme endorsed in Goodman and Quine (1947) goes back as far as 1948; see Mancosu (2008c).
19. There are various reasons telling against the availability of a sharp separation of the vocabularies of scientific theories into observational and theoretical sub-vocabularies. Acknowledging this possibility was crucial for logical positivists, but has been subject to several objections (see Putnam, 1962; Suppe, 1977). We can set these issues aside here, since as far as the discussion on IA is concerned, it is usually granted that the required separation is available for the relevant scientific theories, supposing that the theoretical vocabulary is (or includes) a mathematical vocabulary.
20. A theory is recursively enumerable if its theorems can be effectively enumerated, that is arranged in a succession $\tau_1, \tau_2, \ldots$ whose elements can be effectively obtained one after the other (possibly in an infinite time). Notice that 'recursive' and 'recursively', here and in the next note have a different meaning than that suggested in Chapter 3, note 27. Here, they indicate that some effective decision procedure is available.
21. A theory is recursively axiomatizable if, given any well-formed formula of its language, it is possible to determine whether it is one of the axioms of the theory. The set of the theory's axioms is then said to be recursive.
22. Craig himself implicitly warned against this philosophical use of his theorem, observing that, in his proof, v_Φ is such that its theorems are obtained in a "mechanical and artificial way" and are not "psychologically or mathematically ... more perspicuous" than those of v (1956, p. 49).
23. An example of the necessity of distinguishing between the empirical content of a theory and the content of its observational statements is offered in Melia (2000), on which see § 7.1.
24. Colyvan's discussion of "the role of confirmation theory" in his (2001, pp. 78–81) hints at the relative character of the notion of "preferability". We take our clarification of (in)dispensability to improve on that suggestion.
25. IA apart, this criterion plays a crucial role in ontology: see, among others, Carnap (1950a), Alston (1958), Church (1958), Jackson (1980), Hodes (1990), Melia (1995, 2000), Azzouni (1998), Yablo (1998, 2001), Alspector-Kelly (2001), Eklund (2005), Hofweber (2005a, 2005b, 2007), Chalmers et al. (2009).
26. See note 16 above.
27. Clearly, our choices about which objects to countenance will not be wholly arbitrary, and criteria for this can be thought of. One minimal criterion adopted by Quine is that we should not accept the existence of entities for which no clear identity conditions can be stated: hence his slogan, "no entity without identity" (Quine, 1958, p. 20).
28. Notice that the 'if' direction of this double implication is much less innocent than it appears at first glance, since it depends on admitting the *prima facie* reading of a statement of the form '$\exists x[\mathcal{A}(x)]$', according to which such a statement just asserts that there exists an x such that $\mathcal{A}(x)$: this reading of the quantifiers, sometimes also called 'objectual', might be questioned by alternative conceptions of quantification.

29. Notice, however, that in order to avoid circularity, some basic statements – like those used in reporting results of empirical tests, or, generally, those which Quine calls 'observational statements' – must be to a certain extent independent of the whole theory. See, for details, Quine (1960, § 2.10; 1969b; 1981e, ch. 2; 1993); see also Quine and Ullian (1975, ch. 1).
30. See Haack (2009, ch. 6), on Quine's ambiguity in the use of the term 'science'.
31. This leaves open to be decided what makes a statement part of a theory, and what makes a theory scientific. These are rather complex questions, and different answers might come with different conceptions of naturalism.
32. Stronger conceptions are available. Some see naturalism as the thesis that only non-abstract entities exist. Both the Eleatic Principle discussed by Colyvan (2001, ch. 3), and the naturalism endorsed by Weir (2005) and Armstrong (1997, p. 5) are cases in point (for a discussion, see Cheyne, 2001, ch. 7). Under these readings, naturalism is almost indistinguishable from nominalism. Hence, as argued by Colyvan, this form of naturalism is in principle incompatible with IA.
33. Confirmational holism should be kept distinct from semantic holism, which is roughly the thesis that the meaning of words or statements cannot be determined in isolation, but depends on the role they play in the whole language they belong to (see Fodor, Lepore, 1991). Both theses were suggested by Quine as consequences of his objections to the "dogmas" of logical empiricism (Quine, 1951a), though the former was originally advanced in Duhem (1906, especially p. 225), and is also known as the Duhem-Quine thesis.
34. On occasions, Quine states his view in more moderate terms, claiming for instance that "it is an uninteresting legalism . . . to think of our scientific system of the world as involved *en bloc* in every prediction. More modest chunks suffice" (1981b, p. 71).
35. A noticeable exception, besides Putnam, is Azzouni (2009). As mentioned in § 6.2.2, Colyvan himself (2001, p. 12) observes that the 'only' implication in premise (*i*) of argument [3] is not required, although he includes it in his argument "for the sake of completeness and also to help highlight the important role naturalism plays in questions of ontology". Whatever this role may be, acknowledging that this implication is not required suggests that argument [3] can be freed by naturalism.
36. Some have accepted this unpalatable conclusion nonetheless (see Colyvan, 2000; 2001, pp. 134–40).
37. Quine accepted this conclusion. He claimed that the objects of unapplied mathematical theories are "without ontological rights" and considered those theories as "mathematical recreation" (1986, p. 400; in Quine 1990, pp. 94–5 and 1995, pp. 55–6 this position is partially revised but the general attitude remains the same). This has engendered a wide debate: see § 7.3.1, and Parsons (1983b), Maddy (1992), Leng (2002), Colyvan (2007).

7 The Indispensability Argument: The Debate

1. Since the set of theorems of a theory could itself be taken as a system of axioms.

2. Notice that if this is so, the presence, in these theories, of statements that are neither nominalistic nor mathematical does not prevent the conclusion that, supposing that it is conservative over consistent bodies of non-mathematical statements, acceptable mathematics is dispensable from these theories according to a specification of [*(In)dispensability*] in terms of *o*-equivalence (and of some appropriate criterion of virtuosity). Let s be a consistent, axiomatized, deductively closed, first-order scientific theory, the set of whose axioms is divided into three complementary sets A_N, A_M and A_H, formed, respectively, by nominalistic axioms, by M-loaded axioms for some acceptable mathematical theory M, and by axioms that are neither nominalist nor mathematical. From the conservativeness of acceptable mathematics over any consistent body of non-mathematical statements it follows that a nominalistic statement φ is a theorem of s if and only if it is derivable from A_{N+H}. Thus the sub-theory S_{N+H}, whose axioms are those belonging to A_{N+H}, includes all the nominalistic consequences of s and no other nominalistic consequence. It is then *o*-equivalent to s (granted that *o*-equivalence is defined as said).
3. Pincock (2004) offers a criticism to IA for platonism that somehow recalls Field's use of representation theorems. He suggests what he calls a 'mapping account', according to which applicability of mathematics in science is due to the possibility of correlating (mapping) physical with mathematical models. This would allow him to claim that even if mathematics is indispensable, it does not follow that mathematical objects exist (for discussion, see Bueno and Colyvan, 2011; Batterman, 2010; Pincock, 2011).
4. Representation theorems that must be proved in order to specify the second among the mentioned equivalence relations will have to be necessarily proved in a platonist metatheory. We have already touched upon this issue in § 4.1.
5. The piecemeal character of this argument is criticised in Dummett, 1993, p. 435.
6. We slightly modify Melia's notation for the sake of exposition.
7. Being concrete, a mereological sum is usually identified with the region of space it occupies.
8. These assumptions are not mutually equivalent, not even if one assumes, as Quine does, that statements of the form '$\exists x[\mathcal{A}(x)]$' of a first-order predicate language render the unique sense in which it can be said of an object that it exists. First, it might be that some of the statements considered in (*b*) are resistant to a reduction into a first-order predicate language. Second, one might think that expressions like 'exists', 'there are', etc. occurring in those statements are just idiomatic, and do not indicate the existence of objects at all, or at least do not indicate the existence of objects satisfying the first-order open formula '$\mathcal{A}(x)$'.
9. And leaving aside issues of ontological relativity discussed in Quine (1968). As already mentioned (see note 11, Chapter 6) we abstract here from some of Quine's theses, giving a somewhat simplified picture of Quine's ideas on the relations between language and ontology, which is, however, faithful enough to the way in which Quine's views are usually treated in connection to IA.
10. Wiles' proof is not internal to Peano arithmetic (either first- or second-order), since it employs various other mathematical theories, and nothing

guarantees, to the best of our knowledge, that an internal proof is possible, that is, that the axioms of Peano arithmetic are strong enough to make such a proof possible. Quite to the contrary, in order to prove ¬*Fermat* it would be enough to prove, within Peano arithmetic or any other appropriate version of arithmetic, a formula of the form $ap + bp = cp$ where a, b, c, and p are given natural numbers.

11. One may see also assumption (*a*) as called into question by the claim that all that the acknowledgement that there are numbers allows one to say is that some modal facts concerning progressions hold. Analogously, one could think that the idea that statements of the form '$\exists x[\mathcal{A}(x)]$' of a first-order predicate language should be interpreted in terms of objectual quantification (see note 28, Chapter 6) is contradicted by the idea that ¬*Fermat* and '□[$AX^* \Rightarrow $ ¬*Fermat**]' have the same 'mathematical content'. The way Putnam understands the expressions 'for mathematics' and 'mathematical content' seems however to leave room for different conclusions. In any case, what seems to be beyond discussion (given the overall content of Putnam, 1967, and also Putnam, 2004, ch. 2) is rather that his argument puts pressure on assumption (*b*).
12. For more on Putnam's notion of equivalence involved here, see Putnam (1983). Criticism of Putnam's view of equivalent descriptions is in Burgess and Rosen 1997, pp. 200–1. A recent discussion of equivalent descriptions is in Field (Forthcoming). Maddy (2011) advances a view on the ontology of mathematics that, despite important differences, echoes Putnam's idea of equivalent descriptions.
13. On the evolution of Putnam's thought on the issue, see also Liggins (2007).
14. Of course, (s)he can do that, but it is not mandated. This and related issues for fictionalists are discussed in Sainsbury (2010); see also Azzouni (2010).
15. The two would still be different, since Field gives nominalist reformulations for scientific theories using mathematics, but not for mathematical theories themselves, whereas Yablo would offer them for mathematical theories too.
16. We shall suppose, for the sake of the argument, that the statement 'The average family has 2.3 children' is not understood as meaning that, given a certain particular number of families, and a certain particular number of children, the average family has n children average. Rather, it should be understood as a statement concerning what happens generally, independently of the particular number of families and children under consideration.
17. For reference on the relevance of Quine's criterion in ontological debates, see Chapter 6, note 25.
18. See Chapter 4, note 31.
19. For objections to this argument, see Chihara (2006).
20. Maddy (2005, 2011) examines the differences between her and Burgess' and Rosen's position. These differences do not seem to us to be as clear cut as Maddy seems to suggest.
21. This view echoes Putnam's conception of equivalent descriptions; see § 7.2.2.
22. The debate on constructive empiricism is huge: see Churchland and Hooker (1985) and Monton (2007), for example.
23. It is controversial whether it is allowed, or even mandated, to van Fraassen to reject IBE only when it would lead to justifying the existence of either

unobservable entities or mathematical objects, or whether he should rather reject it in general as a non-valid inference. See Psillos (1999, pp. 211–15).

24. It is unclear what Sober is meaning when he speaks of background assumptions. It is natural to assume that they include bridge-laws (see § 4.1). Should this be the case, however, it would be implausible to think that two scientific theories to which two different mathematical theories are indispensable could be confronted with experience relative to the same background assumptions. In order to oppose Lobacewski, who thought hyperbolic geometry could be experimentally verified, Poincaré (1902) gives an argument showing that no experimental evidence can favour this geometry over Euclid's. His argument rests on the very idea that those geometries cannot be confronted with experience if not through different bridge-laws.
25. For other objections to Sober's view, see Resnik (1995), Hellman (1999), Colyvan (1999; 2001, pp. 126–34). Another objection to IA, somehow related to Sober's views, is raised by Vineberg (1996). For other discussions of IBE as related to IA, see Saatsi (2007, 2011), Busch (2011).
26. Leng (2010) argues against confirmational holism on similar lines, and combines her argument with a fictionalist interpretation of idealizations in empirical sciences.
27. See also § 6.7.
28. Quite independently of IA, Resnik (1997, ch. 7) defends confirmational holism from numerous objections (on this, see Peressini, 2008).
29. The argument is presented in slightly different terms in Resnik (1995, pp. 169–71) and in (1997, pp. 46–8). We stick to the original formulation.
30. Though it has similarities with Burgess' and Rosen's naturalistic argument; see § 7.3.1.1.
31. Resnik's argument in (1997, pp. 46–8) stops with conclusion 7(*vi*); however, conclusion 7(*vii*) is present in the formulation of the argument in (1995, pp. 169–71).
32. See also Leng, 2010, pp. 255–8.

References

Where not otherwise indicated, references in the main text are to the original edition; otherwise, we have indicated with an '*' the edition to which references in the main text refer. As far as texts in languages other than English, and especially Greek texts, are concerned, translations have been occasionally modified with respect to the English translations referenced, through examination of the original texts and confrontation of other available translations in English and other languages.

Alspector-Kelly, M. (2001), "On Quine on Carnap on Ontology", *Philosophical Studies*, 102, 93–122.

Alston, W. (1958), "Ontological Commitments", *Philosophical Studies*, 9, 8–16.

Anderson, R. L. (2008), "Comments on Wayne Martin, *Theories of Judgment*", *Philosophical Studies*, 137, 91–108.

Andrews, P. (2002), *An Introduction to Mathematical Logic and Type Theory: To Truth Through Proof*, 2nd edition (1st edition 1986), Applied Logic Series, 27, Kluwer Academic Publishers, Dordrecht.

Annas, J. (1976), *Aristotle's Metaphysics: Books M And N*, Clarendon Press, Oxford and New York.

Aristotle, (AO), *Aristotelis opera ex recensione I. Bekker*, 11 vols, Oxford, 1837 (original edition *Aristoteles Grecae ex recensione I. Bekker*, vols. I and II of *Aristotelis Opera*, edidit Academia Regia Borussica, 5 vols, Berolini 1831–70).

—— (Metaph), *Metaphysica*, in Aristotle (AO), vol. VIII, (English Translation: *Metaphysica*, by W.D. Ross, 2 vols, Clarendon Press, Oxford, 1924, rev. 1954, in W.D. Ross (ed.), *The Works of Aristotle*, Clarendon Press, Oxford, vol. 8, 1954).

—— (Phys), *Naturalis Auscultationis Libri VIII*, in Aristotle (AO), vol. II, pp. 1–210 (English translation: *Physica*, by R.P. Hardie and R.K. Gaye, in W.D. Ross, ed., *The Works of Aristotle*, Clarendon Press, Oxford, vol. 2, 1930).

Armstrong, D. (1997), *A World of States of Affairs*, Cambridge University Press, Cambridge. Atten, M. van (2004), *On Brouwer*, Thomson/Wadsworth, Belmont (CA).

—— (2006), *Brouwer Meets Husserl. On the Phenomenology of Choice Sequences*, Springer, Dordrecht.

Atten, M. van and Kennedy, J. (2003), "On the Philosophical Development of Gödel", *The Bulletin of Symbolic Logic*, 9, 425–72.

Ayer, A.J. (1946), *Language, Truth and Logic*, Dover, New York.

Azzouni, J. (1997a), "Applied Mathematics, Existential Commitment and the Quine–Putnam Indispensability Thesis", *Philosophia Mathematica* (III), 5, 193–209.

—— (1997b), "Thick Epistemic Access: Distinguishing the Mathematical from the Empirical", *The Journal of Philosophy*, 94, 472–84.

—— (1998), "On 'On What There Is'", *Pacific Philosophical Quarterly*, 79, 1–18.

—— (2004), *Deflating Existential Consequence. A Case for Nominalism*, Oxford University Press, Oxford, New York.

—— (2008), "A Cause for Concern: Standard Abstracta and Causation", *Philosophia Mathematica* (III), 16, 397–401.
—— (2009), "Evading Truth Commitments: The Problem Reanalyzed", *Logique & Analyse*, 206, 139–76.
—— (2010), *Talking About Nothing*, Oxford University Press, Oxford, New York.
Baker, A. (2001), "Mathematics, Indispensability and Scientific Progress", *Erkenntnis*, 55, 85–116.
—— (2003a), "The Indispensability Argument and Multiple Foundations for Mathematics", *The Philosophical Quarterly*, 53, 49–67.
—— (2003b), "Does the Existence of Mathematical Objects Make a Difference?", *Australasian Journal of Philosophy*, 81, 246–64.
—— (2005), "Are there Genuine Mathematical Explanations of Physical Phenomena?", *Mind*, 114, 223–38.
—— (2009), "Mathematical Explanation in Science", *The British Journal for the Philosophy of Science*, 60, 611–33.
Balaguer, M. (1996a), "Towards a Nominalization of Quantum Mechanics", *Mind*, 105, 209–26.
—— (1996b), "A Fictionalist Account of the Indispensable Applications of Mathematics", *Philosophical Studies* 83, 291–314.
—— (1998), *Platonism and Anti-Platonism in the Philosophy of Mathematics*, Oxford University Press, Oxford, New York.
—— (2011), "Fictionalism in the Philosophy of Mathematics", *The Stanford Encyclopedia of Philosophy* (Fall 2011 Edition), Edward N. Zalta (ed.), URL = <http://plato.stanford.edu/archives/fall2011/entries/fictionalism-mathematics/>
—— (2009), "Fictionalism, Theft, and the Story of Mathematics", *Philosophia Mathematica* (III), 17, 131–62.
Barkley Rosser, J. (1936), "Extensions of Some Theorems of Gödel and Church", *Journal of Symbolic Logic*, 1, 87–91.
Baron-Cohen S. and Leslie A.M., Frith U. (1985), "Does the Autistic Child Have a 'Theory of Mind'?", *Cognition*, 21, 37–46.
Batterman, R.W. (2010), "On the Explanatory Role of Mathematics in Empirical Science", *British Journal for the Philosophy of Science*, 61, 1–25.
Benacerraf, P. (1965), "What Numbers Could not Be", *The Philosophical Review*, 74, 47–73*; also in Benacerraf, Putnam (1964), 272–94.
—— (1973), "Mathematical Truth", *The Journal of Philosophy* 70, 661–79*; also in Benacerraf, Putnam (1964), 403–20.
—— (1996a), "What Mathematical Truth Could Not Be – I" in A. Morton and S. P. Stich (eds), *Benacerraf and His Critics*, Blackwell, Oxford, 9–59.
—— (1996b), "Recantation or Any Old ω–Sequence Would Do After All", *Philosophia Mathematica* (III), 4, 184–89.
Benacerraf, P. and Putnam, P. (1964) (eds), *Philosophy of Mathematics. Selected Readings*, Prentice-Hall, Englewood Cliffs (NJ), 1964; 2nd edition 1983*, Cambridge University Press, Cambridge.
Berto, F. (2009), *There's something about Gödel!*, Wiley Blackwell, Oxford.
Blanchette, P.A. (1996), "Frege and Hilbert on Consistency", *The Journal of Philosophy*, 93, 317–36.
Bolzano, B. (1851), *Paradoxien des Unendlichen*, Meiner & Reclam, Leipzig.
Boolos, G. (1975), "On Second-order Logic", *The Journal of Philosophy*, 72, 509–27; also in Boolos (1998), Chapter 3.

—— (1984), "To Be Is To Be a Value of a Variable (or to Be Some Values of Some Variables)", *The Journal of Philosophy*, 81, 430–50; also in Boolos (1998), Chapter 4.

—— (1985), "Nominalist Platonism", *The Philosophical Review*, 94, 327–44; also in Boolos (1998), Chapter 5.

—— (1986–7), "Saving Frege from Contradiction", *Proceedings of the Aristotelian Society*, 87, 137–51; also in Boolos (1998), 171–82.

—— (1987), "The Consistency of Frege's *Foundations of Arithmetic*", in J. Thomson (ed.), *On Being and Saying: Essays in Honor of Richard Cartwright*, MIT Press, Cambridge (MA), 1987, 3–20; also in Boolos (1998), Chapter 12.

—— (1990), "The Standard of Equality of Numbers", in Boolos, G. (ed.), *Meaning and Method: Essays in Honour of Hilary Putnam*, Cambridge, Cambridge University Press, 261–77; also in Boolos (1998), Chapter 13.

—— (1993), "Whence the Contradiction?", *Aristotelian Society Supplementary Volume*, 67, 213–33; also in Boolos (1998), Chapter 14.

—— (1997), "Is Hume's Principle Analytic?", in Heck, R. Jnr. (ed.), *Language, Thought and Logic*, Oxford University Press, Oxford, New York, 1997, 245–62; also in Boolos 1998, Chapter 19.

—— (1998), *Logic, Logic, and Logic*, Harvard University Press, Cambridge (MA), 1998.

Boolos, G. and Heck, R. Jnr. (1998), "*Die Grundlagen der Arithmetik* §§ 82–3", in Schirn (1998), 407–28; also in Boolos (1998), 315–38.

Boolos, G. and Jeffrey, R. (1989), *Computability and Logic*, Cambridge University Press, Cambridge; 1st edition, 1974; 5th edition, Boolos, G., Jeffrey, R., Burgess, J. (eds), 2007.

Bostock, D. (2009), *Philosophy of Mathematics: An Introduction*, Oxford University Press, Oxford, New York.

Bourbaki, N. (EM), *Éléments de mathématiques*. This collective work was written in 1939 by a group of several mathematicians under the pseudonym of 'Nicolas Bourbaki'. At present, it consists of ten volumes, published by Dunod, Paris; some of these have been translated in English and published by Springer-Verlag, Heildelberg. More details can be found at the website of the Association de Collaborateurs de Nicolas Bourbaki (by the École Normale Supérieure de Paris): www.bourbaki.ens.fr/

Boyd, R. (1991) (ed.), *The Philosophy of Science*, MIT Press, Cambridge (MA).

—— (2002), "Scientific Realism", *The Stanford Encyclopedia of Philosophy* (Fall 2008 Edition), Edward N. Zalta (ed.), URL = <http://plato.stanford.edu/archives/fall2008/entries/scientific-realism/>

Brouwer, J.E.J. (1912), *Intuitionisme en formalisme*, Noordhoff, Groningen, 1912; also in Brouwer J.E.J. (1913), *Wiskunde, waarheid, werkelijkheid*; English translation "Intuitionism and formalism", *Bulletin of the American Mathematical Society*, 20; also in Brouwer (CW)*, vol. I, 123–38; also in Benacerraf, Putnam (1964), 77–89.

—— (CW), (edited by A. Heyting), *L. E. J. Brouwer: Collected Works*, vol. I, *Philosophy and Foundations of Mathematics*, 1975; vol. II, *Geometry, Analysis, Topology and Mechanics*, (1976), North Holland, Amsterdam/American Elsevier, New York.

Brown, J.R. (1999), *Philosophy of Mathematics: An Introduction to a World of Proofs and Pictures*, Routledge, London and New York.

Bueno, O. (2003), "Quine's Double Standard: Undermining the Indispensability Argument via the Indeterminacy of Reference", *Principia*, 7, 17–39.

Bueno, O. and Colyvan, M. (2011), "An Inferential Conception of the Application of Mathematics", *Noûs*, 45, 345–74.
Bueno, O. and French, S. (2012), "Can Mathematics Explain Physical Phenomena?", *British Journal for the Philosophy of Science*, 63, 85–113.
Burgess, A. and Burgess, P. (2011), *Truth*, Princeton University Press, Princeton (NJ), 2011.
Burgess, J. (1983), "Why I Am Not a Nominalist", *Notre Dame Journal of Formal Logic*, 24, 93–105.
—— (2004), "Mathematics and *Bleak House*", *Philosophia Mathematica* (III), vol. 12, 18–36.
—— (2005), *Fixing Frege*, Princeton University Press, Princeton (NJ), 2005.
—— (2008), "Review of Charles Parsons, Mathematical Thought and its Objects", *Philosophia Mathematica* (III), 16, 402–20.
Burgess, J.P., Hazen, A. and Lewis, D. (1991), "Appendix on Pairing", in D. Lewis, *Parts of Classes*, Blackwell, Oxford, 121–49.
Burgess, J. and Rosen, G. (1997), *A Subject With No Object. Strategies for Nominalistic Interpretation of Mathematics*, Oxford University Press, Oxford, New York.
—— (2005), "Nominalism Reconsidered", in Shapiro (2005), 515–36.
Burnyeat, M.F. (1987), "Platonism and Mathematics: A Prelude to Discussion", in A. Graeser (ed.), *Mathematics and Metaphysics in Aristotle*, Haupt, Bern and Stuttgart, 213–40.
Busch, J. (2011), "Scientific Realism and the Indispensability Argument for Mathematical Realism: A Marriage Made in Hell", *International Studies in the Philosophy of Science*, 25, 307–25.
Busch, J. and Sereni, A. (2012), "Indispensability Arguments and Their Quinean Heritage", *Disputatio*, 4, 343–60.
Callard, B. (2007), "The Conceivability of Platonism", *Philosophia Mathematica* (III), 15, 347–56.
Cantor, G. (1872), "Über die Ausdehnung eines Satzes aus der Theorie der trigonometrichen Reihen", *Matematische Annalen*, 5, 123–32.
—— (1874), "Über eine Eigenschaft des Inbegriffs aller reellen algebraischen Zahlen", *Journal für die reine und angewandte Mathematik*, 77, 1874, 258–62; also in *Gesammelte Abhandlungen*, 115–18; English translation "On a property of the set of real algebraic numbers", in Ewald (1966), vol. 2, 839–43.
—— (1879–84), "Über unendliche, lineare Punktmannigfaltigkeiten", *Mathematische Annalen* 15(1879), 1–7; 17(1880), 355–58; 20(1882), 113–21; 21(1883), 51–8 and 545–86; 23(1884), 453–88; also in Cantor (GA), 139–145, 145–148, 149–157, 157–164, 165–208, 210–244, reprinted in *Über unendliche, lineare Punktmannigfaltigkeiten. Arbeiten zur Mengenlehre aus den Jahren 1872–84*, edited and with a foreword and commentary by G. Asser, *Teubner–Archiv zur Mathematik*, 2, Teubner, Leipzig, 1984.
—— (1883) *Grundlagen einer allgemeinen Mannigfaltigkeitslehre. Ein mathematisch–philosophischer Versuch in der Lehre des Unendlichen*, Teubner, Leipzig, 1883; also in Cantor (GA), 165–208; English translation "Foundation of the theory of manifolds", *The Campaigner* (The Theoretical Journal of the National Caucus of Labor Committees), 9, 1976, 69–96.
—— (1895–1897), "Beiträge zur Begründung der transfiniten Mengenlehre", *Mathematische Annalen*, 46(1895), 481–512; 49(1897), 207–246; also in Cantor

(GA), 282–311; English translation in *Contributions to the Founding of the Theory of Transfinite Numbers*, Dover, New York, 1952.

—— (GA), *Gesammelte Abhandlungen*, (ed.) by Ernst Zermelo, Springer, Berlin, 1932.

Carnap, R. (1931), "Die logizistische Grundlegung der Mathematik", *Erkenntnis*, 2, 91–105; English translation "The Logicist Foundations of Mathematics", in Benacerraf and Putnam (1964), 41–52.

—— (1939), "Foundation of Logic and Mathematics", in *International Encyclopedia of Unified Science*, vol. I/3, The University of Chicago Press, Chicago.

—— (1950a), "Empiricism, Semantics, and Ontology", *Revue International de Philosophie*, 4, 20–40; revised version in *Meaning and Necessity*, The University of Chicago Press, Chicago, 1956, 205–21; this latter version reprinted in Benacerraf, Putnam (1964)*, 241–57.

—— (1950b), *Logical Foundations of Probability*, University of Chicago Press, 1950, 2nd edition 1965.

—— (1952), *The Continnum of Inductive Methods*, University of Chicago Press, Chicago, 1952.

Cassirer, E. (1918–21), *Kant Leben und Lehre*, Cassirer, 1918 and 1921; English translation *Kant's Life and Thought*, Yale University Press, 1981.

Cassou–Noguès (2005), "Gödel and 'The Objective Existence' of Mathematical Objects", *History and Philosophy of Logic*, 26, 211–28.

Cavaillès, J. (1938), *Remarques sur la formation de la théorie abstraite des ensembles*, Hermann, Paris, 1938, 2nd edition in Cavaillès, J., *Philosophie mathématique*, Herman, Paris, 1962, 23–174, republished by PUF, Paris, 1994; also in Cavaillès, J. *Œuvres complètes de philosophie des sciences*, Hermann, Patis, 1994, 205–374.

Chakravartty, A. (2011), "Scientific Realism", *The Stanford Encyclopedia of Philosophy* (Summer 2011 Edition), Edward N. Zalta (ed.), URL: <http://plato.stanford.edu/archives/sum2011/entries/scientific-realism/>.

Chalmers, D., Manley D. and Wasserman R. (eds) (2009), *Metametaphysics*, Oxford University Press, Oxford,– New York.

Chang C.C. and Keisler, H.J. (1973), *Model Theory*, Elsevier, Amsterdam, 1973, 3rd rev. edition 1990.

Changeux, J.P. and Connes, A. (1989), *Matière à pensée*, Odile Jacob, Paris, 1989; English translation *Conversations on Mind, Matter, and Mathematics*, Princeton University Press (NJ), 1998.

Cheyne, C. (2001), *Knowledge, Cause, and Abstract Objects: Causal Objections to Platonism*, Kluwer, Dordrecht.

Chihara, C.S. (1973), *Ontology and the Vicious-Circle Principle*, Cornell University Press, Ithaca (NY).

—— (1981), "Quine and the Confirmational Paradoxes", in P.A. French, T. Uehling, H. Wettstein (eds), *The Foundations of Analytic Philosophy*, University of Minnesota Press, Minneapolis, 425–52.

—— (1990), *Constructibility and Mathematical Existence*, Clarendon Press, Oxford.

—— (2004), *A Structural Account of Mathematics*, Oxford University Press, Oxford, New York.

—— (2005), "Nominalism", in Shapiro (2005), 483–514.

—— (2006), "Burgess's 'Scientific' Arguments for the Existence of Mathematical Objects", *Philosophia Mathematica (III)*, 14, 318–37.
—— (2008), *The Worlds of Possibility: Modal Realism and the Semantics of Modal Logic*, Clarendon Press, Oxford.
Church, A. (1958), "Ontological Commitment", *The Journal of Philosophy*, 55, 1008–14.
Churchland, P. and Hooker, C. (eds) (1985), *Images of Science: Essays on Realism and Empiricism. With a reply from Bas C. van Fraassen*, University of Chicago Press, Chicago.
Coffa, A.J. (1991), *The Semantic Tradition from Kant to Carnap. To the Vienna Station*, Cambridge University Press, Cambridge, 1991.
Cohen, P.J. (1963), "The Independence of the Continuum Hypothesis", *Proceedings of the National Academy of Sciences of the United States of America*, 50, 1143–48.
—— (1966), *Set Theory and the Continuum Hypothesis*, W.A. Benjamin, New York.
Colyvan, M. (1998), "In Defence of Indispensability", *Philosophia Mathematica* (III), 6, 39–62.
—— (1999), "Contrastive Empiricism and Indispensability", *Erkenntnis*, 51, 323–32.
—— (2000), "Conceptual Contingency and Abstract Existence", *Philosophical Quarterly*, 50, 87–91.
—— (2001), *The Indispensability of Mathematics*, Oxford University Press, Oxford, New York.
—— (2002), "Mathematics and Aesthetic Considerations in Science", *Mind*, 111, 69–74.
—— (2007), "Mathematical Recreation versus Mathematical Knowledge", in Leng *et al.* (2007), 109–22.
—— (2010), "There's No Easy Road to Nominalism", *Mind*, 119, 285–306.
—— (2012), *An Introduction to the Philosophy of Mathematics*, Cambridge University Press, Cambridge.
—— (Forthcoming), "Road Work Ahead: Heavy Machinery on the Easy Road", *Mind*.
Colyvan, M. and Zalta, E.N. (1999), "Mathematics: Truth and Fiction? Review of Mark Balaguer *Platonism and Anti-Platonism in Mathematics*", *Philosophia Mathematica* 7, 336–49.
Cook, R. (ed.) (2007), *The Arché Papers on the Mathematics of Abstraction*, Springer, Dordrecht.
Coquand, T. (2010) "Type Theory", *The Stanford Encyclopedia of Philosophy* (Spring 2010 Edition), Edward N. Zalta (ed.), URL: <http://plato.stanford.edu/archives/spr2010/entries/type-theory/>.
Craig, W. (1953), "On Axiomatizability within a System", *Journal of Symbolic Logic*, 18, 30–32.
—— (1956), "Replacement of Auxiliary Expressions", *Philosophical Review*, 65, 38–55.
Dalen, D. van (1999–2005), *Mystic, Geometer, and Intuitionist. The Life of L.E.J. Brouwer*, Clarendon Press, Oxford, 2 vols., Vol. I 1999, Vol. II 2005.
De Caro, M. and Macarthur, D. (eds) (2004), *Naturalism in Question*, Harvard University Press, Cambridge (MA).
—— (eds) (2010), *Normativity and Naturalism*, Columbia University Press, New York.

Dedekind, R. (1872), *Stetigkeit und irrationale Zahlen*, Vieweg, Brunswick, 1872; also in Dedekind, R. (GMW); English translation "Continuity and Irrational Numbers", in Dedekind (2001), 1–27.

—— (1888), *Was sind und was sollen die Zahlen?*, Vieweg, Brunswick, 1888; also in Dedekind, R. (GMW), English translation "The Nature and Meaning of Numbers", in Dedekind (1901), 31–115.

—— (1901), *Essays on the Theory of Numbers*, Open Court, Chicago, 1901, reprinted by Dover, New York, 1963.

—— (GMW), *Gesammelte mathematische Werke*, 3 vols, (ed.) R. Fricke, E. Noether, Ø. Ore, Vieweg, Brunswick, vol. I 1930, vol. II 1931, vol. III 1932.

Dehaene, S. (1997), *The Number Sense: How the Mind Creates Mathematics*, Oxford University Press, New York, 1997; revised and updated edition (2011).

Demopoulos, W. (2005), "Our Knowledge of Numbers as Self-Subsistent Objects", *Dialectica*, 59, 2005, 141–59.

—— (Forthcoming), "Generality and Objectivity in Frege's *Foundations of Arithmetic*", in A. Miller (ed.), *Logic, Language and Mathematics: Essays in Honour of Crispin Wright*, Oxford Unversity Press, Oxford.

Detlefsen, M. (1986), *Hilbert's Program*, Reidel, Dordrecht.

Dieveney, P. (2007), "Dispensability in the Indispensability Argument", *Synthese*, 157, 105–28.

Divers, J. and Miller, A. (1999), "Arithmetical Platonism: Reliability and Judgement–Dependence", *Philosophical Studies*, 95, 277–310.

Doridot, F. and Panza, M. (2004), "À propos des apports des sciences cognitives à la philosophie des mathématiques", *Intellectica*, 39, 263–87.

Duhem, P. (1906), *La theorie physique. Son object, sa structure*, Chevalier & Rivière, Paris, 1906; 12th revised edition, Marcel Riviere et Cie, 1914; English translation *The Aim and Structure of Physical Theory*, Princeton University Press, Princeton (NJ), 1991.

—— (1908), *SOZEIN TA PHAINOMENA. Essai sur la notion de theorie physique*, A. Hermann, Paris, 1908; English translation *To Save the Phenomena*, University Of Chicago Press, Chicago, 1969.

Dummett, M. (1956), "Nominalism", *Philosophical Review*, 65, 491–505; also in Dummett (1978a), 38–49.

—— (1973), *Frege: Philosophy of Language*, Duckworth, London, 1973; 2nd edition 1981*.

—— (1978a), *Truth and Other Enigmas*, Duckworth, London.

—— (1978b), "Realism", in Dummett (1978a), 145–65.

—— (1991), *Frege: Philosophy of Mathematics*, Harvard University Press, Cambridge (MA).

—— (1993), "What is Mathematics About?", in *The Seas of Language*, Oxford University Press, Oxford, New York, 1993, 429–45.

—— (1994), "Chairman's Address: Basic Law V", *Proceedings of the Aristotelian Society*, 94, 43–251.

Earman, J. and Salmon, W. (1999), "The Confirmation of Scientific Hypothesis", in Salmon, H.M. (ed.), *Introduction to the Philosophy of Science*, Hackett Publishing Co., 1st edition 1992, Chapter 2.

Ebbinghaus, H.-D., Flum, J. and Thomas, W. (2004), *Mathematical Logic*, Springer-Verlag, New York, 2004, 2nd edition; 1st edition, 1994.

Eisler, R. (1930), *Kant Lexikon*, E.S. Mittler, Berlin.

Eklund, M. (2005), "Fiction, Indifference, and Ontology", *Philosophy and Phenomenological Research*, 71, 557–79.
—— (2011), "Fictionalism", The Stanford Encyclopedia of Philosophy (Fall 2011 Edition), Edward N. Zalta (ed.), URL: <http://plato.stanford.edu/archives/fall2011/entries/fictionalism/>.
Engel, P. (1995), "Platonisme mathématique et antiréalisme", in M. Panza and J.M. Salanskis (eds), *L'objectivité mathématique. Platonismes et structures formelles*, Masson, Paris, 133–46.
Errera, A. (1953), "Le problème du continu", *Atti dell'Accademia Ligure di Scienze e Lettere*, 9, 176–83.
Ewald, W.B. (1996), (ed.), *From Kant to Hilbert. A Source Book in the Foundations of Mathematics*, Oxford University Press, Oxford, New York, 2 vols.
Feferman, S. (2005), "Predicativity", in Shapiro (2005), 590–624.
Ferreirós, J. (1999), *Labyrinth of Thought: A History of Set Theory and Its Role in Modern Mathematics*, Birkhäuser, Basel, Boston, Berlin, 2nd edition, 2007.
Field, H. (1974), "Quine and the Correspondence Theory" Philosophical Review, 83, 200–28, reprinted with a Postscript in H. Field, *Truth and The Absence of Fact*, Oxford University Press, Oxford, New York, 2001, 199–221.
—— (1980), *Science Without Numbers*, Blackwell, Oxford.
—— (1982), "Realism and Anti-realism about Mathematics", *Philosophical Topics*, 13, 45–69; also in Field (1989), 53–78.
—— (1984), "Is Mathematical Knowledge just Logical Knowledge", *Philosophical Review*, 93, 509–52; also in Field (1989), 79–124.
—— (1985), "On Conservativeness and Incompleteness", *The Journal of Philosophy*, 82, 239–60; also in Field (1989), 125–46.
—— (1989a), *Realism, Mathematics and Modality*, Blackwell, Oxford.
—— (1989b), "Realism, Mathematics and Modality", in Field (1989a), 227–81 (largely modified version of the same-titled original in *Philosophical Topics*, 19, 1988, 57–107).
—— (1998), "Mathematical Objectivity and Mathematical Objects" in C. MacDonald and S. Laurence (eds), *Contemporary Readings in the Foundations of Metaphysics*, Blackwell, Oxford, 387–403.
—— (Forthcoming), "Mathematical Undecidables, Metaphysical Realism and Equivalent Descriptions", in R.E. Auxier (ed.), *The Philosophy of Hilary Putnam*, Library of Living Philosophers, Open Court Press, Chicago (IL).
Fine, K. (2002),*The Limits Of Abstraction*, Oxford University Press, Oxford, New York.
—— (2005), "The Varieties of Necessity", in K. Fine, *Modality and Tense. Philosophical Papers*, Clarendon Press, Oxford, 2005, 235–60.
Fodor, J. and Lepore, E. (1991), *Holism, A Shopper's Guide*, Wiley Blackwell, Oxford.
Føllesdal, D. (1995), "Gödel and Husserl", in J. Hintikka (ed.), *From Dedekind to Gödel*, Kluwer, Dordrecht, Boston, 1995, 427–46.
Fraassen, B. van (1980), *The Scientific Image*, Oxford University Press, Oxford, New York.
Frege, G. (1879), *Begriffsschrift, eine der Arithmetischen Nachgebildete Formalsprache de reinen Denkens*, Verlag Von L. Nebert, Halle, 1879; English translation "Begriffsschrift, a Formula Language, Modeled upon that of Arithmetic, for Pure Thought", in van Heijenoort (1967), 5–82.

—— (1884), *Die Grundlagen der Arithmetik: eine logische mathematische Untersuchung über den Begriff der Zahl*, Koebner, Breslau, 1884; English translation by J. Austin, in G. Frege, *The Foundations of Arithmetic. A Logico–Mathematical Enquiry into the Concept of Number*, Blackwell, Oxford, 1974.

—— (1893–1903), *Grundgesetze der Arithmetik*, H. Pohle, Jena, 1893–1903 (2 vols.); partial English translation in G. Frege, *The Basic Laws of Arithmetic*, (ed.) M. Furth, University of California Press, Berkeley, 1967.

—— 1919, *Aufzeichnungen für Ludwig Darmstaedter*, in *Nachgelassene Schriften und wissenschaftlicher Briefwechsel*, vol. I, (ed.) H. Hermes *et al.*, Felix Meiner Verlag, Hamburg 1969, pp. 273–7; English translation *Notes for Ludwig Darmstaedter*, in G. Frege, *Posthumous Writings*, Blackwell, Oxford 1979, pp. 253–7*.

—— (WB), *Wissenschaftlicher Briefwechsel*, Felix Meiner, Amburgo, 1976; English translation in Frege (PMC).

—— (PMC), *Philosophical and Mathematical Correspondence*, (ed.) G. Gabriel *et al.*, Blackwell, Oxford, 1980.

Friedman, J. (2005), "Modal Platonism: An Easy Way to Avoid Ontological Commitment to Abstract Entities", *Journal of Philosophical Logic*, 34, 227–73.

Friedman, M. (1992), *Kant and the Exact Sciences*, Harvard University Press, Cambridge (MA), London.

Garavaso, P. (2005), "On Frege's Alleged Indispensability Argument", *Philosophia Mathematica* (III), 13, 160–73.

Gentzen, G. (1936), "Die Widerspruchfreiheit der reinen Zahlentheorie", *Mathematische Annalen*, 112, 493–565; English translation "The Consistency of Elementary Number Theory", in Gentzen (CW), 132–213.

—— (1938), "Neue Fassung des Widerspruchsfreiheitsbeweises fuer die reine Zahlentheorie", *Forschungen zur logik und zur Grundlegung der exacten Wissenschaften, neue Folge*, 4, 19–44; English translation "New version of the consistency proof for elementary number theory", in Gentzen (CW), 252–86.

—— (CW), *The Collected Works of Gerhard Gentzen*, (ed.) M.E Szabo, North-Holland, Amsterdam, 1969.

George, A. and Velleman, D.J. (2001), *Philosophies of Mathematics*, Wiley-Blackwell, Oxford.

Gettier, E. (1963), "Is Justified True Belief Knowledge?", *Analysis*, 23, 121–23.

Giaquinto, M. (1983), "Hilbert's Philosophy of Mathematics", *British Journal for the Philosophy of Science*, 34, 119–32.

—— (2002), *The Search for Certainty: A Philosophical Account of Foundations of Mathematics*, Clarendon Press, Oxford.

Glymour, C. (1980), *Theory and Evidence*, Princeton University Press, Princeton (NJ).

Gödel, K. (CW1), *Collected Works, Vol. I, Publications 1929–1936*, (ed.) S. Feferman, Oxford University Press, Oxford, New York, 1986.

—— (CW2), *Collected Works, Vol. II, Publications 1938–1974*, (ed.) S. Feferman, Oxford University Press, Oxford,New York, 1990.

—— (CW3), *Collected Works, Vol. III*, Unpublished Essays and Lectures, (ed.) S. Feferman *et al.*, Oxford University Press, Oxford,New York, 1995.

—— (UPE), Unpublished Philosophical Essays, (ed.) F.A. Rodriíguez-Consuegra, Birkäuser Verlag, Basel-Boston-Berlin.

—— (1930), "Die Vollständigkeit der Axiome des logischen Funktionenkalküls", *Monatshefte für Mathematik und Physik*, 37; English translation "The completeness of the axioms of the functional calculus of logic", in Gödel (CW1), 103–23 and in van Heijenoort (1967), 582–91.

—— (1931), "Über formal unentscheidbare Sëtze der *Principia Mathematica* un verwandter Systeme I", *Monatshefte Für Mathematick un Physik*, 38, 173–98; English translation "On formally undecidable propositions of *Principia Mathematica* and related systems I", in Gödel (CW1), 145–95.

—— (1939a), "Consistency Proof for the Generalized Continuum Hypothesis", *Proceedings of the National Academy of Sciences*, 25, 220–24; also in Gödel (CW2), 28–32.

—— (1939b), "The Consistency of the Generalized Continuum Hypothesis", *Bulletin of the American Mathematical Society*, 45, 93; also in Gödel (CW2), 27.

—— (1940), "The Consistency of the Axiom of Choice and of the Generalized Continuum Hypothesis with the Axioms of Set Theory", *Annals of Mathematics Studies*, 3, Princeton University Press, Princeton (notes of Gödel's lectures at Princeton in fall 1938, by G.W. Brown); also in Gödel (CW2), 33–101.

—— (1944), "Russell's Mathematical Logic", in Schlipp P.A. (ed.), *The Philosophy of Bertrand Russell*, Northwestern University Press, Evanston (IL), 1944, 125–53; also in Benacerraf, Putnam (1964), 447–69 (1st ed.) and 447–69 (2nd ed. 1983)*, and in Gödel (CW2), 119–41.

—— (1947), "What is Cantor's Continuum Problem?", *The American Mathematical Monthly*, 54, 515–25, and in Gödel (CW2), 176–87.

—— (1964), "What is Cantor's Continuum Problem?", revised and expanded version of Gödel (1947), in Benacerraf, Putnam (1964), 258–73 (1st ed.) and 470–85 (2nd 1983)*; also in Gödel (CW2), 254–70.

Goodman, N. (1951), *The Structure of Appearance*, Harvard University Press, Cambridge (MA); 2nd edition, Bobbs-Merrill, Indianapolis, 1966; 3rd edition Reidel, Boston, 1977.

—— (1954), "The New Riddle of Induction", in N. Goodman, *Fact, Fiction and Forecast*, Harvard University Press, Cambridge (MA); 4th edition, 1979.

Goodman, N. and Quine, W.V. (1947), "Steps Toward a Constructive Nominalism", *Journal of Symbolic Logic*, 12, 105–22.

Haack, S. (2009), *Evidence and Inquiry*, Prometheus Books, Amherst (NY), 1st edition Blackwell, Oxford, 2003.

Haddock, G.R. and Hill, C.O. (2000), *Husserl or Frege?*, Open Court, La Salle (IL).

Halbach, V. and Horsten, L. (2005), "Computational Structuralism", *Philosophia Mathematica* (III), 13, 174–86.

Hale, B. (1987), *Abstract Objects*, Blackwell, Oxford.

—— (1994a), "Dummett's Critique of Wright's Attempt to Resuscitate Frege", *Philosopia Mathematica* (III), 2, 122–47; also in Hale and Wright (2001), 189–213.

—— (1994b), "Singular Terms (2)", in B. McGuinnes and G.L. Oliveri (eds), *The Philosophy of Michael Dummett*, Kluwer, Dordrecht, Boston, London, 1994, 17–44; also in Hale, Wright (2001), 48–71.

—— (1996a), "Singular Terms (1)", in M. Schirn (ed.), *Frege: Importance and Legacy*, Walter de Gruyter, Berlin, New York, 1996, 438–57; also in Hale, Wright (2001), 31–47.

—— (1996b), "Structuralism's Unpaid Epistemological Debts", *Philosophia Mathematica* (III), 4, 124–47.

—— (1999), "Modality", in B. Hale and C. Wright (eds), *A Companion to the Philosophy of Language*, Blackwell, Oxford, 487–514.

—— (2000), "Reals by Abstraction", *Philosophia Mathematica* (III), 8, 100–24; also in Hale and Wright (2001), 399–420.
Hale, B. and Wright, C. (2000), "Implicit Definition and the A Priori", in P. Boghossian and C. Peacocke (eds), *New Essays on the A Priori*, Clarendon Press, Oxford, 286–319; in Hale, Wright (2001), 117–50.
—— (2001), *The Reason's Proper Study. Essays towards a Neo–Fregean Philosophy of Mathematics*, Clarendon Press, Oxford.
—— (2002), "Benacerraf's Dilemma Revisited", *European Journal of Philosophy*, 10, 101–29.
Hallet, M. (1984), *Cantorian Set Theory and Limitation of Size*, Clarendon Press, Oxford.
Hardy, G.H. (1940), *A Mathematician's Apology*, Cambridge University Press, Cambridge.
Hart, W.D. (1996), *The Philosophy of Mathematics*, Oxford University Press, Oxford, New York.
Hebb, D. (1949), *The Organisation of Behavior*, John Wiley and Sons, New York.
—— (1980), *Essay on Mind*, Lawrence Erlbaum Associates, Hillsdale.
Heck, R. Jr. (1993), "The Development of Arithmetic in Frege's *Grundgesetze der Arithmetic*", *Journal of Symbolic Logic*, 58, 579–601; also in W. Demopoulos (ed.), *Frege's Philosophy of Mathematics*, Harvard University Press, Cambdridge (MA), 257–94.
—— (2011), *Frege's Theorem*, Oxford University Press, Oxford, New York.
Heijenoort, J. van, (ed.), (1967), *From Frege to Gödel: A Source Book in Mathematics, 1879–1931*, Harvard University Press, Cambridge (MA).
Heinzmann, G. (1985), *Entre intuition et analyse: Poincaré et le concept de prédicativité*, A. Blanchard, Paris.
Hellman, G. (1989), *Mathematics without numbers*, Oxford University Press, Oxford, New York.
—— (1990), "Modal-Structural Mathematics", in E. Irvine (ed.), *Physicalism in Mathematics*, Dordrecht, Kluwer, 1990, 307–29.
—— (1994), "Real Analysis without Classes", *Philosophica Mathematica* (III), 2, 228–50.
—— (1996), "Structuralism without Structures", *Philosophia Mathematica* (III), 4, 100–23.
—— (1998), "Maoist Mathematics? Review of John P. Burgess and Gideon Rosen, *A Subject With No Object*", *Philosophia Mathematica* (III), 6, 334–45.
—— (1999), "Some Ins and Outs of Indispensability: A Modal Structural Perspective", in A. Cantini, E. Casari, and P. Minari (eds), *Logic and Foundations of Mathematics*, Kluwer Academic Press, Dordrecht, 1999, 25–39.
—— (2005), "Structuralism", in Shapiro (2005), 535–62.
Hempel, G. (1945), "Studies in the Logic of Confirmation", *Mind*, 54, 1–26 and 97–121.
Hermite, C. and Stieltjes, J.T. (1905) *Correspondance d'Hermite et de Stieltjes*, (ed.) B. Baillaud and H. Bouget, with a preface by Émile Picard, Gauthier–Villars, Paris.
Heyting, A. (1930) "Die formalen Regeln der intuitionistischen Logik", *Sitzungsberichte der Preussischen Akademie der Wissenschaften*, 1930, 42–71 and 158–69; partial English translation in Mancosu (1998), 311–27.
—— (1956) *Intuitionism, an Introduction*, North-Holland, Amsterdam.

Hilbert, D. (1899), *Grundlagen der Geometrie*, Teubner, Leipzig; English translation *The Foundations of Geometry*, Open Court, Chicago, 1902.

—— (1922), "Neubegründung der Mathematik: Erste Mitteilung", *Abhandlungen aus dem Seminar der Hamburgischen Universität*, 1, 1922, 157–77; also in Hilbert (GA), vol. 3, 157–77; English translation in Ewald (1996), 1115–34 and Mancosu (1998), 198–214.

—— (1926), "Über das Unendliche", *Mathematische Annalen*, 95, 161–90; also in the 7th edition of Hilbert (1899), 1930; English translation *On the Infinite*, in van Heijenoort (1967), 367–92.

—— (GA), *Gesammelte Abhandlungen*, Springer, Berlin, 1932–35, 3 vols., I vol. 1932, II vol. 1933, III vol. 1935.

Hilbert, D. and Bernays, P. (1934–39), *Grundlagen der Mathematik*. Springer, Berlin, 2 vols., vol. I 1934, vol. II 1939.

Hintikka, J. (1968), "The Varieties of Information and Scientific Explanation", in B. van RootSelaar and J.F. Staal (eds), *Logic, Methodology and Philosophy of Science*, III, North Holland Publishing Company, Amsterdam, 311–31.

—— (1996), *Lingua Universalis vs. Calculus Ratiocinator: An Ultimate Presupposition of Twentieth-century Philosophy*, Springer, Berlin.

Hodes, H.T. (1990), "Ontological Commitments: Thick and Thin", in G. Boolos (ed.), *Meaning and Method: Essays in Honor of Hilary Putnam*, Cambridge University Press, Cambridge, 347–407.

Hodges, W. (1993), *Model Theory*, Cambridge University Press, Cambridge, 1993, abridged version: *A Shorter Model Theory*, Cambridge University Press, Cambridge, 1997.

Hofweber, T. (2005a), "A Puzzle About Ontology", *Noûs*, 39, 256–83.

—— (2005b), "Number Determiners, Numbers, and Arithmetic", *The Philosophical Review*, 114, 179–225.

—— (2007), "Innocent Statements and Their Metaphysically Loaded Counterparts", *The Philosophers' Imprint*, 7, 1–33.

Horwich, P. (1991), "On the Nature and Norms of Theoretical Commitment", *Philosophy of Science*, 58, 1–14.

Howson, C. and Urbach, P. (1989), *Scientific Reasoning. The Bayesian Approach*, Open Court, La Salle.

Hughes, G. and Cresswell, M. (1996), *A New Introduction to Modal Logic*, Routledge, London.

Hume, D. (1739–40), *A Treatise of Human Nature*, (ed.) L.A. Selby-Bigge, 2nd revised edition by P.H. Nidditch, Clarendon Press, Oxford, 1975.

Hussey, E. (1991), "Aristotle on Mathematical Objects", *Apeiron*, 24, 105–34.

Irvine, A. (1989), *Epistemic Logicism and Russell's Regressive Method*, in "Philosophical Studies", 55, 3, pp. 303–27.

—— (1990), (ed.), *Physicalism in Mathematics*, Kluwer A.P., Dordrecht, Boston, London.

Jackson, F. (1980), "Ontological Commitment and Paraphrase," *Philosophy*, 55, 303–15.

Jacob, P. (1980), *L'Empirisme logique*, Editions de Minuit, Paris.

Kalderon, M.E. (2005) (ed.), *Fictionalism in Metaphysics*, Oxford University Press, Oxford, New York.

Kanamori, A. (2004), *The Higher Infinite: Large Cardinals in Set Theory from Their Beginnings*, Springer, Berlin (2nd enlarged and corrected edition; 1st edition 1994).

Kant, I. (AK), *Gesammelte Schriften*, hrgb. von der Königlich Preussischen (Deutschen) Akademie der Wissenschaften, Berlin, Leipzig, 1900–sgg., 29 vols.
—— (KRV), *Kritik der reinen Vernunft*, 1st edition [A] 1781 in Kant (AK), vol. IV, 2nd edition [B] 1787 in Kant (AK), vol. III; also in *Kritik der reinen Vernunft* [editions A and B], hrsg. von Ingeborg Heidemann, Reclam, Stuttgart 1995; English translation *Critique of Pure Reason*, (ed.) P. Guyer and A. Wood, Cambridge University Press, Cambridge, 1997.
—— (LB), *Logik Blomberg*, in Kant (AK), vol. XXIV (1–2), *Vorlesungen über Logik*; English translation *The Logic Blomberg*, in Kant (LL), Part I.
—— (LJ), *Logik. Ein Handbuch zu Vorlesungen (Logik Jäsche)*, in Kant (AK), vol. IX, *Logik*; English translation *The Jäsche Logic*, in Kant (LL), Part IV.
—— (WL), *Wiener Logik*, in Kant (AK), vol. XXIV (1–2), *Vorlesungen über Logik*; English translation *The Vienna Logic*, in Kant (LL), Part II.
—— (LL), *Lectures on Logic*, (ed.) J.M. Young, Cambridge University Press, Cambridge, 1992.
Kasa, I. (2010), "On Field's Epistemological Argument Against Platonism", *Studia Logica*, 96, 141–47.
Kenny, A. (1995), *Frege*, Penguin Books, London.
Keränen, J. (2001), "The Identity Problem of Realist Structuralism", *Philosophia Mathematica* (III), 9, 308–30.
Kitcher, P. (1983), *The Nature of Mathematical Knowledge*, Oxford University Press, Oxford, New York.
Kitcher, P. and Salmon, W. (eds) (1989), *Scientific Explanation, Minnesota Studies in the Philosophy of Science*, Vol. 13, University of Minnesota Press, Minneapolis.
Kleene, S.C. (1967), *Mathematical Logic*, John Wiley and Sons, Inc., New York, reprinted by Dover, New York, 2002.
Klein, J. (1934–6), "Die griechischste Logistik und die Entstehung der Algebra", *Quellen und Studien zur Geschichte der Mathematik, Astronomie und Physik*, Abteilung B, Studien, 3, 1934, 18–105, and 3, 1936, 122–235; modified English translation by E. Brann, *Greek Mathematical Thought and the Origin of Algebra*, MIT Press, Cambridge (MA), 1967*.
Kripke, S. (1980), *Naming and Necessity*. Harvard University Press, Cambridge (MA).
Kuipers, T. (1997), "The Carnap-Hintikka Programme in Inductive Logic," in M. Sintonen (ed.), *Knowledge and Inquiry: Essays on Jaakko Hintikka's Epistemology and Philosophy of Science*, Poznan Studies, vol. 51, Rodopi, Amsterdam, 1997, 87–99.
Künne, W. (2003), *Conceptions of Truth*, Oxford University Press, Oxford, New York.
Lakatos, I. (1968), "Changes in the Problem of Inductive Logic", in I. Lakatos (ed.), *The Problem of Inductive Logic*, North-Holland, Amsterdam.
—— (1970), "Falsification and the Methodology of Scientific Research Programmes", in I. Lakatos, A. Musgrave (eds), *Criticism and the Growth of Knowledge*, Cambridge University Press, Cambridge, reprinted with corrections 1976, 91–196.
—— (1976), "A Renaissance of Empiricism in the Recent Philosophy of Mathematics", *The British Journal for the Philosophy of Science*, 27, 201–23.
Landini, G. (2011), *Russell*, Routledge, London.
—— (2012), *Frege's Notations. What they are and how they mean*, Palgrave Macmillan, Basingstoke.

Largeault, J. (ed.) (1992), *Intuitionnisme et théorie de la démonstration*, Vrin, Paris.
Laudan, L. (1981), "A Confutation of Convergent Realism", *Philosophy of Science*, 48, 19–49.
Lear, J. (1982), "Aristotle's Philosophy of Mathematics", *Philosophical Review*, 91, 161–92.
Leitgeb, H. and Ladyman, J. (2008), "Criteria of Identity and Structuralist Ontology", *Philosophia Mathematica* (III), 16, 388–96.
Leng, M. (2002), "What's Wrong With Indispensability? (Or, The Case for Recreational Mathematics)", *Synthese*, 131, 395–417.
—— (2005a), "Revolutionary Fictionalism: A Call to Arms", *Philosophia Mathematica* (III), 13, 277–93.
—— (2005b), "Mathematical Explanation", in C. Cellucci and D. Gillies (eds), *Mathematical Reasoning, Heuristics and the Development of Mathematics*, King's College Publications, London, 167–89.
—— (2010), *Mathematics and Reality*, Oxford University Press, Oxford, New York.
Leng M., Paseu A. and Potter M. (2007) (eds), *Mathematical Knowledge*, Oxford University Press, Oxford, New York.
Leslie, A. (1987), "Pretence and Representation: The Origins of the 'Theory of Mind'", *Psychological Review*, 94, 412–26.
—— (1994), "Pretending and Believing: Issues in the ToMM", *Cognition*, 50, 211–38.
Lewis, D. (1986), *On the Plurality of Worlds*, Blackwell Press, Oxford.
Liggins, D. (2006), "Is There a Good Epistemological Argument Against Platonism?", *Analysis*, 66, 135–41.
—— (2007), "Quine, Putnam, and the 'Quine-Putnam' Indispensability Argument", *Erkenntnis*, 68, 113–27.
—— (2010a), "The Autism Objection to Pretence Theories", *Philosophical Quarterly*, 60, 764–82.
—— (2010b), "Epistemological Objections to Platonism", *Philosophy Compass*, 5, 67–77.
Linnebo, Ø. (2003), "Plural Quantification Exposed", *Noûs*, 37, 71–92.
—— (2006), "Epistemological Challenges to Mathematical Platonism", *Philosophical Studies*, 129, 545–74.
—— (2008), "Plural Quantification", *The Stanford Encyclopedia of Philosophy* (Spring 2009 Edition), Edward N. Zalta (ed.), URL: <http://plato.stanford.edu/archives/spr2009/entries/plural-quant/>
Linsky, B. and Zalta, E. (1995), "Naturalized Platonism versus Platonized Naturalism", *The Journal of Philosophy*, 92, 525–55.
—— (2006), "What is Neo-logicism?", *The Bulletin of Symbolic Logic*, 12, 60–99.
Lipton, P. (2002), *Inference to the Best Explanation*, Routledge, London, 2nd revised edition; 1st edition 1991.
MacBride, F. (1999), "Listening to Fictions: A Study of Fieldian Nominalism", *British Journal for the Philosophy of Science*, 50, 431–55.
MacBride, F. (2003), "Speaking with Shadows: A Study of Neo-Logicism", *British Journal for the Philosophy of Science*, 54, 103–63.
MacFarlane, J. (2002), "Frege, Kant, and the Logic in Logicism", *The Philosophical Review*, 111, 25–65.
Mackie, J.L. (1977), *Ethics: Inventing Right and Wrong*, Penguin.

—— (2006), *How Things Might Have Been: Individuals, Kinds, and Essential Properties*, Oxford University Press, Oxford, New York.
MacLane, S. (1997), *Categories for the Working Mathematician*, Springer, Dordrecht.
Maddy, P. (1980), "Perception and Mathematical Intuition", *Philosophical Review*, 89, 163–96.
—— (1989), "The Roots of Contemporary Platonism", *The Journal of Symbolic Logic*, 54, 1121–44.
—— (1990a), *Realism in Mathematics*, Oxford University Press, Oxford, New York.
—— (1990b), "Physicalistic Platonism", in A. Irvine (ed.), *Physicalism in Mathematics*, Kluwer Academic Publishers, 259–89.
—— (1992), "Indispensability and Practice", *The Journal of Philosophy*, 89, 275–89.
—— (1994), "Taking Naturalism Seriously", in D. Prawitz, B. Skyrms, and D. Westerståhl (eds), *Logic, Methodology and Philosophy of Science*, IX, 383–407.
—— (1997), *Naturalism in Mathematics*, Clarendon Press, Oxford.
—— (1998), "How to be a Naturalist about Mathematics", in H.G. Dales and G. Oliveri (eds), *Truth in Mathematics*, Clarendon Press, Oxford, 161–80.
—— (2005), "Three Forms of Naturalism", in S. Shapiro (2005), 437–60.
—— (2007), *Second Philosophy*, Oxford University Press, Oxford, New York.
—— (2011), *Defending the Axioms*, Oxford University Press, Oxford, New York.
Malament, D. (1982), "Review of Hartry Field's *Science Without Numbers*", *The Journal of Philosophy*, 79, 523–34.
Mancosu, P. (1998) (ed.), *From Brouwer to Hilbert. The Debate on the Foundations of Mathematics in the 1920s*, Oxford University Press, Oxford, New York.
—— (2001), "Mathematical Explanation: Problems and Prospects", *Topoi*, 20, 97–117.
—— (2008a) (ed.), *The Philosophy of Mathematical Practice*, Oxford University Press, Oxford, New York.
—— (2008b), "Mathematical Explanation: Why it Matters", in P. Mancosu (ed.), 2008, *The Philosophy of Mathematical Practice*, Oxford University Press, Oxford, New York.
—— (2008c), "Quine and Tarski on Nominalism", *Oxford Studies in Metaphysics*, vol. IV, 2008, 22–55.
—— (2011) "Explanation in Mathematics", *The Stanford Encyclopedia of Philosophy* (Summer 2011 Edition), Edward N. Zalta (ed.), URL: <http://plato.stanford.edu/archives/sum2011/entries/mathematics–explanation/>.
Martin, W. (2006), *Theories of Judgment: Psychology, Logic, Phenomenology*, Cambridge University Press, Cambridge.
McGee, V. (1997), "How We Learn Mathematical Language", *The Philosophical Review*, 106, 35–68.
McGrath, M. (2007), "Propositions", *The Stanford Encyclopedia of Philosophy* (Fall 2008 Edition), Edward N. Zalta (ed.), URL: <http://plato.stanford.edu/archives/fall2008/entries/propositions/>.
Melia, J. (1995), "On What There's Not", *Analysis*, 55, 223–29.
—— (2000), "Weaseling Away the Indispensability Argument", *Mind*, 109, 455–79.
—— (2002), "Response to Colyvan", *Mind*, 111, 75–9.
—— (2006), "The Conservativeness of Mathematics", *Analysis*, 66, 202–08.
Mendelson, E. (1964), *Introduction to Mathematical Logic*, Van Nostrand Reinhold, New York, 4th edition, 1997.

Meyer, G. (2009), "Extending Hartry Field's Instrumental Account of Applied Mathematics to Statistical Mechanics", *Philosophia Mathematica* (III), 17, 273–312.

Mill, J.S. (1843), *A System of Logic, Ratiocinative and Inductive*. Longmans, London.

Monton, B. (ed.) (2007), *Images of Empiricism: Essays on Science and Stances. With a Reply from Bas C. van Fraassen*, Oxford University Press, Oxford, New York.

Mueller, I. (1970), "Aristotle on Geometrical Objects", *Archiv für Geschichte der Philosophie*, 52, 156– 71.

Nagel, E. and Newman, J.R. (1958), *Gödel's Proof*, NYU Press, New York.

Neumann, von J. (1923), "Zur Einführung der transfiniten Zahlen", *Acta Litterarum ac Scientiarum: Sectio Scientiarum Mathematicarum*, 1, 199–208; English translation "On the Introduction of Transfinite Numbers", in van Heijenoort (1967), 346–54.

—— (1947), "The Mathematician", in R.B. Heywood (ed.), *The Works of the Mind*, University of Chicago Press, Chicago, 1947, 180–96; also in *John von Neumann. Collected Works*, (ed.) A.H. Taub, Pergamon Press, Oxford, 1961*, vol. 1, 1–9.

Panza, M. (1997), "Mathematical Acts of Reasoning as Synthetic *a priori*", in M. Otte M., M. Panza (eds), *Analysis and synthesis in Mathematics*, Kluwer, Dordrecht, Boston, London, 261–326.

—— (2012), "The Twofold Role of Diagrams in Euclid's Plane Geometry", in M., J. Mumma, M. Panza and G. Sandu (eds), *Diagrammatic Reasoning in Mathematics*, special issue of *Synthese*, 186, n. 1, 2012, 55–102.

Papineau, D. (1988), "Mathematical Fictionalism", *International Studies in the Philosophy of Science*, 2, 151–73.

—— (1993), *Philosophical Naturalism*, Basil Blackwell, Oxford.

—— (2007), "Naturalism", *The Stanford Encyclopedia of Philosophy* (Spring 2009 Edition), Edward N. Zalta (ed.), URL: <http://plato.stanford.edu/archives/spr2009/entries/naturalism/>.

Parsons, C. (1964), "Frege's Theory of Number", in M. Black (ed.), *Philosophy in America*, Allen & Unwin, London, 180–203; reprinted in W. Demopoulos (ed.), *Frege's Philosophy of Mathematics*, Harvard University Press, Cambridge (MA), 1995, 182–210.

—— (1979–80), "Mathematical Intuition", *Proceedings of the Aristotelian Society*, New Series, 80, 145–68, in Hart (1996), 95–113.

—— (1982), "Objects and Logic", *The Monist*, 65, 91–516.

—— (1983a), *Mathematics in Philosophy: Selected Essays*, Cornell University Press, Ithaca.

—— (1983b), "Quine on the Philosophy of Mathematics", in Parsons (1983a), Chapter 7; also in L. Hahn, P. Schilpp (eds), *The Philosophy of W.V. Quine*, Open Court, La Salle, (Ill.), 369–95.

—— (1990), "The Structuralist View of Mathematical Objects", *Synthese*, 84, 303–46.

—— (1998), "Finitism and Intuitive Knowledge", in Schirn (1998), 251–70.

—— (2000), "Reason and Intuition", *Synthese*, 125, 299–315.

—— (2004), "Structuralism and Metaphysics", *The Philosophical Quarterly*, 54, 56–77.

—— (2008), *Mathematical Thought and Its Objects*, Cambridge University Press, Cambridge.

Paseu, A. (2007), "Scientific Realism", in Leng *et al.* (2007), 123–49.

Peano (1889), *Arithmetices Principia, Nova Methodo Exposita*, Bocca, Augustæ Taurinorum, 1998; also in G. Peano, *Opere scelte*, Cremonese, Roma, 1957–59, vol. II, 20–55.

Peressini, A. (2008), "Confirmational Holism and its Mathematical (W)holes", *Studies in History and Philosophy of Science*, 39, 102–11.

Petitot, J. (1995), "Pour un platonism transcendantal", in M. Panza and J.M. Salanskis (eds), *L'objectivité mathématique. Platonismes et structures formelles*, Masson, Paris, 147–78.

Pettigrew, R. (2009), "Aristotle on the Subject Matter of Geometry", *Phronesis* 54, 239–60.

Pincock, C. (2004), "A Revealing Flaw in Colyvan's Indispensability Argument", *Philosophy of Science*, 71, pp. 61–79.

—— (2011), "On Batterman's 'On the Explanatory Role of Mathematics in Empirical Science'", *British Journal for the Philosophy of Science*, 62, 211–17.

—— (2012), *Mathematics and Scientific Representation*, Oxford University Press, Oxford, New York.

Plantinga, A. (1974), *The Nature of Necessity*, Clarendon Press, Oxford.

Plato (PO), *Platonis opera, recognovit brevique adnotatione critica instruxit Ioannis Burnet*, 5 Vols., Oxford 1900–07. References in the text follow numeration in Plato, *Platonis opera quae extant omnia, excudebat H. Stephanus*, 3 vols., Geneva 1578, which is indicated in brackets in the following lemmas, together with the corresponding volume).

—— (Epist.), *Epistulae*, in Platone (PO), vol. V (St. III, 309a–363e) (English translation: *Letters, in Plato in Twelve Volumes*, vol. 7, translated by R.G. Bury. Cambridge (MA), Harvard University Press; London, William Heinemann Ltd, 1966).

—— (Gorg.), *Gorgias*, in Platone (PO), vol. III (St. I, 447a–527e) (English translation: *Letters, in Plato* (PTW), Vol. 7, by R.G. Bury, 1966).

—— (Hipp. Maj.), *Hippias Maior*, in Platone (PO), vol. III (St. III, 281a–304e) (English translation: *Gorgias*, in Plato (PTW), vol. 3, by W.R.M. Lamb., 1967).

—— (Men.), *Meno*, in Platone (PO), vol. III (St. II, 70a–100b) (English translation: *Meno*, in Plato (PTW), Vol. 3, by W.R.M. Lamb, 1967).

—— (Phaed.), *Phaedrus*, in Platone (PO), vol. II (St. III, 227a–279c) (English translation: *Phaedrus*, in Plato (PTW), vol. 9, by Harold N. Fowler, 1966).

—— (Rep.), *Respublica*, in Platone (PO), vol. IV (St. II, 327a–621d) (English translation: *The Republic*, in Plato (PTW), vols. 5 and 6, by Paul Shorey, 1969).

—— (Theaet.), *Thaetetus*, in Platone (PO), vol. I (St. I, 142a–210d) (English translation: *Theaetetus*, in Plato (PTW), vol. 12, by Harold N. Fowler, 1921).

—— (PTW), *Plato in Twelve Volumes*, 12 Vols., Harvard University Press, Cambridge (MA), William Heinemann Ltd., London.

Poincaré, H. (1900), *La valeur de la science*, Flammarion, Paris; English translation *The Value of Science*, in Poincaré (1913), 201–55.

—— (1902), *La science et l'hypothèse*, Flammarion, Paris; English translation *Science and Hypothesis*, in Poincaré (1913), 9–197.

—— (1906), *Science et méthode*, Flammarion, Paris; English translation *Science and Method*, in Poincaré (1913), 359–546.

—— (1913), *The Foundations of Science*, The Science Press, Lancaster (PA).

Popper, K. (1934), *Logik der Forschung*, Mohr Siebeck, Tübingen; English translation *The Logic of Scientific Discovery*, Hutchinson, London, 1959.

Potter, M. (2002), *Reason's Nearest Kin: Philosophies of Arithmetic from Kant to Carnap*, Oxford University Press, Oxford, New York.

—— (2004) *Set Theory and Its Philosophy. A Critical Introduction*, Oxford University Press, Oxford, New York.

Proclus (CEM) (1992), *A Commentary on the First Book of Euclid's* Elements, translated and annotated by G.R. Morrow, with a Preface by I. Mueller, Princeton University Press, Princeton NJ).

Psillos, S. (1999), *Scientific Realism: How Science Tracks Truth*, Routledge, London.

Putnam, H. (1962), "What Theories Are Not", in E. Nagel, P. Suppes and P. A. Tarski (eds), *Logic, Methodology and Philosophy of Science; Proceedings of the 1960 International Congress*, Stanford University Press, Stanford (CA), 240–51; also in Putnam (1975b), Chapter 13.

—— (1965), "Craig's Theorem", *The Journal of Philosophy*, 62, 251–60; also in Putnam (1975b), Chapter 14.

—— (1967), "Mathematics without Foundations", *The Journal of Philosophy*, 64, 5–22*; also in Putnam (1975b), Chapter 3; also in Benacerraf, Putnam (1973), 295–311.

—— (1971), *Philosophy of Logic*, Harper & Row, New York; also in Putnam (1975b), Chapter 20*.

—— (1975a), "What is Mathematical Truth?", *Historia Mathematica*, 2, 529–43; also in Putnam (1975b), Chapter 6.

—— (1975b), *Mathematics Matter and Method: Philosophical Papers Vol. 1*, Cambridge University Press, Cambridge, 2nd edition, 1985.

—— (1983), "Equivalence", originally published in Italian as lemma 'Equivalenza' in Enciclopedia Einaudi, Giulio Einaudi Editore, 5, 547–64; also in H. Putnam, *Realism, and Reason, Philosophical Papers*, vol. III, Cambridge University Press, Cambridge, Chapter 2.

—— (2004), *Ethics Without Ontology*, Harvard University Press, Cambridge (MA).

—— (2012), "Indispensability Arguments in the Philosophy of Mathematics", in H. Putnam, *Philosophy in an Age of Science*, (ed.) M. De Caro and D. Macarthur, Cambridge (MA), Harvard University Press, 2012, 181–201.

Quine, W.V.O. (1936), "Truth by Convention", (ed.) O.H. Lee, *Philosophical Essays for A.N. Whitehead*, Longmans, New York, 90–124; also in Quine (1966a), Chapter 1.

—— (1939), "Designation and Existence", *The Journal of Philosophy*, 36, 701–709; also in F. Feigl and W. Sellars (eds), *Readings in Philosophical Analysis*, Appleton-Century-Crofts, Inc., New York.

—— (1948), "On What There Is", *Review of Metaphysics*, 2, 21–38; also in Quine (1953a)*, Chapter 1.

—— (1950), *Methods of Logic*, Harvard University Press, Cambridge (MA), 4th edition 1982.

—— (1951a), "Two Dogmas of Empiricism", *Philosophical Review*, 60, 20–43; also in Quine (1953a)*, Chapter 2.

—— (1951b), "On Carnap's Views on Ontology", *Philosophical Studies*, 2, 65–72; also in Quine (1966a), Chapter 19.

—— (1953a), *From a Logical Point of View*, Harper & Row, New York, 2nd edition 1961.

—— (1953b), "Logic and the Reification of Universals", in Quine (1953a), Chapter 6.

―― (1954), "Carnap and Logical Truth", first published in *Synthese*, 12, 1960, 350–74; also in Quine, 1966a*, Chapter 12.
―― (1955), "Posits and Reality", in S. Uyeda (ed.), *Basis of Contemporary Philosophy*, Waseda University Press, Tokyo; also in Quine (1966a), Chapter 23.
―― (1958), "Speaking of Objects", *Proceedings and Addresses of the American Philosophical Association*, 31, 5–22; also in Quine, 1969a, Chapter 1.
―― (1963), *Set Theory and its Logic*, Harvard University Press, Cambridge (MA), second edition 1969.
―― (1960), *Word and Object*, MIT Press, Cambridge (MA).
―― (1966a), *The Ways of Paradox and Other Essays*, Columbia University Press, New York, 2nd revised edition, 1976.
―― (1966b), "The Scope and Language of Science", in Quine (1966a)*, Chapter 22.
―― (1968), "Ontological Relativity", *Journal of Philosophy*, 65, 185–212 in Quine (1969a), Chapter 2.
―― (1969a), *Ontological Relativity and Other Essays*, Columbia University Press, New York and London.
―― (1969b), "Epistemology Naturalized", in Quine (1969a), Chapter 3.
―― (1969c), "Existence and Quantification", *L'Âge de la Science*, 1, 151–64; also in Quine, 1969a*, Chapter 4.
―― (1970), *Philosophy of Logic*, Harvard University Press, Cambridge (MA), 2nd edition 1986.
―― (1975), "On Empirically Equivalent Systems of the World", *Erkenntins*, 9, 313–28.
―― (1981a), "Things and Their Place in Theories", in Quine (1981e), 1–23.
―― (1981b), "Five Milestones of Empiricism", in Quine (1981e), 67–72.
―― (1981c), "Success and Limits of Mathematization", in Quine (1981e)*, 148–55.
―― (1981d), "Reply to Chihara", in P.A. French, T. Uehling and H. Wettstein (eds), *The Foundations of Analytic Philosophy*, University of Minnesota Press, Minneapolis, 453–5.
―― (1981e), *Theories and Things*, Harvard University Press, Cambridge (MA).
―― (1986), "Reply to Charles Parsons", in L. Hahn and P. Schilpp (eds), *The Philosophy of W.V. Quine*, Open Court, La Salle, (IL), 396–403.
―― (1990), *Pursuit of Truth*, Harvard University Press, Cambridge (MA), 2nd revised edition 1992.
―― (1993), "In Praise of Observation Sentences", *The Journal of Philosophy*, 90, 107–16.
―― (1995), *From Stimulus to Science*. Harvard University Press, Cambridge (MA).
Quine, W.V.O. and Ullian, J.S. (1978), *The Web of Belief*, McGraw-Hill, New York, 2nd edition 1970.
Ramsey, F.P. (1926), "The Foundations of Mathematics", *Proceedings of the London Mathematical Society*, II, 338–84; also in F.P. Ramsey, *Philosophical Papers*, Cambridge University Press, Cambridge, 164–224.
Rayo, A., Yablo, S. (2001), "Nominalism through De-nominalization", *Noûs*, 35, 74–92.
Reichenbach, H. (1951), *The Rise of Scientific Philosophy*, University of California Press, Berkeley, Los Angeles, London.
Resnik, M.D. (1981), "Mathematics as a Science of Patterns: Ontology and Reference", *Noûs*, 15, 529–50.

—— (1983), "Review of Hartry Field's *Science without Numbers*", *Noûs*, 17, 514–19.
—— (1985), "How Nominalist is Hartry Field's Nominalism", *Philosophical Studies* 47, 163–81.
—— (1992), "Applying Mathematics and the Indispensability Argument", in J.Echeverria, A. Ibarra, and T. Morman (eds), *The Space of Mathematics*, Walter de Gruyter, Berlin, 115–31.
—— (1995), "Scientific vs Mathematical Realism: The Indispensability Argument", *Philosophia Mathematica* 3, 166–74.
—— (1997), *Mathematics as a Science of Patterns*, Clarendon Press, Oxford.
—— (2005), "Quine and the Web of Belief', in Shapiro (2005), 412–37.
Richardson, A. and Übel, T. (eds) (2007), *The Cambridge Companion to Logical Empiricism*, Cambridge University Press, Cambridge, New York.
Ritchie, J. (2009), *Understanding Naturalism*, Acumen Publishing, Durham.
Rosen, G. (2001), "Abstract Objects", *The Stanford Encyclopedia of Philosophy* (Fall 2009 Edition), Edward N. Zalta (ed.), URL: <http://plato.stanford.edu/archives/fall2009/entries/abstract–objects/>.
Ross, S. (1979), *A First Course in Probability Theory*, Prentice Hall, Upper Saddle River (NJ), 8th edition 2008.
Rothmaler, P. (2000), *Introduction to Model Theory*, Gordon and Breach, Amsterdam.
Russell, B. (1903), *The Principles of Mathematics*, Allen & Unwin, London, 2nd edition 1938.
—— (1905), "On Denoting", *Mind*, 14, 479–93.
—— (1906), *Les paradoxes de la logique,* in "Revue de Métaphysique et de Morale", XIV, pp. 627–50 (English translation *On "insolubilia" and Their Solution by Symbolic Logic,* in Russell, 1973, Chapter 9).
—— (1907), "The Regressive Method of Discovering the Premises of Mathematics", in Russell (1973, Chapter 13).
—— (1908), "Mathematical Logic as Based on the Theory of Types", *American Journal of Mathematics*, 30, 222–62.
—— (1918–19), "The Philosophy of Logical Atomism", *The Monist*, 28 (1918)–29 (1919); also in B. Russell, *The Collected Papers of Bertrand Russell*, vol. III, *The Philosophy of Logical Atomism and Other Essays*, (ed.) J.G. Slater, Allen & Unwin, London, 1914–19.
—— (1973), *Essays in Analysis*, Allen & Unwin, London.
Russell, B. and Whitehead, A.N. (1910–13), *Principia Mathematica*, 3 vols, Cambridge, Cambridge University Press, 2nd edition 1925 (vol. 1), 1927 (vols 2, 3).
Saatsi, J. (2007), "Living in Harmony: Nominalism and the Explanationist Argument for Realism", *International Studies in the Philosophy of Science*, 21, 19–33.
—— (2011), "The Enhanced Indispensability Argument: Representational versus Explanatory Role of Mathematics in Science", *British Journal for the Philosophy of Science*, 62, 143–54.
Sainsbury, R.M. (2010), *Fiction and Fictionalism*, Routledge, London, New York.
Salmon, W. (1989), *Four Decades of Scientific Explanation,* University of Minnesota Press, Minneapolis.
Schirn, M. (ed.) (1998), *Philosophy of Mathematic Today*, Clarendon Press, Oxford.
Shapiro, S. (1983), "Conservativeness and Incompleteness", *The Journal of Philosophy*, 80, 521–31.

—— (1997), *Philosophy of Mathematics. Structure and Ontology*, Oxford University Press, Oxford, New York.
—— (2000a), *Thinking about Mathematics: The Philosophy of Mathematics*, Oxford University Press, Oxford, New York.
—— (2000b), "Frege Meets Dedekind: A Neo-logicist Treatment of Real Analysis", *Notre Dame Journal of Formal Logic*, 41, 335–64.
—— (2005a) (ed.), *Oxford Handbook of Philosophy of Mathematics and Logic*, Oxford University Press, Oxford, New York.
—— (2005b), "Higher-order Logic", in S. Shapiro (2005), Chapter 25.
—— (2006a), *Structure and Identity*, in F. MacBride (ed.), *Identity and Modality*, Oxford University Press, Oxford, New York, 109–45.
—— (2006b), "The Governance of Identity", in F. MacBride (ed.), *Identity and Modality*, Oxford University Press, Oxford, New York, 164–73.
—— (2008), "Identity, Indiscernibility, and *ante rem* Structuralims: The Tale of *i* and *–i*", *Philosophia Mathematica* (III), 16, 285–309.
—— (2009) "A Scientific Enterprise? Penelope Maddy's Second Philosophy", *Philosophia Mathematica (III)*, 71, 247–71.
Shapiro, S. and Weir, A. (2000), "'Neo-logicist' Logic is not Epistemologically Innocent", *Philosophia Mathematica* (III), 8, 160–89.
Shoenfield, J.R. (1967) *Mathematical Logic*, Addison-Wesley Publishing Co., Reading (MA), London, Don Mills, Ont., 1967, reprinted by AK Peters, Ltd, Natick (MA), 2001.
Skolem, T. (1923), "Begründung der elementaren Arithmetik durch die rekurrierende Denkweise ohne Anwendung scheinbare Veränderlichen mit unendlichem Ausdehnungsbereich", *Videnskapsselkapet skrifter Kristiania, I. Mathematisk – naturvidenskabelig klasse*, 6; English translation in van Heijenoort (1967), 302–33.
Smullyan, R. (1992), *Gödel's Incompleteness Theorems*, Oxford University Press, Oxford.
Smith, P. (2007), *An Introduction to Gödel's Theorem*, Cambridge University Press, Cambridge, 2007, reprinted with corrections (2008).
Sober, E. (1993), "Mathematics and Indispensability", *The Philosophical Review*, 102, 35–57.
Stadler, F. (1997), *Studien zum Wiener Kreis. Ursprung, Entwicklung und Wirkung des Logischen Empirismus im Kontext*, Suhrkamp, Frankfurt a. M.; English translation *The Vienna Circle. Studies in the Origins, Development and Influence of logical Empiricism*, Springer, Vienna, New York, 2001.
Stanley, J. (2001), "Hermeneutic Fictionalism", *Midwest Studies in Philosophy*, XXV, 2001, 36–71.
Steiner, M. (1978), "Mathematics, Explanation, and Scientific Knowledge", *Noûs*, 12, 17–28.
—— (1989), "The Application of Mathematics to Natural Science", *The Journal of Philosophy*, 86, 449–80.
—— (1998), *The Applicability of Mathematics as a Philosophical Problem,* Harvard University Press, Cambridge (MA).
—— (2005), "Mathematics: Application and Applicability", in Shapiro (2005), 625–50.
Stigt, W.P. van (1990), *Brouwer's Intuitionism*, Studies in the History and Philosophy of Mathematics, vol. 2, North Holland.

Stolz, O. (1885), *Vorlesungen über Allgemeine Arithmetik*, Teubner, Leipzig.
Strawson, P. (1966), *The Bounds of Sense: An Essay on Kant's Critique of Pure Reason*, Methuen, London.
Suppe, F. (1977), "The Search for Philosophical Understanding of Scientific Theories", in F. Suppe (ed.), *The Structure of Scientific Theories*, University of Illinois Press, 2nd edition revised and enlarged, 3–241, 1st edition 1974.
Suppes, P. (1972), *Axiomatic Set Theory*, Dover, New York.
Tait, W.W. (1986), "Truth and Proof: The Platonism of Mathematics", *Synthese*, 69, 341–70; also in Tait W.W., *The Provenance of Pure Reason. Essays in Philosophy of Mathematics and its History*, Oxford University Press, Oxford, New York, 2005, 61–88.
Thomas, R. (2000), "Mathematics and Fiction I: Identification", *Logique et Analyse*, 43, 171–2, 301–40.
—— (2002), "Mathematics and Fction II: Analogy", *Logique et Analyse*, 45, 177–8, 185–228.
Tieszen, R. (1992), "Kurt Gödel and Phenomenology", *Philosophy of Science*, 59, 176–94.
—— (1995), "Mathematics", in B. Smith and D.W. Smith (eds), *The Cambridge Companion to Husserl*, Cambridge University Press, Cambridge.
—— (2005), *Phenomenology, Logic, and the Philosophy of Mathematics*, Cambridge University Press, Cambridge.
Varzi, A. (2009), "Mereology", *The Stanford Encyclopedia of Philosophy* (Summer 2009 Edition), Edward N. Zalta (ed.), URL: <http://plato.stanford.edu/archives/sum2009/entries/mereology/>.
Vineberg, S. (1996), "Confirmation and the Indispensability of Mathematics to Science", *Philosophy of Science*, 63: Supplement, S256–S263.
Walton, K. (1990), *Mimesis As Make-believe: On the Foundations of the Representational Arts*, Harvard University Press, Harvard.
—— (1993), "Metaphor and Prop Oriented Make-believe" *European Journal of Philosophy* 1, 39–57; also in Kalderon, 2005, 65–87.
Watkins, J. (1958), "Confirmable and Influential Metaphysics", *Mind*, 67, 344– 65.
Weir, A. (2005), "Naturalism Reconsidered", in S. Shapiro (2005), 460–82.
White, N. (1974), "What Numbers Are", *Synthese*, 27, 111–24; also in D. Giaquette (ed.), *Philosophy of Mathematics, An Anthology*, Wiley Blackwell, Oxford, 2001, 91–8.
Wiles, A. (1995), "Modular Elliptic Curves and Fermat's Last Theorem", *Annals of Mathematics*, 141, 443–551.
Woodin, H. (2001a), "The Continuum Hypothesis, Part I", *Notices of the AMS*, 48, 567–76.
—— (2001b), "The Continuum Hypothesis, Part II", *Notices of the AMS*, 48, 681–90.
Woodward, J. (2009), "Scientific Explanation", *The Stanford Encyclopedia of Philosophy* (Spring 2009 Edition), Edward N. Zalta (ed.), URL: <http://plato.stanford.edu/archives/spr2009/entries/scientific-explanation/>.
Wright, C. (1983), *Frege's Conception of Numbers as Objects*, Aberdeen University Press, Aberdeen.
—— (1990), "Field and Fregean Platonism", in A. Irvine (1990), 73–93; also in Hale and Wright (2001), 153–68.

—— (1997), "On the Philosophical Significance of Frege's Theorem", in R. Heck Jr. (ed.), *Language, Truth and Logic: Essays in Honour of Michael Dummett*, Clarendon Press, Oxford, 201–44; also in Hale and Wright (2001), 272–306.

—— (1998a), "Response to Dummett", in Schirn (1998), 389–405; also in Hale and Wright (2001), 256–71.

—— (1998b), "On the Harmless Impredicativity of $N^=$", in Schirn (1998), 339–68; also in Hale and Wright (2001), 229–55.

—— (1998c), "Why Frege Does Not Deserve his Grain of Salt: A Note on the Paradox of "The Concept Horse" and the Ascription of *Bedeutungen* to Predicates", in *Grazer Philosophische Studien*, 55, *New Essays on the Philosophy of Michael Dummett*, (ed.) J. Brandl and P. Sullivan, Rodopi, Amsterdam, Atlanta, 1998, 239–63; also in Hale and Wright (2001), 72–90.

—— (1999), "Is Hume's Principle Analytic?", *Notre Dame Journal of Formal Logic*, 40, 6–30; also in Hale and Wright (2001), 307–32.

—— (2000), "Neo-Fregean Foundations for Real Analysis", *Notre Dame Journal of Formal Logic*, 41, 317–34.

Yablo, S. (1998), "Does Ontology Rest on a Mistake?", *Proceedings of the Aristotelian Society*, Supp. vol. 72, 229–61; also in Yablo (2010), 117–44.

—— (2001), "Go Figure: A Path Through Fictionalism", *Midwest Studies in Philosophy*, XXV, 72–102; also in Yablo (2010), 177–99.

—— (2002), "Abstract Objects: A Case Study", in A. Bottani, M. Carrara and P. Giaretta (eds) *Individuals, Essence and Identity. Themes of Analytic Metaphysics*, Topoi Library, vol. 4, Kluwer, Dordrecht, 2002*, 163–88; also in E. Sosa and E. Villanueva (eds), *Philosophical Issues 12: Realism and Relativism*, Blackwell, Oxford, 220–40; also in Yablo (2010), 200–20.

—— (2005), "The Myth of the Seventh", in Kalderon (2005), 88–115; also in Yablo (2010), 221–45.

—— (2006), "Non-catastrophic Presupposition Failure", in J.J. Thomson and A. Byrne (eds), *Content and Modality*, Oxford University Press, Oxford, New York, 164–90; also in Yablo (2010), 269–95.

—— (2009), "Must Existence-questions Have Answers?", in D. Chalmers, D. Manley and R. Wasserman (eds), *Metametaphysics*, Oxford University Press, Oxford, New York, 507–25; also in Yablo (2010), 296–314.

—— (2010), *Things*, Oxford University Press, Oxford, New York.

Zalta, E. (1983), *Abstract Objects: An Introduction to Axiomatic Metaphysics*, D. Reidel, Dordrecht.

—— (1999), "Natural Numbers and Natural Cardinals as Abstract Objects: A Partial Reconstruction of Frege's *Grundgesetze* in Object Theory", *Journal of Philosophical Logic*, 28, 619–60.

—— (2000), "Neo-logicism? An Ontological Reduction of Mathematics to Metaphysics", *Erkenntnis*, 53, 219–65.

—— (PM), *Principia Metaphysica*, draft available online at http://mally.stanford.edu/principia.pdf.

Zermelo, E. (1908), "Untersuchungen über die Grundlagen der Mengenlhere I", *Mathematische Annalen*, 65, 261–81; also in van Heijenoort (1967), 199–215.

Index

Notes: This index only points to the most salient passages where the relevant arguments of the listed authors are discussed. This is not meant, therefore, as an exhaustive index of names or of the occurrences of listed lemmas.

abstract structure of progression, 104
abstraction
   abstraction principles 157–8, 164–5 (*see also* Hume's Principle)
   in Aristotle, 29, 31
   of Dedekind, 192–3
   linguistic, 185
analysis (real and complex), 45–7 (*see also* real numbers, definition)
analytic philosophy, x–xi
Annas, J., 27–9
applicability of mathematics, ix, 113, 117–18, 121, note 1 ch. 6, note 3 ch. 7
   and conservativeness, 112–25
argument from pessimistic meta-induction, 241
Aristotle, 27–32, 33–4, 38
   on ideas, 27–8
   Metaphysics, 27–30, note 10, ch. 1
   Physics, 29, note 11 ch. 1
arithmetic
   applied, 130, 133
   finitary, 81–6 (*see also* mathematics, finitary)
   first-order, 139–40, 193, 227–8, 245, note 23 ch. 2, note 10 ch. 7
   Frege arithmetic (FA) *see* Frege
   primitive recursive (PRA), 193–4
   pure, 25
   second-order, 139, 158, note 33 ch. 2, note 10 ch. 7
arithmetization of analysis, 45–7
assert, xiii
atoms
   atomic theory, *see* theory
   mereological, 140–3, 222–3, note 11 ch. 4 (*see also* mereology)
autism objection, *see* Stanley

automorphism, 186–7
axioms, xv, 48 (*see also* Frege, Peano)
Azzouni, J., 230, 247, note 6 ch. 3, note 35 ch. 6, note 14 ch. 7

Baker, A., 199, 203, 209, note 13 ch. 6 (*see also* indispensability argument; explanation)
Balaguer, M., 135
   objection to Field, 127
   on easy road, *see* easy road
   on fictionalism, 125
Bedeutung, *see* reference
Benacerraf, P., viii, ch. 3, 112, 136, 162, 164, 176, 180, 181, 228, notes 2, 3, 6 ch. 3
Benacerraf's (1965) argument, 101–7
Benacerraf's (1973) argument, *see* Benacerraf's dilemma
Benacerraf's dilemma, chs 3, 4, 5
Bernays, P., 81, note 40 ch. 2.
Bolzano, B., 70
Boolos, G., 65, 141, 157, 160, 162, note 15 ch. 2, note 28 ch. 2
Bourbaki, N., 184–5
Brouwer, L. E. J., 87–90, 90–2, 190, 225, note 45 ch. 2 (*see also* intuitionism; pure two-oneness)
Bueno, O., 230, notes 11, 13 ch. 6, note 3. ch 7
Burgess, J., 122, 137, 141–2, 161, 192, 200, 230, 237–8, notes 12, 13, 14, 16 ch. 4, note 12, 20, 30 ch. 7
   objections to fictionalism, 135
   hermeneutic/revolutionary distinction, 128–9
Burnyeat M. F., 22–3, 29

296

Index 297

calculus of individuals, *see* mereology
canonical notation, 210
Cantor, G., 63, 68–9, 94, 98 (*see also* transfinite cardinals; sets; Continuum Hypothesis; real numbers; transfinite ordinals)
  Cantor's paradox, 68
  definition of real numbers, 63
Carnap, R., x, 73–5, 80–1, 87, 91, 134–5, 196, 225–6, 230, 234, 243, note 38 ch. 2 (*see also* existence, internal and external; linguistic *framework*)
Cauchy, A.L., 46 (*see also* arithmetization of analysis)
characteristic marks (of a concept), note 17 ch. 2
Chihara, C., 69, 121, 137, 215, 225, 242, note 38 ch. 2, note 13 ch. 4, note 19 ch. 7
class
  equivalence class, xiii, 79,
  class L of constructible sets, note 48 ch. 2
  no-class theory, 69, 92
  "class as many" vs "class as one", note 20 ch. 2
Cohen, P., 71, 93, note 36 ch. 2
Colyvan, M., 198, 201, 202–3, 204, 207–8, 215, 230, 247, note 13, 24, 32, 35, 36, ch. 6, note 3, 25 ch. 7 (*see also* indispensability argument; naturalism; holism)
concepts (*see also* Frege; Kant)
  concept ⌜(being a) natural number⌝, 62, 232
  conditions of application and identity (for objects supposed to fall under a concept), 57, 59, 155
  of level n, 51
  equinumerous, 55–8, 101, 154, 157
  numbers of concepts, *see* numbers
  pure, 42–3
  sortal, 5, 52, 57, 60, 62, 100–1, 152, 155, 163, 174
consistency proof in Hilbert, *see* formalism, Hilbert's

constraint
  application constraint or Frege's constraint, 163–5
  of conservativeness, 85
conditions of identity,
  and HP, 57
  for quasi-concrete objects, 190–1
  for sortal concepts, *see* concepts, sortal
confirmation, 229–30, 241–9, note 24 ch. 6,
  probabilistic conception, 243
confirmational holism, 214–15, 215–16, 243, note 33 ch. 6, notes 26, 28 ch. 7
  hypothetico-deductive model, 241–2
consequence of a theory, xv
  false nominalistic consequences, 114
  necessarily true nominalistic consequences, 114–15, note 3 ch. 4
conservativeness, 112–25, 217–24, note 2 ch. 7 (*see also* theory, conservative)
  Field's definition, 115–16, note 7 ch. 4
  according to Hilbert, 85
  Shapiro's objection, 121–2
  semantics vs syntactic, 121
consistent body of nominalistic / non-mathematical statements, 116–17, 217, 221, note 2 ch. 7
constants
  functional, 159, note 17 ch. 2, notes 8, 14 ch. 6
  individual, 12, 105, 192, 222–3, note 22 ch 6
  logical / non-logical, 48–9, 52–3, 61, 67, 77, 94, 174, 190, 211, note 15 ch. 2, note 14 ch. 6 (*see also* Frege, fundamental functions)
  predicate, 77, 106, 131, 166–8, 171, 174, 204, 222–3, note 8 ch. 5, note 16 ch. 6
content
  literal, 129–30, 135, 232, 234
  mathematical, 228–9, note 11 ch. 7
  real, 128, 130–4, 233–5, notes 20, 21 ch. 4

Continuum Hypothesis (CH), 7l–2, 93–5, 98, 127, note 48 ch. 2
contradiction in Frege's formal system, 64–6
Craig, W., *see* theorem, Craig's criterion of ontological commitment, 210–12, 224–35
criterion of virtuosity for a theory, 208–10, 212, 218, note 2 ch. 7

Dedekind, R., 63, 79–80, 137, 192–3 (*see also* abstraction, of Dedekind)
  definition of real numbers, 63
  Dedekind-infinite, 70
deductive rules, xiv, 47–8, 50, 83, note 5 ch. 2
  in GA, 53–4, 65
definite description symbol, 60, 168
definitions
  contextual, 101, 189, 206–7, 232
  creative, 75–6
  explicit, 48, 54–5, 59–60, 101, 152, 171–2, 180, 206
  impredicative, 69, 74, 91–2, note 38 ch. 2
  implicit, 56, 81–2, 120, 155, 161, 167, 180, 185, note 17 ch. 2
Demopoulos, W., 164
Detlefsen, M., 87, note 40 ch. 2
diagrams, 23, 32, 190
dianoia, 21–2, 33–6, 38, 194, note 3 ch. 1
  objects of (*dianoēta*), 34
discourse, figurative or metaphorical, 233–4
figurative or metaphorical language, 129, 134
double negation elimination, 89, note 46 ch. 2
doxa, 22
Dummett, M., 24, 61, 149, 163, note 2, 14, 20, 26, ch. 1
  definition of realism, 24, note 3 Introduction, note 9 ch. 6, note 5 ch.7
  on Russell's paradox, 64, 66, note 28 ch. 1

easy road for rejecting IA, 230
ε-equivalence, 209, 218–19
eikasia, 21–2
empirical sciences, ix, xiv, 142–3, 235, note 26 ch. 7
empiricism, 10–11
  contrastive, 242–5
  constructive, 242–3, note 22 ch. 7
  logical, 73–90, note 33 ch. 6
encoding, 166–72, 174, 176
entities
  saturated / unsaturated, 50
  theoretical, vii–viii, 118, 203, 235
epistēmē, 17
equinumerosity, *see* concepts, equinumerous
equivalence
  observational (*o*-equivalence), 207–9
  observational as different from empirical equivalence, 208–9
  and paraphrase (*p*-equivalence), 205–7
equivalence relation, note 3 ch. 5
equivalent descriptions, 227–30
ermēneia, see interpretation
Errera, A., 93
existence
  internal and external, 73–90, 134, 234
  of abstract objects, *see* objects, abstract
explanation (*see also* indispensability; inference to the best explanation)
  intrinsic and extrinsic, 221
extension (*Umfang*), 50, 57–61, 63, 65–7, 107, 156–8, note 18 ch. 2
Euclid, 32, 37, 45, 76, 83, 94–5, 190, note 6 ch. 1, note 24 ch. 7
  Elements, 11, 23, 26, 32, 37, 45, notes 6, 12 ch. 1
  (non-)Euclidean geometry, 63, 76, 78–9, 81, 94, 120, 123–4, 187, 190, 245
Euclid's fifth postulate, 94
Euler, L., 46, 50

façon de parler, 10, 57, 206
False (as an object), 51

fictionalism, 13–14, 125–35
  Field's fictionalism, 125–9
  hermeneutic vs revolutionary, 129, 135
  meta-fictionalism, 130
  as an answer to IA, 231–5
  object, 130–1
  Yablo's, 129–35
Field, H., 11–14, 106, 110, 112–25, 125–35 (*see also* conservativeness; fictionalism; nominalism; space-time points)
  on Benacerraf's dilemma, 109–10
  objections to the indispensability premise in IA, 217–24
field of reals, 81, 120
formal system, 47–54, 67, 90, note 5 ch. 2
formalism, Hilbert's, 81–6
  ideal / finitary distinction, 206, 85–7, 89
  metamathematics, 81–6, 191
  consistency proofs, 78–9, 81, 84–7, 90
formula (of a language), xii
formulation (of a statement), xiii
Fraassen, B. Van, 203, 240, 242–3, note 23 ch. 7 (*see also* constructive empiricism; scientific realism, objections; inference to the best explanation)
Fraenkel, A., 71 (*see also* sets)
Frege, G., 45–66, 149–65 (*see also* assertion; *Bedeutung*; definitions; False; Russell's paradox; Peano, axioms; reference; context principle; formal system; True)
  Frege arithmetic (FA), 161–3, 206, 232, note 3 ch. 2.
  axioms of GA, 52–4 (*see also* law, of *Grundgesetze*)
  Begriffsschrift, 47
  on concepts, 51
  on definitions against Hilbert, 78–80
  definition of natural numbers, 54–62
  definition of real numbers, 62–4
  concept/object distinction, 54
  psychological/logical distinction, 54
  fundamental functions, 52–3
  GA (Grundgesetze Arithmetic), 47, 47–54
  Grundgesetze der Arithmetik, 47, 49, 59–63, 65, 75–6, 78, 163–4, note 4 ch. 2
  Grundlagen der Arithmetik, 47, 54, 59–60, 62, 76, 78, 81, 101, 119, 153, 156, 158
  formal system, 47–54
fruitfulness (of new axioms), 95, 98 (*see also* successful applications)

games of make-believe, 134, 234
Gauss, C. F., 46 (*see also* arithmetization of analysis)
geometry (*see also* Plato, Aristotle, Proclus):
  Cartesian, 37, 45
  Euclidean, *see* Euclid
  Hilbert axiomatic foundation, 75–80
  hyperbolic, note 24 ch. 7
  non-Euclidean, 94, 245
Gentzen, G., 87
Gettier, E., 17
Gödel, K., x, 71, 86–7, 90–8, 109, 121, 145, 194–6, note 43, 44, 48 ch. 2, note 2 ch. 3 (*see also* Continuum Hypothesis; understanding; intuition; platonism; theorem, Gödel's incompleteness)
Goodman, N., 132, 207, 244, note 18 ch. 6 (*see also* calculus of individuals)

Hale, B., 110–11, 149–65 (*see also* neo-logicism; responses to Benacerraf's dilemma)
  definition of real numbers, 164–5
Hazen, A., 141
Hebb, D., 146
Hellman, G., 110, 112, 138–43, 200, 225, 229, 242, note s13, 25, 26, 27 ch. 4, note 25 ch. 7 (*see also* modal eliminative structuralism)
  on the hermeneutic / revolutionary distinction, note 16 ch. 4

Hempel, C. G., 242
Heyting, A., 90
Hilbert, D.
   on definitions, 78–80
   axiomatic foundation of geometry (*see* geometry)
   Grundlagen der Geometrie, 76, 78
   formalist programme, *see* formalism, Hilbert's
holism
   confirmational, *see* confirmation, confirmational holism
   delivers sufficient conditions for the existence of mathematical objects / for the truth of mathematical theories in IA, 215–16
   semantic, note 33, ch. 6
Horwich, P., 135
Husserl, E., 5, 90–1, 191, note 47 ch. 2
Hussey, E., 31
hierarchy of types, *see* type-theory
hyperouranios, 19
hypothetico-deductive model, *see* confirmation

identity (*see also* Frege, fundamental functions; conditions of identity)
   of indiscernibles, 124, 186–8
   problem of identity for structuralism, 186–7
   tout court, concrete, and among monadic properties, 168
imagination (*phantasia*), 34–6, 38–9, 43–4, 169, 252
imperscrutability of reference, 107
implication (*see also* Frege, fundamental functions)
   expressed by "if" and "only if", 202 (*see also* Colyvan's IA, confirmational holism, naturalism)
(in)dispensability, 203–10
   arguments against the indispensability premise in IA, 217–24
   general clarification, 209–10
   and observational equivalence, 207–9
   for explanation, 209
   and paraphrase, 205–7

   s make indispensable use of T, 204–5
indispensability argument (IA), chs 6, 7
   IA as a transcendental argument, note 7 ch. 6
   IA enhanced (Baker version), 203
   IA Colyvan's version, 202
   IA epistemic for platonism, 198
   IA epistemic for semantic realism, 199–200
   IA non-epistemic for platonism, 197
   IA non-epistemic for semantic realism, 199
   objections to IA, ch. 7
   IA pragmatic, 248
   IA Quine-Putnam, 201
indispensability premise, 217–24, 231–2, note 4 ch. 7
ineliminable metaphors, 233–4
inference to the best explanation (IBE), 147, 203, 203, 240, 243, notes 23, 25 ch. 7
integral, Lebesgue's, Radon's, Stieltjes', 6
interpretation
   of axioms, 80–1
   ermēneia, 18
   of modality, *see* modality
   of a statement, 48–9
intuition, 37–43, 76, 78, 80, 82–3, 87–91, 110, note 18 ch. 1, note 40 ch. 2 (*see also* intuitionism)
   mathematics, 90–1, 94–8, 145 (*see also* Gödel)
   for Parsons, 190–5
   pure, 42–4, 83, 87–8, 91, 191
intuitionism, 87–90, 225 (*see also* Brouwer; Heyting)
   pure two-oneness, 88

judgement
   analytic / synthetic, 40–1
   a priori / *a posteriori*, 40–1
   for Kant, 38–9
   mathematical as synthetic *a priori*, 38
   judgement-stroke ("⊢"), 53, note 13 ch. 2

Kant I., 36–44 (*see also* concepts; judgement; intellect; intuition; schema; sensibility)
  Critique of Pure Reason, 39, 57
  Logik Blomberg, note 18 ch. 2
  Logik Jäsche, note 18 ch. 2
  sphaera conceptus (or extension of a concept), 57, note 18 ch. 2
  space as form of outer sense, 42, 87
  time as form of inner sense, 42
  Vienna Logic, 57
Keränen, J., 186
Kitcher, P., 148, 203
knowledge
  a posteriori, 39–41, 43, 125
  a priori, 39
  hypothetical, 21, 44
  as justified true belief/opinion, 17–18
  propositional / non-propositional, 18

Ladyman, J., 187–8
Lagrange, J.-L., 46, 50
Lakatos, I., 11, 240, 243
language
  formal, 47–50
  nominalistic, xv, 119, 125, 220-2
  predicate, 77, 131, 174, 225, note 8, 11 ch 7
  first-order predicative language and canonical notation, 210–13
law
  of excluded middle, 89
  Leibniz's law *see* identity of indiscernibles
  bridge-laws, 114, 117, 222, note 2 ch. 4, note 24 ch. 7
law (in *Grundgesetze*), 53–4, 59–60, note 16 ch. 2
  v (o *Basic Law* v, BLV), 59, 61, 64–5, 156, 158, notes 22, 25 ch. 2
  vi, 60, 156
Leitgeb, H., 187–8
Leng, M., 135, notes 23, 29 ch. 4, note 37 ch. 6, note 26 ch. 7
linguistic conventions, 10 (*see also* linguistic *framework*)
linguistic *framework*, 73–4, 91, 134, 135, 225

Linsky, B., 149, 165–77 (*see also* Object Theory)
logistic, 21, 25–6, note 2 ch. 1

MacBride, F., 121, note 1, 2 ch. 5
Mackie, J. L., 125
Maddy, P., 11, 111–12, 212, 225, note 31 ch. 4, note 37 ch. 6, notes 12, 20 ch. 7
  on atomic theory, *see* atomic theory
  cognitive basis of set theory, 144–8
  compromise platonism, 145
  against Quine's naturalism, 235–9
  scientific practice objection to IA, 245–7
Malament, D., 119, 123, 222
Mancosu, P., ix, 92, note 40 ch. 2, notes 13, 18 ch. 6
mathematics
  current mathematics, 85
  contentual and ideal, *see* formalism, Hilbert's
  finitary, 82–7, 190–1, 193, note 40 ch. 2
  good, according to Field, 113–14, 116–18, 217
  as modal logic, 227–9
  pre-modern, 183
  recreational, *see* mathematical recreation
  as set theory, 227–9
  structuralist, 188–90, 192–3, 194–5
Melia, J., 117, 222–4, 230, note 23 ch. 6, note 6 ch. 7
mereological sums, *see* mereology
mereology, 141, 222, note 11 ch. 4
meta-fictionalism, *see* fictionalism
metamathematics, 81, 84, 95, 191
method of paraphrase, 9–10, 205–7 (*see also* paraphrase)
Mill, J. S., 11
modality, 140, 143–4, note 29 ch. 4
Mueller, I., 31

naturalism, 214–15, 215–16
  delivers necessary conditions for the existence of mathematical objects / for the truth of mathematical recreation, 235, note 37 ch. 6

mathematical theories in IA, 215–16
  Maddy's, 235–9
  ontological, 213, 237, 245
  semantic, 213, 237, 239
negation, *see* Frege, fundamental functions
neo-logicism, 149–65
  definition of natural numbers, 158–60
  definition of real numbers, 164–5
  application constraint, *see* constraint, Frege's
Neumann, J. von, *see* sets, von Neumann's
Newton, I., 37, 69, 120
  Principia (*Philosophiae naturalis principia mathematica*), 37, 69
  nominalistic reformulation of Gravitational Theory, 119–21, 123–4, 205, 220–1
noēsis, 21–2
nominalism, 9
  Field's, 112–25
  global, 9
  hermeneutic vs revolutionary, 128–9
nous, 22, 33–6, 38, 97, notes 3, 5 ch. 1
numbers
  of concepts, 155, 159, 162–3, 181
  naturals, definition, *see* Frege; neo-logicism; Russell
  reals, definition, *see* Cantor; Dedekind; Frege; neo-logicism
numerical quantifiers, 119, 130, 206, 233, 252, note 9 ch. 4

objectivity
  of mathematics, 7–8, 127, 200, 227–30
  of mathematical objects, 16, 92, 97, 188–9
objects
  abstract, 2–3
  copies of an idea, 18, 20, 28, 44
  intermediate, 27–8, 33–4
  mathematical, 3–4
  mathematical as characters in a fiction, 13–14, 126, 129
  ordinary, 167–8

"quasi-concrete", 190–3, 195, note 42 ch. 2
  of a theory T ($O_{[T]}$), 197
Object Theory, 165–77
o-equivalence (*see* equivalence)
one-one correspondence, 55, 70, 102, 157
ontological commitment, *see* criterion of ontological commitment
ontological relativity, note 11 ch. 2, note 9 ch. 7
Ω, 79, 81

paraphrase, xv, 9–10, 136, 205–7, 220, 233, 247, 252 (*see also* p-equivalence)
Parsons, C., 11, 106, 111, 136–8, 142–3, 147, 149, 156, 177–8, 181, 183, 186–96, note 42 p. 2, notes 27, 30 ch. 4, note 37 ch. 6 (*see also* eliminative structuralism; intuition; Reason)
parts or portions of a theory, xiv
pattern recognition, 185
Peano, G.,
  Peano axioms, 60, 72, 81, 139, 156, 174, 228, notes 23, 34, 35 ch. 2
  derivation of Peano axioms in Frege's formal system, 60
p-equivalence, *see* equivalence
Peressini, A., note 28 ch. 7
Pettigrew, R., 32
pistis, 21–2
Plato, 17–26 (*see also eikasia*; *epistēmē*; *ermēneia*; geometry; hyperuranios; logistic; *nous*; *pistis*; recollection)
  Gorgias, 25–6
  Hippias Major, 55
  on ideas, 18, 22, 27–8
  Meno, 18–20, 23, 25, note 1 ch. 1
  Phaedrus, 19–20, notes 1, 3 ch. 1
  Philebus, 25
  Republic, 20, 44
  divided line, in *Republic*, 20–2, note 5 ch. 1
  Seventh Letter, 20, 24, 26, 44
  Theaetetus, 17–20, 23, 25, note 1 ch. 1

platonism
  arithmetical, 4
  arrogant, 107–8, 161–2
  compromise platonism,
    see Maddy
  empiricist, 11
  physicalist, 11
  sober, 107
Poincaré, H., x, 89, 91, 246, note 38 ch. 2, note 24 ch. 7
Popper, K., 240
posit, 213, 240, 242
PRA, *see* arithmetic
principle
  Vicious Circle principle 69, 91
  context principle, 54, 61–2, 149, 167, note 31 ch. 2
  Eleatic principle, note 32 ch. 6
  Hume's Principle (HP), 55–6, 60–1, 152–65, 177, note 17 ch. 2
  Likelihood principle, 244–5
problem
  of access, 17, 110, 125, 149, 154, 162, 250
  Caesar problem, 56–8, 101, 155, 162–3, 164, 180
  of identity, *see* identity
  Plato's problem, vii–ix, 1, 26–8, 37, 39, 74, 99, 101–2, 105, 110, 181–2, 195, 254
  neo-logicist solution to Caesar problem, 162–4
Proclus, 32–6 (*see also* geometry, *phantasia*, projections)
  Commentary to the First Book of Euclid's Elements, 32–3, note 13 ch. 1
  on ideas, 32–5
  logoi, 34–6, 38, 44, 97
progression, note 8 ch. 1
projections (*probolai*), 35–6, 44
proposition, xiii, note 7 ch. 5
  (*see also* encoding)
propositional attitude, 128, 134–5, 234, note 15 ch. 4
prospective places-are-objects, 179
prospective places-are-roles, 179
Psillos, S., 239, 241, note 23 ch. 7
  (*see also* scientific realism)

Putnam, H, 142, 196, 197–201, 202, 215, 227–30, note 27 ch. 4, notes 8, 19, 35 ch. 6, notes 11, 12, 13, 21 ch. 7

quantification
  and IA, 204–5
  and ontological commitment, 210–12, 224–35
  objectual, note 28 ch. 6, note 11 ch. 7 (*see also* existential statements *and* ontological commitment)
quasi-assertions, 130
quasi-concrete, *see* objects
Quine, W.V.O, viii, 107, 109, 134 (*see also* criterion of ontological commitment; IA; imperscrutability of reference; indeterminacy of translation; objectual interpretation of existential statements; second-order logic; naturalism; holism; quantification and ontological commitment, ontological relativity; canonical reformulation of a theory)

Ramsey, F.P., 92
realism
  ontological, 8, 201
  scientific, 8, 203, 235–41, 248, note 10 ch. 6
  objections to scientific realism, 239–41
  semantic, 8, 113, 199, 201, 229, 250 (*see also* indispensability argument)
Reason
  distinct from reasons, 194
recollection, 19–20
reconceptualization, 152, 160
recursive, note 27 ch. 2, (*see also* recursively enumerable sets; recursively axiomatizable theories)
reference
  sense/reference distinction in Frege, 53
  of names of functions, 61

reformulation
 of statements, xiii, xv, 9–10
 canonical and ontologically minimal of a theory, 211–13, 230, 233–4
relation
 ancestral, 56, 159
 of equivalence, *see* equivalence
 of set membership, 67–8
responses to Benacerraf's dilemma, 110–11
 conservative, ch. 5
 non-conservative, ch. 4
 intellectual and intuitional, 110
Resnik, M., 105, 123–4, 136–7, 187, 215, notes 25, 28, 29, 31 ch. 7 (*see also* pragmatic IA)
Rosen, G., 122, 128–9, 135, 200, 230, 237–8, note 1 Introduction, notes 12, 13, 14, 16 ch. 4, notes 12, 20, 30 ch. 7
 hermeneutic / revolutionary distinction, 128–9
Russell, B., 66–9, 71, 73–4, 81, 87, 91–3, 99, note 29 ch. 2 (*see also* Russell's paradox; Vicious Circle principle; type theory)
 no-class theory, *see* class
 definition of natural numbers, 67–8
 Principia Mathematica, 69, 74, 91
 The Principles of Mathematics, 66–7, 69, 99
 Russell's paradox, 64–6
 incomplete symbols, 69

schema
 of axioms, 157–8, 222–4, notes 23, 35 ch. 2, note 3 ch. 4, notes 25, 26 ch. 4
 comprehension schema, 140–3, 157–60, 167–8, 222–3
 in Kant, 39, 43–4, note 19 ch. 1
 of statements, 131, 228
 of theorems, 83
 translation schema, 139–40, note 24 ch. 4
scientific practice objection, *see* Maddy
second order logic, 47–8, 50, 54, 61, 66, 72, 131, 160

nominalistic interpretation of, 140–2
modal, 139–40, note 25 ch. 4 (*see also* mathematics as modal logic)
as set theory or mathematics, 131, 140–1, 160, note 15 ch. 2
semantics, 48–9
sense of a name, *see* reference
sensibility, 34, 38, 42–3, 58, note 15 ch. 1
sentence, xiii–xiv, 47–8.
separation conditions between nominalistic and mathematical statements, 115
sets
 Cantor and the development of set theory, *see* Cantor
 equinumerous, 67–9
 and the foundation of arithmetic, *see* Benacerraf's (1965) argument
 natural numbers as sets of equinumerous sets, 67–8
 recursively enumerable, 208, note 20 ch. 6
 set theory, 69–72
 Von Neumann's, 71, 81, 102–3, 178, note 3 ch 2
 Zermelo-Fraenkel's, 71–2
 ZF, 71–2, 93–5, 98, 127, 222–4, note 48 ch. 2
Shapiro, S., 30, 74–5, 80, 85–6, 111, 115, 121–3, 137, 143–4, 149, 160, 164, 177–87, 188–9, 192, 217–8, 239, 252, note 15 ch. 2, note 1 ch. 3, note 10 ch. 4 (*see also* structuralism; conservativeness; structure theory)
sheaf of properties, 168–9, note 6 ch. 5
simulation, 134–5, 234
Skolem, T., 193 (*see also* Löwenheim-Skolem theorem)
Sobe,r E., 242–5 (*see also* contrastive empiricism)
Socrates, 17–20, 25
space-time points, 11, 120, 122–4, 142
Stanley, J., 129
 autism objection, 135
statement, xiii–xv, 48

Index 305

existential statement and ontological commitment, 210–12, 224–5
mathematical / of mathematics, xiv–xv
nominalistic, xv
observational, 207–9, 219, 239, note 29 ch. 6
M-loaded, 217, 219, 232, 235, note 2 ch. 7
T-loaded, 205–9 (*see also* M-loaded)
vacuously true, note 4 Introduction, 12–13, 112, 117, 138, 140
Steiner, M., notes 1, 4 ch. 6
stipulations, in a formal system, 48–9
Stolz, O., 75, 78
subtheories, xiv
structuralism
  ante rem, 111, 149, 177–87, 188
  eliminative, 136–44
  modal eliminative, Hellman's, 138–44
  modal eliminative as a form of neutralism, 143
  non-eliminative, 149, 177–87, 187–95
  non-eliminative, Parsons', 149, 187–95
  suggested by Benacerraf (1965), *see* abstract structure of progression
structure
  grammatical, 6–7, 12, 110, 131
  semantical, 7, 12, 106
successful applications (o mathematics), 145 (*see also* fruitfulness)
successor, 56, 139, 159, 187, 192–3, 259–60, note 3 ch. 4
syntax, 48–9
systems or bodies of statements, xiv

terms
  natural numerical, 161
  role of representational aids, 133–4
  role of things represented, 134
  empty singular, 13
theorem
  Craig's, 208

  Frege's, 157, 160–1, 164
  Gödel's incompleteness theorem, 86–7, 121–2, notes 43, 44 ch. 2
  Löwenheim-Skolem theorem, note 35 ch. 2
  last theorem of Fermat, 75, 227–9, notes 10, 11 ch. 7
  representation theorems, 120, 220–1, notes 3, 4 ch. 7
theory
  atomic, 246
  category theory, 199
  conservative, 85, 116–23, 217–18, 220, 221, note 7 ch. 4, note 2 ch. 7 (*see also* conservativeness)
  as a deductively close set of statements, xiv–xv
  empirically adequate, 240, 242
  error theory, 125
  of functions, or analysis, *see* analysis
  indispensable, *see* indispensability
  impure mathematics, xv
  nominalistically impure mathematics, 114–17, note 2 ch. 4
  mature scientific, 240–1
  model theory, 260
  natural numbers theory, note 2 Introduction
  Newton's Gravitational Theory, *see* Newton
  nominalist, *see* nominalistic language
  platonic theory of ideas, *see* Plato
  preferable, 208
  pure mathematics, xv
  Q-justified, 246–7
  reasonably attractive, 208
  recursively axiomatizable, 208, note 21 ch. 6
  scientific, xiv–xv
  scientific theory under-determined by empirical evidence, 240
  set-theory, *see* sets
  set-theory in disguise, *see* second-order logic
  structure theory, 183, 185
  type theory, 69

thesis
  Frege's, 150–2, 154, 162, 177, 189
  Hilbert's, 193
  syntactic priority thesis, 151–2, 154–5, 159, 161, 162
thought
  in Frege, 53
  as understanding in Gödel, see Gödel, understanding
token / type, 83, 190
transfinite cardinals, 70
transfinite ordinals, 70
True (as object), 51
true-in-a-fiction/-in-the-story, 126–7, 129–33, 135, 231
truth *tout court*, 14, note 4 Introduction

understanding
  for Gödel, 96–7
  for Kant, 38, 42, 97

value-range of a function, 59–62, 64–5, notes 10, 25 ch. 2
Vineberg, S., note 25 ch. 7

vocabulary, 204–7, 217, 220, 223, 228, 233, 263, notes 14, 16, 19 ch. 6
  observational vs theoretical, 207–9

Walton, K., 134 (*see also* games of make-believe)
Weir, A., 160, note 32 ch. 6
White, N., 106, 137
Whitehead, A. N., 69, 91–2
Principia Mathematica, *see* Russell
Wiles, A., 75, 227, note 10 ch. 7
Wright, C., 107, 110–12, 144, 149–65, note 26 ch. 2, notes 1, 2 ch. 5 (*see also* neo-logicism; responses to Benacerraf's dilemma)

Yablo, S., 14, 111–12, 125, 129–35, 230, 232–4, 240, notes 17, 18, 20, 21, 23, ch. 4, note 15 ch. 7 (*see also* fictionalism)

Zalta, E., 111, 149, 165–77 note 28 ch. 4, notes 4, 8 ch. 5 (*see also* Object Theory)
Zermelo, E., 71, 81, 102–3, 163, 178, note 3 ch. 3
ZF e ZFC, *see* set theory

The manufacturer's authorised representative in the EU is Springer Nature Customer Service Centre GmbH, Europaplatz 3, 69115 Heidelberg, Germany. If you have any concerns regarding our products, please contact ProductSafety@springernature.com

Printed and bound by CPI Group (UK) Ltd, Croydon, CR0 4YY

23/03/2026

02076447-0020